MATHEMATICS IN HEALTH ADMINISTRATION

Robert W. Broyles, Ph.D.
Colin M. Lay, Ph.D.

University of Ottawa

AN ASPEN PUBLICATION®
Aspen Systems Corporation
Rockville, Maryland
London
1980

Library of Congress Cataloging in Publication Data

Broyles, Robert W.
Mathematics in health administration.

Includes bibliographical references and index.

1. Health facilities—Administration—
Mathematics. 2. Health services administration—
Mathematics. I. Lay, Colin M., joint author.
II. Title. [DNLM: 1. Mathematics. 2. Public
health administration. WA950 B885m]
RA971.B88 510 80-19451
ISBN: 0-89443-297-4

Library of Congress Catalog Card Number: 80-19451
ISBN: 0-89443-297-4

Printed in the United States of America

1 2 3 4 5

GREEK ALPHABET

Α	α	alpha	Ν	ν	nu
Β	β	beta	Ξ	ξ	xi
Γ	γ	gamma	Ο	o	omicron
Δ	δ	delta	Π	π	pi
Ε	ϵ	epsilon	Ρ	ρ	rho
Ζ	ζ	zeta	Σ	σ	sigma
Η	η	eta	Τ	τ	tau
Θ	θ	theta	Υ	υ	upsilon
Ι	ι	iota	Φ	ϕ	phi
Κ	κ	kappa	Χ	χ	chi
Λ	λ	lambda	Ψ	ψ	psi
Μ	μ	mu	Ω	ω	omega

Table of Contents

Preface

That the management of the health care facility has increasingly required an understanding and application of mathematics and mathematical techniques is well recognized. In this regard, an understanding of mathematics enhances the ability of the administrator to employ the principles of "scientific management" when reaching decisions or resolving problems that confront the institution. In addition, mathematical techniques are commonly employed in health care research and in formulating national or regional policies that influence the health care delivery system. In order to evaluate health policies and the results of health care research, the administrator frequently requires an understanding of mathematics.

We do not present arithmetic or mathematical tricks that can be used in a cookbook fashion in particular situations, as has been done in some books for the health care field. Rather we have focused on the *principles* of *basic mathematics,* showing how those principles can be used, by means of numerous examples. It is far better to learn to speak and use the basics of a language than to rely on a phrase book. Each chapter presents a group of principles that are closely related to each other. In general, the principles covered in early groups of chapters (sections) must be understood before proceeding to later sections.

Even though the importance of mathematical techniques is well recognized, students and practitioners frequently approach the study of mathematics with fear and trepidation. However, it should be recognized that mathematics is simply a language that, unlike English, is unambiguous and governed by well-defined rules and conventions. For example, when used in isolation, the word "can" is somewhat ambiguous since it might refer to

1. a drinking cup,
2. the ability to perform a given task,

3. the act of discharging or dismissing an employee,
4. a case or vessel formed from sheet metal; or
5. the process of preserving certain foodstuffs.

Similarly, the letter "a" not only is the first letter of the alphabet but also is an indefinite article. Just as the symbol "a" is the first letter of the English alphabet, the symbol "Σ" is a letter of the Greek alphabet. Unlike the symbol "a," however, the Greek letter Σ is unambiguous since it pertains to a specific mathematical operation.

The primary purpose of this text is to present mathematics in a language that is understandable and useful to managers and students of health administration. As such, the book is devoted to an examination of those aspects of mathematics that may be used in managing the health care facility and to a discussion of the application of mathematical techniques to the problems confronting the institution. As a result, it does not consider the more esoteric aspects of mathematics that are of greater interest to mathematicians.

A DESCRIPTION OF THE BOOK

This text consists of five major sections. The first section examines the number system as well as the tabular and graphic presentation of data. Part II addresses the fundamentals of algebra, including relationships, functions, linear equations and functions, the mathematics of finance, and the process of solving a system of equations. On the basis of the preceding material, Part III explores differential and integral calculus as mathematical tools that may be used by the health care administrator.

Part IV focuses on set theory and the concepts of probability as applied to discrete and continuous variables. We employ the fundamental principles of integral calculus when examining the concept of probability as applied to a continuous variable; a portion of this discussion is simply an extension of the analysis in Part III.

Finally, Part V covers the fundamentals of matrix algebra—beginning with the operations of addition, subtraction, and multiplication as applied to matrices. In addition, we examine the matrix algebra counterpart to division as well as linear transformations, operations on partitioned matrices, and the transpose of a matrix.

This volume assumes that the reader has little or no knowledge of mathematics. To accommodate such a user, we begin our discussion with an examination of fundamental principles and proceed in later chapters to an analysis of the more complex aspects of mathematics.

We anticipate that this book will be of interest to a wide variety of readers. To maximize the usefulness of this material, we have characterized each section in terms of three dimensions. First, each section is marked as being either required or optional reading. The mathematical novice should review sections identified as required reading with considerable care, while those classified as optional may be omitted without a loss of continuity.

The second dimension attempts to capture the extent to which the discussion in various sections is devoted to the generic (theoretical) aspects of mathematics or to the application of mathematical techniques to the problems confronting the health administrator. Mathematics may be presented in varying degrees of sophistication and in almost any balance between theory and application. In this work, we not only attempt to develop a basic understanding of the generic aspects of mathematics but we also emphasize the usefulness of mathematical techniques in the management of the health facility.

The final dimension involves the relative complexity of the material in each section. We distinguish sections in which fundamental principles are discussed from those that consider more sophisticated or complex areas of mathematics.

At the beginning of each chapter is a map that summarizes each section in terms of these three dimensions. In addition, the following code is used to characterize each section:

CODE	*EXPLANATION*
R	Required reading
O	Optional reading
G	Presentation of generic material
A	Application of mathematical techniques
F	Development of fundamental principles
C	Development of more complex or sophisticated material

To illustrate, consider the code represented by (R, G, A, F) *where*:

R indicates that the material in the section should be regarded as required reading,

G,A indicates that the section is devoted to a discussion of generic and applied mathematics,

F indicates that this discussion is of a less complex or sophisticated nature.

In addition to these letter codes, a set of study objectives and problems for solution accompanies each chapter, followed at the end of the book by solutions to selected problems that we hope will assist readers in verifying their mastery of the material in each chapter.

Acknowledgments

We wish to express our gratitude to our wives, Rita and Marjorie, and to our children, Erin, Steven, and Andrew, for their patience and understanding during the preparation of this book.

In addition, we are in debt to the students who used earlier versions of this text and whose suggestions resulted in an improved presentation. Further, to the reviewers of the manuscript, we express our sincere gratitude for the many suggestions that we have incorporated.

We also wish to express our sincere appreciation to Micheline LeBlanc, our secretary, whose diligence, patience, and dedication defy description. And finally, we express our gratitude to the many mathematicians on whose work we have relied heavily. We, of course, assume the responsibility for any errors or oversights.

R.W.B.
and
C.M.L.
Ottawa, Ontario
October 1980

Part I

Introduction to Numerical Information and the Presentation of Data

Chapter 1

Importance of Mathematics to Health Care Managers

1.1 USES OF MATHEMATICS IN HEALTH CARE MANAGEMENT

The importance of applying mathematical techniques in the management of the health care facility has grown in recent years, and administrators who are unfamiliar with the fundamentals of mathematics now find themselves at an increasing disadvantage.

This work is predicated on the premise that a fundamental understanding of mathematics is required in order to discharge many, if not most, of management's responsibilities. For example, health care administrators frequently are required to present and interpret a large volume of numerical information. To do so, managers should understand the number system as well as the various methods of presenting data. Two-dimensional tables, histograms, line charts, polygons, bar charts, frequency distributions, and pie charts are among the methods health care managers may use to rearrange data and to present numerical information in a more useable or understandable form.

Mathematical techniques also may be used when discharging the managerial functions of planning, monitoring, and controlling operational activity. In planning the future course of organizational activity, one of the first tasks confronting management is the development of projections concerning the future workload of the institution or one of its units. As will be seen, the principles of mathematical expectation, algebra, and matrix algebra are techniques that may be employed when developing forecasts on the volume and type of services that will be required by the community.

Once the workload of the institution or organizational unit has been projected, management is in a position to estimate the resources that will be required. In developing such an estimate, it is useful to specify the functional relationship between anticipated rates of activity and the corresponding volume of resources that will be needed. For example, suppose that the care our institution provides is limited to stay-specific services, which we measure by the number of days of care, and several ancillary services such as urinalysis, blood chemistry examinations, etc. For each component of care, the functional relation between the rate of activity and factor inputs such as labor, capital equipment, and consumable supplies provides the basis for projecting resource requirements. In order to specify and employ such a functional relationship, however, an understanding of fundamental algebra is required.

The relation between the rate of activity and the various factor inputs also may be employed when reaching decisions as to the use of an additional unit of a specific resource. In such a situation, management is interested in the additional care that results from using an additional unit while holding other inputs constant. In this case, the partial derivative is a mathematical technique that may be employed to obtain the information on which a rational decision might be based.

Consider next the use of mathematics in controlling and evaluating the results of operational activity. That governmental authorities and other third party payers are exerting increasing pressure to contain health care costs or to reduce their rate of increase is well known. In response to these pressures, health care administrators frequently are required to examine the behavior of costs in relation to differing rates of activity and, on the basis of this investigation, formulate and implement policies designed to restrain expenses.

Of particular importance in this regard is the use of flexible budgets in the control process. A flexible budget simply expresses the different cost implications associated with differing rates of activity and, as such, represents the standard against which actual performance is compared. As will be seen later, the flexible budget might be expressed in the form

$$TC = a + bV \qquad \textbf{(1.1)}$$

where

TC represents the total costs of operation,
a represents the fixed cost component of total costs (i.e., costs that remain invariant with respect to changes in the volume of care provided),
b represents the change in total cost resulting from a change in the volume of services provided (i.e., the variable cost component of total cost), and
V represents the volume of service.

To illustrate the usefulness of a flexible budget in health care administration, as expressed by Equation 1.1, suppose that actual costs of $70,000 are incurred in providing 5,000 units of service during the period. Assume further that the expected costs of providing those 5,000 units, as expressed by the flexible budget, is $40,000. A comparison of actual with expected outcomes alerts management to the presence of a problem that may require remedial action to redress the imbalance between the standard of performance and the actual performance of the unit. This illustration demonstrates that the ability of management to control cost is enhanced by an understanding of basic algebra and a functional relationship in which the dependence of one variable on another is specified.

As a second illustration, suppose that management wishes to evaluate the desirability of maintaining the status quo, in which 200 patients would be treated on an inpatient basis, or adopting a new proposal in which the 200 patients would receive care on an ambulatory basis. Assume further that the quality or the effectiveness of the care provided under the two approaches is identical and that the following information is available to management:

Approach	Days of Care	Outpatient Visits	Total Costs
Status Quo	1,200	500	$62,000
Proposal	900	1,000	$57,000
Change	(300)	500	($5,000)

An inspection of these data reveals that, by implementing the new proposal, management is able to realize a net saving of $5,000 without compromising the quality of the care rendered to the 200 patients. In evaluating the desirability of implementing the new approach, also notice that management required information depicting the responsiveness of costs to changes in operational activity. As will be seen later, these data may be obtained by employing the techniques of differential calculus.

As a final illustration, an element of uncertainty is present in almost every facet of managing a health care facility. For example, in scheduling the operating room, management might be interested in determining the chances that a given surgical procedure will require more than two hours to perform. In this situation, a decision concerning an aspect of operational activity must be reached in a world of uncertainty. As will be seen later, however, the uncertainty in this situation may be captured quantitatively by applying the techniques of integral calculus to an appropriate probability density function. As shown in this example, an application of mathematical techniques can provide the manager with the data required to reach a well-informed decision.

An understanding of mathematical techniques also may enhance management's ability to use high-speed electronic computers productively. Matrix algebra and electronic data processing equipment should be regarded as complementary tools that managers may employ to reduce the time and energy required to complete computational tasks. Once a well-defined set of mathematical equations has been expressed in the notation and operations of matrix algebra, programs may be developed that enable management to use computers to perform required calculations. As a result, the use of matrix algebra and high-speed computers allows the health care administrator to devote more time and energy to other areas of managerial responsibility.

It thus is clear that mathematics can help the health care executive in reaching decisions and in discharging managerial responsibilities. It also should be noted that the use of mathematics in health care is by no means limited to the areas described here; rather, the application is constrained only by the needs and imagination of the manager.

Chapter 2

The Number System and the Tabular Presentation of Data

Objectives

After completing this chapter, you should be able to:

1. Describe the family tree of numbers;
2. Describe the types of tables used to display numeric information;
3. Describe the components of a table;
4. Construct a table.

Chapter Map

The sections comprising this chapter may be summarized as follows:

Section Number	Required Reading	Optional Reading	Generic Development	Application to Management	Fundamental Principles	Complex Material
	(R)	(O)	(G)	(A)	(F)	(C)
2.1	x			x	x	
2.2	x		x	x	x	
2.3	x			x	x	
2.4	x			x	x	
2.4.1	x			x	x	
2.4.2	x			x	x	
2.5	x		x	x	x	
2.5.1	x		x		x	
2.5.2	x			x	x	
2.5.3	x		x	x	x	

2.1 INTRODUCTION(R,A,F)

In this chapter and the next, methods of presenting data in a more useable or understandable form are considered. As noted earlier, health care administrators are required to present and interpret large volumes of data. An ability to present data in tabular form, therefore, is of considerable value when preparing the budget, reviewing the performance of the institution, engaging in labor negotiations, preparing medical records, and reporting the results of the most recent utilization review. In this chapter, methods of presenting data in tabular form are examined, with various techniques of displaying information graphically considered later. Before discussing the construction of tables, however, it is useful to describe the types of numerical information that might be displayed in tabular form.

2.2 SYSTEMS OF NUMBERS(R,G,A,F)

The family tree of numbers includes integers, fractions, rationals, irrationals, reals, and imaginaries. With the exception of imaginary numbers, the health care administrator should be familiar with all of these.

Integers, which also are called "whole" numbers, were defined originally by counting 1, 2, 3, 4, 5, $\cdots$ where the notation "$\cdots$" means "and so on." In this system, zero and negative integers (i.e., $\cdots$, -3, -2, -1) were added when the concepts of "nothing" and less than zero were recognized. Integers are used extensively in health care management. For example, patients in a ward are counted to obtain the census. A patient's stay is counted in days. A waiting list has a certain number of patients. Department heads must know the number of employees for whom they are responsible. Integers are used to address problems in which frequencies or counts are required. Negative integers arise when there is no natural zero level, or when some point other than "absolute" zero is used as the reference point. Temperature and elevation are physical examples where the commonly used zero is arbitrarily defined. On the Celsius scale, zero is the freezing point for water, and -18 degrees is regarded as being "very" cold by most people in the United States and Canada (zero on the old Farenheit scale), but those living in the Arctic or the Antarctic believe temperatures below -40 degrees are "very" cold. Since 37°C represents the "normal" body temperature, hypothermia and fever are measured relative to that point, and temperatures are termed above and below "normal."

When dealing with integers, the operations of addition, subtraction, and multiplication always result in other integers. However, the division of one integer by another does not always result in another whole number and, as a result, fractions must be defined in order to accommodate the results of division. In this regard, the average length of stay (ALOS) of a group of patients frequently requires fractions that customarily are expressed in decimal form. For example, a group of 30 patients using a total of 220 days of care has an ALOS of 220/30 or $7\frac{1}{3}$ days, which usually is expressed as 7.33 days. In this case, simply ignore the remaining digits of the *decimal fraction* that results from dividing one by three.

Note that some decimal fractions "terminate" but that others do not, and still others repeat in a definite cycle as indicated by dots that appear above the repeating figures (or groups). For example:

$$\begin{aligned}
1/3 &= .33\dot{3}\cdots \\
1/8 &= .125 \\
1/9 &= .11\dot{1} \\
1/11 &= .0909\dot{0}\dot{9}\cdots \\
1/7 &= .142857\dot{1}4285\dot{7}
\end{aligned}$$

Often fractions are expressed in a decimal form that has been rounded to 2, 3, 4, or 5 *decimal places,* depending on the accuracy required. To illustrate the process of rounding, suppose a manager wants to retain three digits in a decimal fraction of the form .333 x. In this case, if the value of x is equal to or greater than 5, simply add .001 to the decimal fraction .333 and obtain .334. Similarly, to retain four significant digits in the decimal fraction .3478921, it can be seen that the fifth digit (i.e., 9) is greater than 5. In this case add .0001 to .3478 and obtain .3479. Rounding also may be accomplished by

1. adding a decimal fraction consisting of zeros up to the last decimal place to be retained and a 5 as the next digit; and
2. truncating after the last digit to be retained.

Applying this approach, we find that

$$\begin{aligned}
1/8 &= .125 \text{ (exactly)} \\
1/3 &= .3333 + .0005 = .333\not{8} = .333 \\
1/11 &= .0909 + .0005 = .091\not{4} = .091 \\
1/7 &= .1428 + .0005 = .143\not{3} = .143
\end{aligned}$$

In this case, the decimal fractions have been rounded to three digits.

A fraction is the *ratio* of two integers (e.g., 220/30), which implies that any integer may be viewed as a ratio of itself and one (e.g., 7/1 = 7). However, zero may not be used as a divisor and a ratio of the form 7/0 is *not defined.* Since a fraction is the ratio of two integers, fractions and integers are called rational numbers. In this case, the term "rational" refers to a mathematical rather than a psychological concept.

Certain numbers cannot be expressed as ratios of integers and, since they are not rational, they are called "irrational." Two examples of great theoretical importance are the number π (the Greek letter pi), which is the ratio of the circumference to the diameter of the circle and has a value of approximately 3.14159, and e, which is the base of natural logarithms and has a value of approximately 2.71828.

Together, the rational and irrational numbers are called *real* numbers. Health care management is concerned almost exclusively with real numbers. Any of the regular arithmetic operations performed on a pair of real numbers produces a real number as the result. Only division by zero is not defined and not allowed. The square roots of positive real numbers are defined (e.g., $\sqrt{4} = 2$, $\sqrt{5} = 2.23606\cdots$, etc.). However, the square root of negative numbers is not defined in the real number system. Therefore, mathematicians have defined *imaginary* numbers that involve $\sqrt{-1}$. In this book, we ignore imaginary numbers and deal only with reals.

The relationships among the various types of numbers may be presented in the form of a "family tree" (Figure 2-1). In this figure, the most elementary types of numbers appear on the left and are grouped into more complete systems toward the right.

Figure 2-1 The Family Tree of Numbers

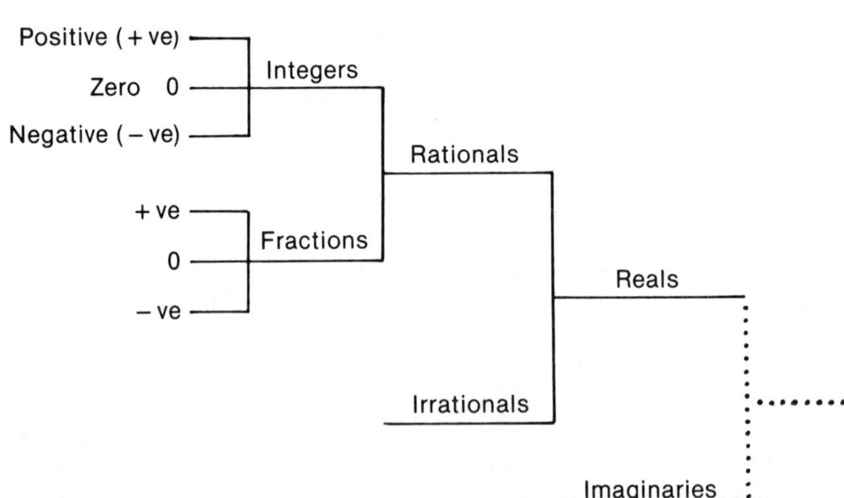

In Figure 2-2, a real number can be represented as a point on the line called the *real number line.* This figure shows it is conceptually possible to extend the real line forever in each direction. The term "forever" is represented in mathematics by the concept of infinity (∞) and, at least conceptually, the real number line extends from $-\infty$ to $+\infty$.

We can compare the sizes of numbers appearing on the real line; by definition, numbers to the left are smaller than numbers to the right. Three basic comparisons are identifiable:

1. One number may be less than another; such a comparison is represented by the symbol "$<$," which means "less than."
2. One number may be equal to another, which is represented by the symbol "$=$."
3. One number may be greater than another; such a comparison is represented by the notation "$>$," which means "greater than."

The three comparisons may be combined in symbolic form:

1. The symbol "$\leq$" means "less than or equal to," while
2. the symbol "$\geq$" means "greater than or equal to."

The basic comparisons also may be negated. In this case, the symbols "$\not<$, $\neq$ and $\not>$" mean "not less than," "not equal to," and "not greater than," respectively.

To illustrate the use of the three comparison symbols, we find that

$$\begin{aligned} -10 &< -6 \\ -6 &< 5 \\ 3 &\neq 2 \\ 16 &> 0 \\ -5 &> -17 \\ -3 &\not> 1 \end{aligned}$$

Similarly, 2 degrees of fever is a higher temperature than 7 degrees hypothermia, while a hospital stay that is one day below average is shorter than a stay that is one day above average.

Figure 2-2 The Real Number Line

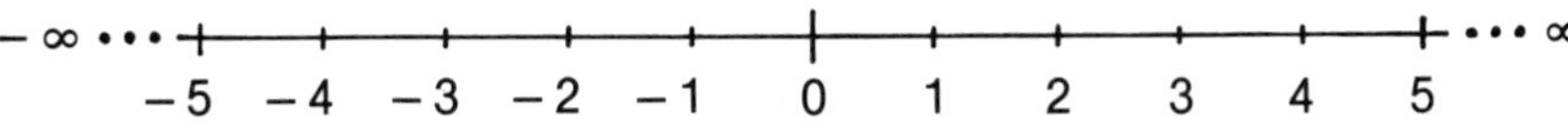

2.3 THE NEED FOR NUMERICAL INFORMATION[R,A,F]

An understanding of the number system described in the previous section provides the basis for interpreting and presenting data in tabular and graphic form. As implied earlier, the numeric information of interest to the health care manager usually assumes the form of a real number. This section considers the tabular presentation of data as a means of rearranging numeric information into a more useable and understandable form.

All health care organizations employ numeric information in the process of planning, evaluating, and controlling operational activity. Any such institution requires information concerning the volume of service provided to patients and to other clients. The manager also must be interested in obtaining and organizing resources required to provide service. Ensuring that activities are well organized and respond appropriately to changing conditions requires the collection, interpretation, and analysis of large amounts of data on the volume, type, and characteristics of the patient population, the services they require, and the resources needed (both human and nonhuman), as well as the organization and operation of various ancillary and other service departments.

More specifically, hospital administrators require data and information on patient loads, staffing, supplies, equipment, volume of services, activities, costs, etc., to determine how effectively and efficiently these services are being provided to patients. Most management decisions are based, directly or indirectly, on the interpretation of data. In addition, numerical information or the interpretation of data are frequently employed when evaluating managerial decisions.

Managers of smaller health care organizations are interested in identical kinds of information but face volumes of data far smaller than those confronting the hospital administrator. Smaller health care organizations usually consist of a few departments and, as a result, the range of services they provide is limited. Consequently, the problems associated with the collection, presentation, and analysis of data are less complex in small facilities than in the larger institutions. It should be noted, however, that the basic principles of data presentation and interpretation are applicable irrespective of the scale or scope of operation.

In addition, the principles of data presentation also are applicable to qualitative information. Qualitative information usually is based on data that have been recorded and reported on an informal basis. Even though insufficient data may prevent a formal study, judgments may be improved by presenting the known facts in a useable or understandable form. In such a situation, an application of the principles of data presentation may highlight the dimensions of a problem or the relationship that is being examined by management.

2.4 TYPES OF TABLES(R,A,F)

Numerical information that has been accumulated and recorded by management frequently is reported in tabular form. The purpose of the table is to present:

1. raw data in the form in which it was collected originally,
2. a summary of the raw data represented by a total of the original data,
3. more complex summaries,
4. the process and results of analysis, and
5. generalized guidelines for analyzing future data as well as making decisions.

2.4.1 Classification of Tables by Function(R,A,F)

Tabular presentations may be classified in terms of the function or purpose served by the table or in terms of the nature of the data contained in the table. Employing the classification scheme developed by Clover and Balsley (1974, Chapter 12), we refer to tabular presentations that contain raw or summarized data as *repository tables.* For example, Table 2-1 summarizes information concerning the bed complement, the number of patients discharged, and the number of patient days of care associated with each nursing unit in a hypothetical hospital. In addition, tabular presentations that contain inferences drawn from raw data or illustrate operations required to determine relationships are called *analytical tables.* Table 2-2 is an example of this category since three sets of operations are performed on the data in Table 2-1. Average length of stay (ALOS, col. 4) is calculated by dividing patient days by the number of patients discharged. Occupancy (col. 5) is calculated by dividing patient days by available bed days (beds × days in month). Patients per bed month is a "turnover" ratio that is calculated by dividing the number of patients discharged by the bed complement.

A table that contains information based on previous research or a theoretical development that can be used in a specific problem-solving situation can be regarded as a *reference table.* For example, tables containing logarithms, square roots, and mathematical functions such as sine, cosine, and tangent have been available for many years. These functions are routinely

Table 2-1 Patients Discharged and Patient Days by Service for September: Example of a Repository Table

Service and Floor	Bed Complement	Patients Discharged	Patient Days
Medicine-1	30	174	857
Medicine-2	27	139	731
Medicine-3	30	165	826
Medicine-4	25	110	642
Medicine-5	20	89	528
Surgery-1	28	109	798
Surgery-2	30	120	783
Surgery-3	25	108	638
Surgery-4	25	112	690
Gynecology	25	154	668
Pediatrics	30	212	705
Obstetrics	18	61	243
Nursery	20	65	330
Total	333	1,618	8,439

provided now on all but the simplest of hand calculators, and the need for such tables is diminishing. The analysis of waiting lines (queues) is an example of a common management problem in which the theoretical results expressed by complex equations have been summarized. As an example, Table 2-3 summarizes information that emanates from queuing theory. Notice that by letting

$$r = \text{arrival rate/service rate}$$

and

$$m = \text{the number of providers,}$$

the manager can determine from the table certain crucial aspects of the behavior to be expected from the queuing service system. These include average values for the number of patients receiving service, the number waiting for service (queue length), the time an individual spends waiting, and the proportion (fraction) of time the service providers are busy. For example, if an emergency room is required to provide care to 18 patients

Table 2-2 Patient Length of Stay and Bed Utilization for September: Example of an Analytical Table

Service and Floor	Bed Comple- ment (1)	Patients Discharged (2)	Patient Days (3)	ALOS (days) (4)	Occu- pancy (%) (5)	Patients Per Bed Month (6)
Med 1	30	174	857	4.93	95.2	5.80
Med 2	27	139	731	5.26	90.2	5.15
Med 3	30	165	826	5.01	91.8	5.50
Med 4	25	110	642	5.84	85.6	4.40
Med 5	20	89	528	5.93	88.0	4.45
Surg 1	28	109	798	7.32	95.0	3.89
Surg 2	30	120	783	6.53	87.0	4.00
Surg 3	25	108	638	5.91	85.1	4.32
Surg 4	25	112	690	6.16	92.0	4.48
GYN	25	154	668	4.34	89.1	6.16
Ped	30	212	705	3.33	78.3	7.07
Obstet	18	61	243	3.98	45.0	3.39
Nursery	20	65	330	5.08	55.0	3.25
Total	333	1,618	8,439	5.22	84.5	4.86

per hour (arrival rate, λ) and if each member of the medical staff can treat an average of six cases per hour (service rate, μ), the parameter appearing on the left hand side of the table is $r = \lambda/\mu = 18/6 = 3$.

Referring to the row specified by $r = 3$, we find five columns in the feasible region for M values 4, 5, 6, 7, or 8. M indicates the number of providers (medical staff), which is a decision variable under management control. The values in the table indicate the average length of the queue. For example, the average length of the queue associated with four and eight providers is 1.5282 and .0077, respectively. The latter value indicates that a patient rarely, if ever, would be forced to wait to see a provider. Special formulas are used to calculate waiting time and the other factors. This information can be used to evaluate the effect of choices available to the manager.

Finally, there is a class of tables called "statistical" tables that are useful in judging the extent to which the variation in a set of data represents random chance or the operation of some nonrandom factor. The most com-

Table 2-3 Queuing System—Average Length of Queue: Example of a Reference Table

VALUES OF L_q FOR $M = 1 - 15$, AND VARIOUS VALUES OF $r = \lambda/\mu$.
POISSON ARRIVALS, NEGATIVE EXPONENTIAL SERVICE TIMES

	Number of Service Channels, M														
r	1	2	3	4	5	6	7	8	9	10	11	12	13	14	15
0.10	0.0111														
0.15	0.0264	0.0008													
0.20	0.0500	0.0020													
0.25	0.0833	0.0039													
0.30	0.1285	0.0069													
0.35	0.1884	0.0110													
0.40	0.2666	0.0166													
0.45	0.3681	0.0239	0.0019												
0.50	0.5000	0.0333	0.0030												
0.55	0.6722	0.0449	0.0043												
0.60	0.9000	0.0593	0.0061												
0.65	1.2071	0.0767	0.0084												
0.70	1.6333	0.0976	0.0112												
0.75	2.2500	0.1227	0.0147												
0.80	3.2000	0.1523	0.0189												
0.85	4.8166	0.1873	0.0239	0.0031											
0.90	8.1000	0.2285	0.0300	0.0041											
0.95	18.0500	0.2767	0.0371	0.0053											
1.0		0.3333	0.0454	0.0067											
1.2		0.6748	0.0904	0.0158											
1.4		1.3449	0.1778	0.0324	0.0059										
1.6		2.8444	0.3128	0.0604	0.0121										
1.8		7.6734	0.5320	0.1051	0.0227	0.0047									
2.0			0.8888	0.1739	0.0398	0.0090									
2.2			1.4907	0.2770	0.0659	0.0158									
2.4			2.1261	0.4305	0.1047	0.0266	0.0065								
2.6			4.9322	0.6581	0.1609	0.0426	0.0110								
2.8			12.2724	1.0000	0.2411	0.0659	0.0180								
3.0				1.5282	0.3541	0.0991	0.0282	0.0077							
3.2				2.3856	0.5128	0.1452	0.0427	0.0122							
3.4				3.9060	0.7365	0.2085	0.0631	0.0189							
3.6				7.0893	1.0550	0.2947	0.0912	0.0283	0.0084						
3.8				16.9366	1.5184	0.4114	0.1292	0.0412	0.0127						
4.0					2.2164	0.5694	0.1801	0.0590	0.0189						
4.2					3.3269	0.7837	0.2475	0.0827	0.0273	0.0087					
4.4					5.2675	1.0777	0.3364	0.1142	0.0389	0.0128					
4.6					9.2885	1.4867	0.4532	0.1555	0.0541	0.0184					
4.8					21.6384	2.0708	0.6071	0.2092	0.0742	0.0260					
5.0						2.9375	0.8102	0.2786	0.1006	0.0361	0.0125				
5.2						4.3004	1.0804	0.3680	0.1345	0.0492	0.0175				
5.4						6.6609	1.4441	0.5871	0.1779	0.0663	0.0243	0.0085			
5.6						11.5178	1.9436	0.6313	0.2330	0.0883	0.0330	0.0119			
5.8						26.3726	2.6481	0.8225	0.3032	0.1164	0.0443	0.0164			
6.0							3.6828	1.0707	0.3918	0.1518	0.0590	0.0224			
6.2							5.2979	1.3967	0.5037	0.1964	0.0775	0.0300	0.0113		
6.4							8.0768	1.8040	0.6454	0.2524	0.1008	0.0398	0.0153		
6.6							13.7692	2.4198	0.8247	0.3222	0.1302	0.0523	0.0205		
6.8							31.1270	3.2441	1.0533	0.4090	0.1666	0.0679	0.0271	0.0105	
7.0								4.4471	1.3471	0.5172	0.2119	0.0876	0.0357	0.0141	
7.2								6.3135	1.7288	0.6521	0.2677	0.1119	0.0463	0.0187	
7.4								9.5102	2.2324	0.8202	0.3364	0.1420	0.0595	0.0245	0.0097
7.6								16.0379	2.9113	1.0310	0.4211	0.1789	0.0761	0.0318	0.0129
7.8								35.8956	3.8558	1.2972	0.5250	0.2243	0.0966	0.0410	0.0168
8.0									5.2264	1.6364	0.6530	0.2796	0.1214	0.0522	0.0220
8.2									7.3441	2.0736	0.8109	0.3469	0.1520	0.0663	0.0283
8.4									10.9592	2.6470	1.0060	0.4288	0.1891	0.0834	0.0361
8.6									18.3223	3.4160	1.2484	0.5286	0.2341	0.1043	0.0459
8.8									40.6824	4.4806	1.5524	0.6501	0.2885	0.1298	0.0577
9.0										6.0183	1.9368	0.7980	0.3543	0.1603	0.0723
9.2										8.3869	2.4298	0.9788	0.4333	0.1974	0.0899
9.4										12.4189	3.0732	1.2010	0.5287	0.2419	0.1111
9.6										20.6160	3.9318	1.4752	0.6437	0.2952	0.1367
9.8										45.4769	5.1156	1.8165	0.7827	0.3588	0.1673
10.0											6.8210	2.2465	0.9506	0.4352	0.2040

Under-Utilization Region (upper right of table)

Over-Utilization (Impossible) Region (lower left of table)

Source: E. S. Boffa, Modern Production Management. Wiley/Hamilton, 1977 (Reprinted with permission of John Wiley and Sons, Inc.)

monly used among these tables are the "normal" (see Table 2-4), the "*t*," the "chi-square" (the "*ch*" is pronounced as a "*k*"), and the "*F*" tables.

In summary, we have described repository and analytical tables as those containing data that result from performing a study, taking a census, or conducting research. Reference tables were described as those containing information that results from a theoretical analysis of a given class of situations such as queuing or statistics. The next section considers the classification of tables in terms of the nature of data that are presented.

2.4.2 Classification of Tables by Nature of Data[R,A,F]

In providing an alternate approach to the classification of tables, Clover and Balsley identify qualitative, quantitative, geographical, and time series data as different types of information. In particular, data that are subdivided in qualitative terms reflect differences in *kind.* For example, the categories used in the construction of Tables 2-1 and 2-2 are qualitative in nature and represent differences in kind. Other categorical groupings include males and females, nurses, doctors, lab technicians, secretaries, etc.; X-rays, lab tests, operative procedures, prescriptions, etc.; labor costs, supplies costs, depreciation charges, contracted service costs, etc. All of these groupings represent qualitative distinctions that might appear in reports prepared for hospital administrators.

Data subdivided in *quantitative* terms reflect differences in *amount* or degree. Departments could be classified (for some purposes) by the number of employees; hospitals routinely are classified by number of beds; people in censuses by family income, household size, age, etc.; group practices by the number of physicians, or the number of patients, or patient billings; health maintenance organizations by the number of enrollees or by the cost per enrollee. All of these are examples of quantitative data that are used to classify and describe members of relevant populations.

Data subdivided in *geographical* terms reflect differences in *place.* For example, the manager of an outpatient clinic requires information concerning the characteristics of the population at risk. The required information may be obtained from an appropriate census tract or from enumeration area tabulations of population characteristics. Hospital administrators frequently require information depicting the population residing in their service area. In the United States, many different health care programs have been organized on a geographical basis (e.g., Regional Medical Programs, Medicare/Medicaid, Blue Cross, Blue Shield, and others).

Finally, *time series* data are important when identifying trends in costs, patient characteristics, the number of admissions and visits, the volume of lab tests or radiological examinations, etc. These data are necessary for

Table 2-4 The Normal Distribution Function: Example of a Reference Table

z	.00	.01	.02	.03	.04	.05	.06	.07	.08	.09
0.0	.0000	.0040	.0080	.0120	.0160	.0199	.0239	.0279	.0319	.0359
0.1	.0398	.0438	.0478	.0517	.0557	.0596	.0636	.0675	.0714	.0753
0.2	.0793	.0832	.0871	.0910	.0948	.0987	.1026	.1064	.1103	.1141
0.3	.1179	.1217	.1255	.1293	.1331	.1368	.1406	.1443	.1480	.1517
0.4	.1554	.1591	.1628	.1664	.1700	.1736	.1772	.1808	.1844	.1879
0.5	.1915	.1950	.1985	.2019	.2054	.2088	.2123	.2157	.2190	.2224
0.6	.2257	.2291	.2324	.2357	.2389	.2422	.2454	.2486	.2517	.2549
0.7	.2580	.2611	.2642	.2673	.2704	.2734	.2764	.2794	.2823	.2852
0.8	.2881	.2910	.2939	.2967	.2995	.3023	.3051	.3078	.3106	.3133
0.9	.3159	.3186	.3212	.3238	.3264	.3289	.3315	.3340	.3365	.3389
1.0	.3413	.3438	.3461	.3485	.3508	.3531	.3554	.3577	.3599	.3621
1.1	.3643	.3665	.3686	.3708	.3729	.3749	.3770	.3790	.3810	.3830
1.2	.3849	.3869	.3888	.3907	.3925	.3944	.3962	.3980	.3997	.4015
1.3	.4032	.4049	.4066	.4082	.4099	.4115	.4131	.4147	.4162	.4177
1.4	.4192	.4207	.4222	.4236	.4251	.4265	.4279	.4292	.4306	.4319
1.5	.4332	.4345	.4357	.4370	.4382	.4394	.4406	.4418	.4429	.4441
1.6	.4452	.4463	.4474	.4484	.4495	.4505	.4515	.4525	.4535	.4545
1.7	.4554	.4564	.4573	.4582	.4591	.4599	.4608	.4616	.4625	.4633
1.8	.4641	.4649	.4656	.4664	.4671	.4678	.4686	.4693	.4699	.4706
1.9	.4713	.4719	.4726	.4732	.4738	.4744	.4750	.4756	.4761	.4767
2.0	.4772	.4778	.4783	.4788	.4793	.4798	.4803	.4808	.4812	.4817
2.1	.4821	.4826	.4830	.4834	.4838	.4842	.4846	.4850	.4854	.4857
2.2	.4861	.4864	.4868	.4871	.4875	.4878	.4881	.4884	.4887	.4890
2.3	.4893	.4896	.4898	.4901	.4904	.4906	.4909	.4911	.4913	.4916
2.4	.4918	.4920	.4922	.4925	.4927	.4929	.4931	.4932	.4934	.4936
2.5	.4938	.4940	.4941	.4943	.4945	.4946	.4948	.4949	.4951	.4952
2.6	.4953	.4955	.4956	.4957	.4959	.4960	.4961	.4962	.4963	.4964
2.7	.4965	.4966	.4967	.4968	.4969	.4970	.4971	.4972	.4973	.4974
2.8	.4974	.4975	.4976	.4977	.4977	.4978	.4979	.4979	.4980	.4981
2.9	.4981	.4982	.4982	.4983	.4984	.4984	.4985	.4985	.4986	.4986
3.0	.4987	.4987	.4987	.4988	.4988	.4989	.4989	.4989	.4990	.4990

Source: This table is based on Table 1 of *Biometrika Tables for Statisticians. Volume 1,* 3rd ed. Cambridge: University Press, 1966, by permission of the *Biometrika* trustees.

planning operations in the short run as well as in the near and distant future. For example, staffing on a day-by-day and month-by-month basis requires a knowledge of expected patient volumes and the services required by those patients. On the other hand, planning the acquisition of plant and equipment requires similar but less detailed information for longer periods of time.

Any given table can be classified by the most important subdivision of the data. Some tables are predominantly oriented to one type of information. However, many tables have a number of types of data and several intended uses. Each of the intended uses may be oriented toward a different type of data, which may be represented using a qualitative, quantitative, geographical, and time series classification. For such tables there may be no value in specifying the most important type of data.

2.5 TABLE CONSTRUCTION(R,G,A,F)

Appropriate arrangement of information in tables enhances the ease of comprehension. Although guidelines are suggested here, they are not inviolable, and the overriding principle is always ease of comprehension. However, the writer must assume the position of the reader in order to judge comprehensibility. The United States *Bureau of the Census Manual of Tabular Presentation* (1949) gives a very detailed discussion of the principles of presentation. The following presentation draws heavily upon both it and the Clover and Balsley work.

2.5.1 The Parts of a Table(R,G,F)

Every table in a report or book should contain all the parts indicated by the classification in Table 2-5 and illustrated in Figure 2-3. The *heading* (Age of all persons . . .) is the means by which the table is identified in the table of contents, indicating the table's general content. The heading also is the link between the discussion in the text and the data in the table. By convention, the table is referred to in the text by the table number, and the title should clearly establish the appropriateness to the context in which the data are discussed.

The data under consideration are presented in the *body* of the table. The body also must identify the types and classifications of the data that define both the rows and columns. Informative headings identifying the content or classification of the rows are arranged in the *stub* that appears on the left side of the table. The *block headings* (center headings, if required) indicate major groups of related rows and the line captions provide the de-

Table 2-5 Classification of the Parts of a Table

A. HEADING
   1. Table Number
   2. Major Title
      a. Subtitle, if any
      b. Designation of units
B. BODY
   1. Stub (the classifications of the rows of the table)
      a. Stub Head
      b. Block (captions for a selfcontained group of lines, often a subclassification)
         (i) Block Heading
         (ii) Line Captions
   2. Boxhead (the classifications of the columns of the table)
      a. Captions
         (i) Column Heads
         (ii) Spanner Heads
      b. Panel (captions for a selfcontained group of columns, often a subclassification)
   3. Field
      a. Rows
      b. Columns
      c. Cells
C. NOTATIONS
   1. Footnotes
   2. Source

Sources: U.S. Bureau of the Census: *Bureau of the Census Manual of Tabular Presentation,* by Bruce L. Jenkinson, Washington, D.C.: U.S. Government Printing Office, 1949.
V. Clover and H. Balsley, *Business Research Methods,* Columbus, Ohio: Grid, Inc., 1974, chap. 12, pp. 325–327 (modified to conform to the Bureau of the Census).

tailed categories of the rows. The *boxhead* across the top of the table indicates the column classifications by means of column heads and *spanner heads*. The latter are used to show groupings of several related columns in the same way as block headings. A *panel* consists of a spanner head and its related column heads. The *field*, on the other hand, consists of all the rows and columns of data, excluding stub and boxhead (i.e., only the data). Finally, the *notation* section gives supplementary footnotes about items in the table and/or the source of the information presented.

2.5.2 Tables with Two or More Dimensions[R,A,F]

The table in Figure 2-3 is interesting because it illustrates the presentation of a number of different categories or "dimensions" of the data. The *two basic dimensions* of the presentation of the data are the seven age groups (under 5; 5–14; 15–24; 25–34; 35–44; 45 and over, as well as 21 and over) and sex (male and female), and their respective totals. A *third dimension* is added by grouping the place of residence into urban and rural (and the total). This is accomplished by having three blocks in the stub (one each for Urban, Rural, and Total U.S.) with identical (repeated) stub items for each block. A *fourth dimension*, citizenship, is added by creating a panel for citizens and another for "all persons" (i.e., citizens plus noncitizens). Three of these dimensions are "qualitative" (sex, place of residence, and citizenship) while the fourth is "quantitative" (age group). Such a table is extremely useful for examining variations in the basic age-sex relationship while controlling for place of residence and/or citizenship.

The use of multiple blocks and panels is an easy way to extend the complexity of the relationship portrayed in the table. These techniques can be considered when more than two variables (dimensions) may be involved in the relationship. This method of presenting data is closely related to the topic of causal analysis discussed later when we examine relationships in greater depth. In addition, the use of multiple blocks and panels is related to the construction of contingency tables, also described later.

2.5.3 Guidelines for Table Arrangement[R,G,A,F]

The field consists of rows and columns of data. The first and last rows and/or columns are preferred positions for the emphasis of particular data. Government statistical reports usually present the overall total *first*, followed by the detail (as in Figure 2-3), but other organizations usually prefer to list the overall category last. In this regard, psychological studies have shown that items near the beginning or the end of a list are remembered more easily than those in the middle. Subtotals of blocks or panels in the middle of the table can be emphasized by drawing single and/or double separator lines through the field.

Normally the data are arranged so that the information progresses naturally from top to bottom and left to right. This is facilitated by having more rows than columns. Frequently, the data appearing in rows or columns are presented for comparative purposes and comparisons are facilitated by placing values that are to be compared side by side.

Clover and Balsley suggest a number of different methods of sequencing the rows and columns of a table. Remember that the rows and columns

Figure 2-3 Sample Table Indicating Names and Locations of All Parts of the Table

HEADING →

TABLE 6.—AGE OF ALL PERSONS AND OF CITIZENS BY SEX, FOR THE UNITED STATES, URBAN AND RURAL: 1940

[Age classification based on completed years]

Area and age	All persons			Citizens [1]		
	Total	Male	Female	Total	Male	Female
UNITED STATES						
All ages				769		
Under 5 years				26		
5 to 14 years				115		
15 to 24 years				139		
25 to 34 years				178		
35 to 44 years				205		
45 and over				106		
21 and over	988	475	513	567	302	265
URBAN						
All ages				453		
Under 5 years				15		
5 to 14 years				73		
15 to 24 years				86		
25 to 34 years				104		
35 to 44 years				116		
45 and over				59		
21 and over				328		
RURAL						
All ages				316		
Under 5 years				11		
5 to 14 years				42		
15 to 24 years				53		
25 to 34 years				74		
35 to 44 years				89		
45 and over				47		
21 and over				239		

← BOXHEAD STUB → ← FIELD

[1] Includes both native and naturalized.

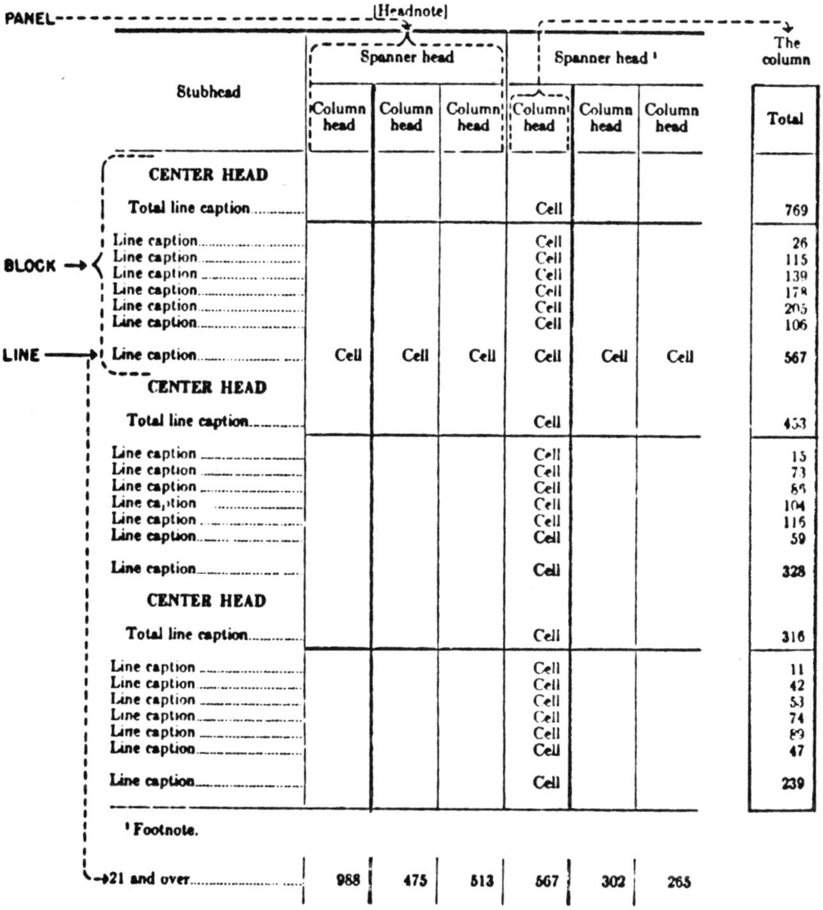

TABLE No.—TITLE OF TABLE

[Headnote]

PANEL — The column

BLOCK — LINE

Stubhead	Spanner head			Spanner head[1]			Total
	Column head	Column head	Column head	Column head	Column head	Column head	
CENTER HEAD							
Total line caption				Cell			769
Line caption				Cell			26
Line caption				Cell			115
Line caption				Cell			139
Line caption				Cell			178
Line caption				Cell			205
Line caption				Cell			106
Line caption	Cell	Cell	Cell	Cell	Cell	Cell	567
CENTER HEAD							
Total line caption				Cell			453
Line caption				Cell			15
Line caption				Cell			73
Line caption				Cell			65
Line caption				Cell			104
Line caption				Cell			116
Line caption				Cell			59
Line caption				Cell			328
CENTER HEAD							
Total line caption				Cell			316
Line caption				Cell			11
Line caption				Cell			42
Line caption				Cell			53
Line caption				Cell			74
Line caption				Cell			89
Line caption				Cell			47
Line caption				Cell			239

[1] Footnote.

21 and over	988	475	513	567	302	265

Source: U.S. Bureau of the Census: *Bureau of the Census Manual of Tabular Presentation*, by Bruce L. Jenkinson, U.S. Government Printing Office, Washington, D.C., 1949, p. 10-11.

Table 2-6 Distribution of Number of Patients by Number of Diagnoses: Example of Offsetting Percentages Diagonally Downward

Group	Number of Diagnoses (NUMDIAG) 1	2	3	4	5	Total
	# Pat. (%)	# Pat. (%)	# Pat. (%)	# Pat. (%)	# Pat. (%)	# Pat. (%)
IRBL	529 (74.7)	98 (13.8)	51 (7.2)	24 (3.4)	6 (.8)	708 (100.0)
ULCR	43 (38.1)	38 (33.6)	19 (16.8)	9 (8.0)	4 (3.5)	113 (100.0)
LIVR	38 (34.9)	37 (33.9)	17 (15.6)	10 (9.2)	7 (6.4)	109 (100.0)
GALL	56 (33.3)	48 (28.6)	33 (19.6)	22 (13.1)	9 (5.4)	168 (100.0)
CANC	27 (32.9)	23 (28.0)	14 (17.1)	9 (11.0)	9 (11.0)	82 (100.0)
BNGN	68 (43.6)	43 (27.6)	31 (19.9)	11 (7.1)	3 (1.9)	156 (100.0)
Total	761 (57.0)	287 (21.5)	165 (12.4)	85 (6.4)	38 (2.8)	1336 (100.0)

Source: Disease Costing in an Ambulatory Clinic—Disease and Physician Profiles and the Selection of Patients for Review, a Ph.D. dissertation by Colin M. Lay, © 1978.

represent subdivisions of basic dimensions. Quantitative data often are arranged in order of size (smallest to largest, or vice versa). Qualitative data subdivisions may be arranged alphabetically, in some customary order (e.g., Medicine, Surgery, Gynecology, · · ·), or with the one for which emphasis is desired placed first, highlighted, with the others following in one of the preceding arrangements. Geographical subdivisions usually are arranged in a customary order for regions of the country, states or provinces, or counties. They also may be arranged alphabetically. Time series data usually are grouped in terms of an appropriate time interval (i.e., days, weeks, months, and years).

Sometimes the number of observations in different groups will be very different—disparities that may obscure a relationship concerning the composition of the groups. The composition of one or more groups is better shown by percentages, but the absolute numbers also may be of interest. In a disease costing study, Lay used the number of diagnoses registered for any patient as one surrogate for the severity of the patient's problem, within each of six problem groups. There were substantial differences in the number of patients in each group and in the distributions of the number of diagnoses within each group. Table 2-6 shows one method of presenting both raw data and percentages in the same table, emphasizing the difference by offsetting the percentage figures along the diagonal and enclosing them in parentheses.

Chapter 3

Graphic Presentation of Data

Objectives

After completing this chapter you should be able to:

1. Describe the types of charts used to display numeric information graphically;
2. Identify and describe the components of charts;
3. Construct line charts, bar charts, histograms, ogives, scattergrams, pie charts, radial charts, and maps.

Chapter Map

The sections comprising this chapter may be summarized as follows:

Section Number	Required Reading	Optional Reading	Generic Development	Application to Management	Fundamental Principles	Complex Material
	(R)	(O)	(G)	(A)	(F)	(C)
3.1	x		x	x	x	
3.2	x		x	x	x	
3.3	x		x	x	x	
3.4	x		x	x	x	
3.5	x		x	x	x	
3.6	x		x	x	x	
3.7	x		x	x	x	
3.8	x		x	x	x	
3.9		x	x		x	

This chapter continues the discussion of methods that may be used to present numerical information. However, unlike the previous chapter, the focus here is on the graphic display of numeric information.

3.1 THE ADVANTAGES OF GRAPHIC PRESENTATION(R,G,A,F)

Although tables are extremely important for storing and reporting information, the more data that are contained in a table, the less comprehensible it is. In such a case, it is useful to turn to a graphic portrayal of the information. The essential relationships may be highlighted in a chart or a graph while the obscuring details are suppressed. The old saying can be updated and paraphrased to read: *A graph is better than a thousand numbers*. Since a table of numbers is frequently the source of data on which a chart is constructed, tabular and graphic displays of information complement one another. Since a chart tends to suppress detail, a well-chosen graphic display appeals to visual senses and usually is easy to comprehend. On the other hand, a table can be used to retain detailed information for reference purposes.

3.2 LINE CHARTS(R,G,A,F)

This section limits the discussion to the simplest and perhaps the most popular approaches to the construction of line charts. In Chapter 4, where we examine logarithms, we augment the following discussion and present an alternate approach to the construction of line charts.

Perhaps the most familiar approach to the construction of line charts involves the use of *arithmetic,* or *Cartesian, graph paper.* This paper is ruled vertically and horizontally and there usually are 10 or 20 lines per inch or per centimeter. Usually every fifth or tenth line is accented (darker). Two axes (X and Y) must be drawn, with scales chosen so that the points to be graphed will fit conveniently on the paper, almost filling the available space.

Each point (observation) requires a pair of values, or *coordinates* (usually labelled x_i and y_i, where the subscript i differentiates between observations), representing the variables of interest in the situation. For example, suppose we are provided with the following information concerning the use of service, as measured by the annual number of doctor office visits by patients of differing ages. Assuming that we represent the use of care on the vertical axis of the arithmetic graph paper and age on the horizontal axis, we might choose to let y_i represent the number of visits used by patient i and x_i represent the patient's age (see Figure 3-1).

Patient Identification Number	Use (Number of Visits)	Age (in Years)
1	1	10
2	2	15
3	4	18
4	5	21
5	6	24
6	7	26
7	9	30
8	10	42

The scales for the X and Y axes are chosen so that the *origin* (point (0,0)) is included and the most negative values (if any) and the most positive values required can be plotted. In the example neither the number of visits, nor the age can be negative, so only the positive portions of the axes are required. The largest value shown on an axis will usually be the next multiple of 5, 10, 50, 100 or 1000 larger than or equal to the largest x or y value to be plotted. In this case the largest age value is 42 and therefore the X axis is marked up to and including 45, while the Y axis is marked to accommodate a maximum of 10 visits. (Since the graph paper originally used for this chart had 10 divisions to the inch, the X axis scale was set to 10 years per inch, and the Y axis scale was set to 2 visits per inch. These scales made the plotting of individual points very easy.)

Each point is plotted by first counting the appropriate number of units along the X axis, and then up a vertical line the appropriate number of units for the y coordinate. For a line chart such as this one all points are plotted, and then straight lines are drawn connecting the first point to the second, the second to the third, and so on.

Using a similar approach, we may employ the data from Table 2-6 to construct the line chart in Figure 3-2. In this figure, the decline in the absolute number of patients presenting an increasing number of diagnoses is evident. For example, in relative terms, patients presenting Irritable Bowel syndrome decline from approximately 75 percent with one coexisting diagnosis to less than one percent with five coexisting conditions. These examples also illustrate the complementary use of tables and graphs.

Figure 3-1 Line Chart Depicting the Use of Service by Age of Patients

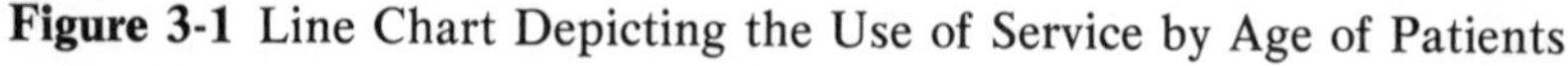

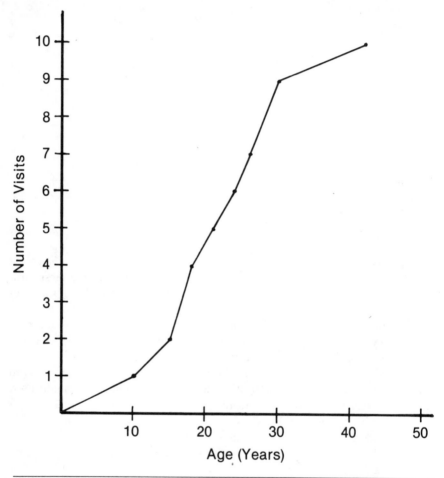

As noted, line charts may be constructed using arithmetic graph paper. However, the background lines tend to obscure the visual clarity of the chart. To avoid this problem, we may construct the line chart on arithmetic paper and then trace the graph onto unlined paper. Such a procedure is illustrated in Figure 3-4, which has been created by tracing Figure 3-3 onto unlined paper. Notice that by eliminating the background lines, the visual confusion in Figure 3-3 is reduced markedly.

Several graphic aids that are available on a commercial basis should be mentioned. In particular, we might have used a graphic tape to construct the

Figure 3-2 Number of Recorded Diagnoses Per Patient by Disease Group

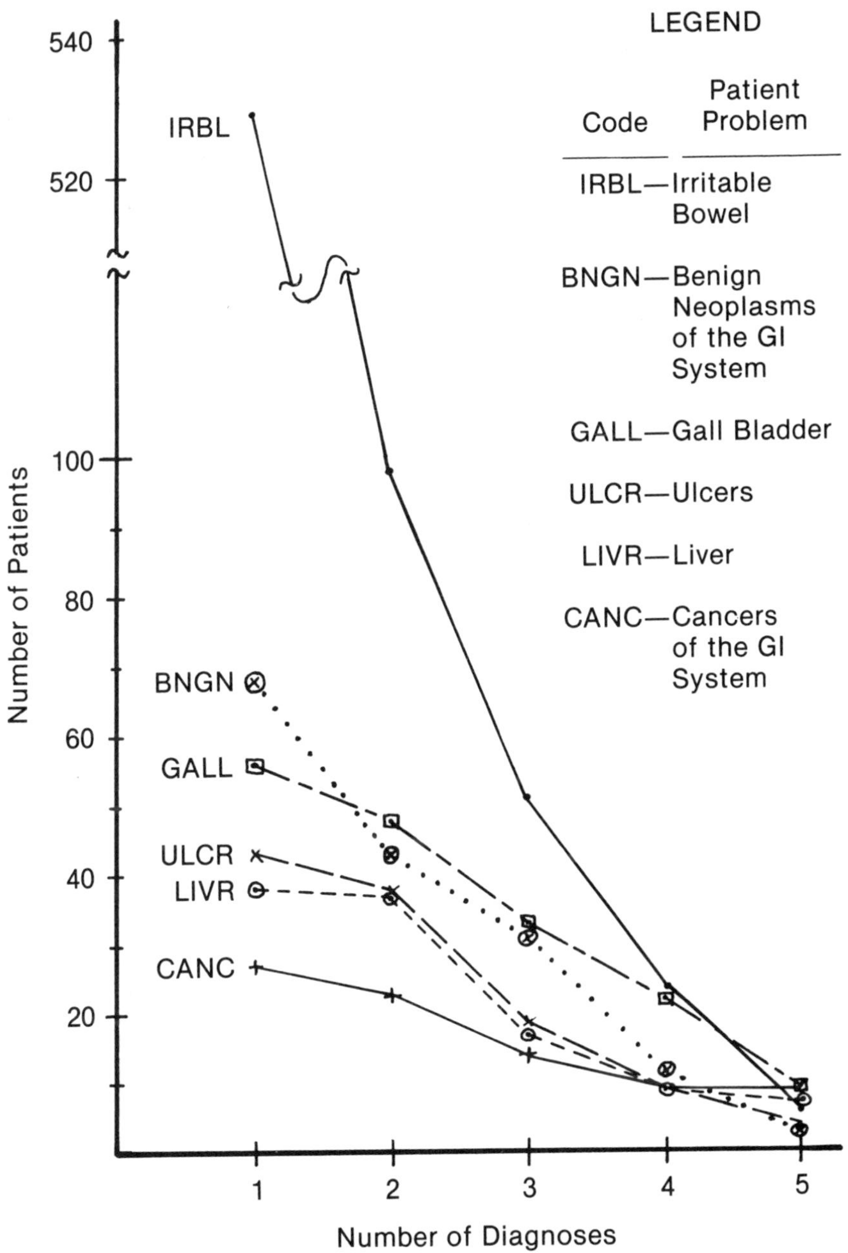

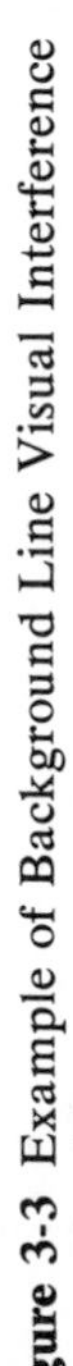

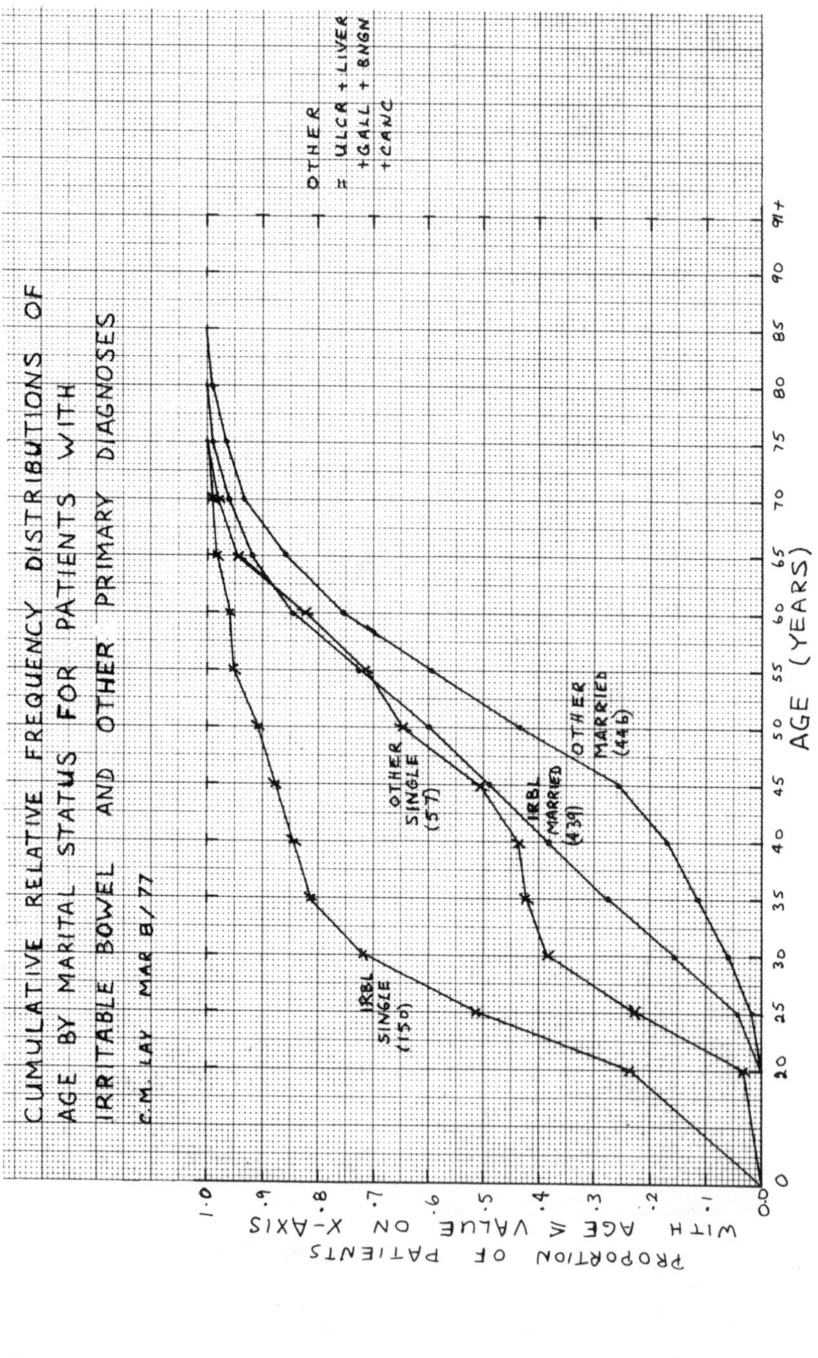

Figure 3-3 Example of Background Line Visual Interference

line charts in Figure 3-4. Figure 3-5 shows examples of the types of tape that might be used to construct line charts. These tapes are available in different widths and designs.

When these tapes are not available, or are not used for whatever reason, the person constructing the chart can distinguish between lines by using a variety of combinations of dots, short and long dashes, solid lines, and point indicators, such as · * x + · ⊙ ⊡ as well as other appropriate symbols similar to those in Figure 3-2. Care should be exercised to ensure that the lines are drawn accurately, sharply, and with good contrast (as black as possible, not with pencil). The use of colored lines in constructing the original chart should be avoided unless all reproductions will also be printed in color. Most photocopying processes reproduce colors poorly as shades of gray, which cannot be distinguished in the copy.

3.3 BAR CHARTS AND HISTOGRAMS[R,G,A,F]

Bar charts are commonly used to display quantitative information (such as numbers, dollars, ages, days of stay, etc.) for various categories of some dimension that may be qualitative or quantitative in nature. For example,

Figure 3-4 Example of Visual Clarity When Background Lines Are Removed

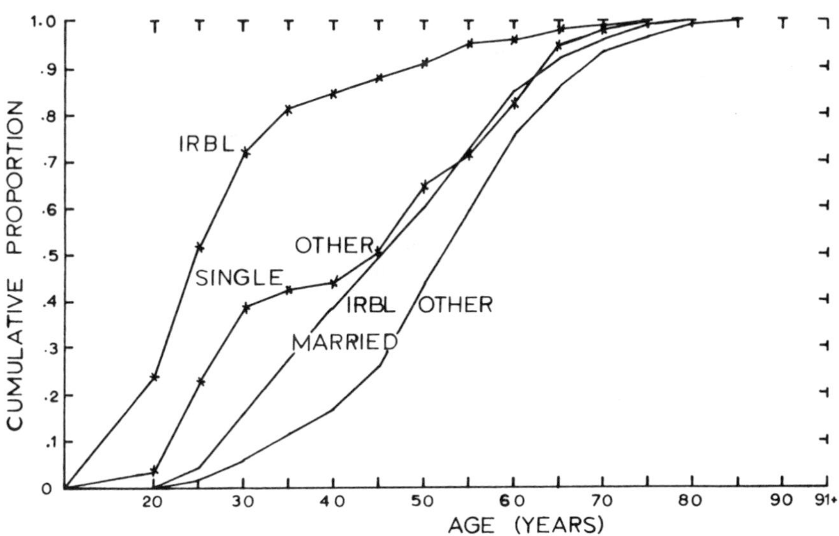

Source: Disease Costing in an Ambulatory Clinic: Disease and Physician Profiles and the Selection of Patients for Review, a Ph.D. dissertation by Colin M. Lay, © 1978.

Figure 3-5 Sample of Graphic Aids

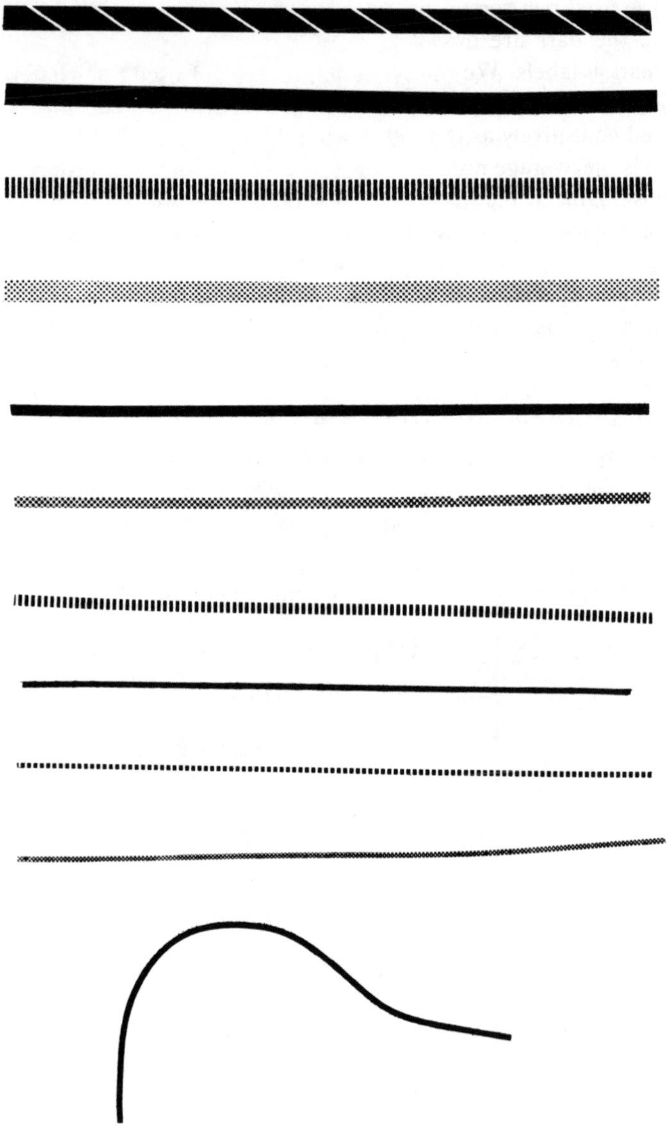

Figure 3-6 shows a hypothetical example of the average number of days of sick leave used per person per year for various groups of employees. In this example the bars are drawn horizontally to facilitate the reproduction of identification labels. We might have constructed the figure by using vertical bars. In both cases, however, the groups are qualitatively different and can be viewed collectively as the "independent" variable, while the "dependent" variable is the average number of days of sick leave. The length of the bar obviously is quantitative, a condition that must be satisfied in any bar chart. In terms of Figure 3-6, the average for each group is given by the length of the bar.

Bar charts also can be constructed to illustrate differences in a given phenomenon relative to zero. For example, consider the problem of presenting graphically the net cash flows of an institution. In this case we define the net cash flow of some period t as the difference between the cash inflows (CI_t) and cash outflows (CO_t) of the period. Hence, the expression

$$NCF_t = CI_t - CO_t$$

Figure 3-6 Example of a Bar Chart: Sick Leave Per Person Per Year

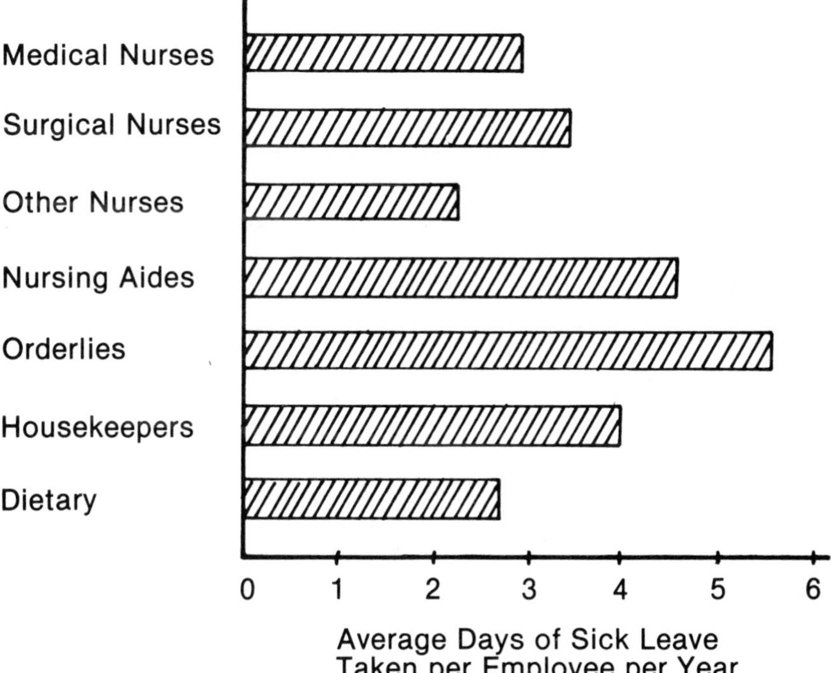

implies that the net cash flows of the period (NCF_t) is: (1) positive when $CI_t > CO_t$, (2) zero when $CI_t = CO_t$, and (3) negative when $CI_t < CO_t$. In Figure 3-7 the zero line corresponds to periods in which cash inflows are equal to cash outflows. Similarly, rectangles a, b, and e represent periods in which a positive net cash flow was realized while rectangles c and d indicate periods of negative net cash flow. Such a bar chart also might be constructed vertically or horizontally.

A *histogram* is a bar chart that usually is based on a frequency distribution. Suppose we are interested in the number of hospitals that reported a per diem charge of \$100–\$119, \$120–\$139, \$140–\$159 and \$160–\$179. Suppose further that we obtain the per diem charges of 200 hospitals and that these data provide the basis for constructing the frequency distribution presented in Table 3-1. In this case, we refer to the number of hospitals in each category as the class frequencies and the values \$100–\$119, \$120–\$139, etc., as the class intervals. We may now employ this information to construct the histogram presented in Figure 3-8. Notice that the height of each rectangle in the histogram is defined by one of the class frequencies appearing in Table 3-1, while the base of each rectangle is defined by the corresponding class interval.

We also may use the data in Table 3-1 to construct a relative frequency distribution. Such a distribution is obtained by dividing each class frequency by the total number of observations. Referring to Table 3-2, we find that the relative frequency of .15 that is associated with the first group of hospitals is obtained by dividing the class frequency of 30 hospitals by the 200 institutions in our study. The other values in Table 3-2 are obtained in a similar fashion. Employing this approach, we may now use the resulting relative fre-

Figure 3-7 Graphic Display of Net Cash Flows

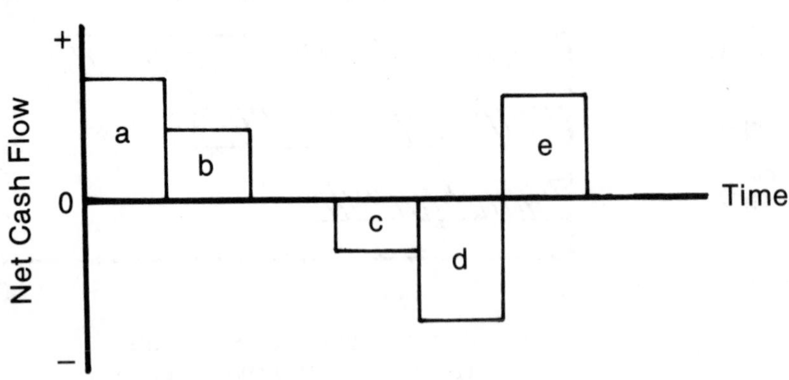

Table 3-1 Distribution of 200 Hospitals in Terms of the Per Diem Charge

Per Diem Charge	Number of Hospitals
\$100–119	30
\$120–139	50
\$140–159	80
\$160–179	40
Total	200

quencies and the corresponding class intervals to construct a histogram that depicts the relative frequency distribution in Table 3-1.

3.4 OGIVES(R,G,A,F)

As noted in the previous section, a frequency distribution indicates the frequency or proportion of observations associated with a given category. Returning to our example, it also might be of interest to determine the number of hospitals that charge less than \$120 per day, less than \$140 per day, and so on. In such a situation, we would develop a cumulative distribution similar to the one in Table 3-3. Table 3-1 shows that 80 of the hospitals (i.e., 50 + 30) charged less than \$140 per day. Similarly, 160 of the hospitals (30 + 50 + 80) charged less than \$160 per day and all the institutions charged less than \$180 per day. These data are recorded in Table 3-3, which is an example of a cumulative frequency distribution. Note that no hospitals charged less than \$100 per day.

We may now employ the data in Table 3-3 to construct a line chart that is called an ogive. As seen in Figure 3-9, the ogive pertaining to our example is obtained by first plotting the coordinates (\$100,0), (\$120,30), (\$140,80), (\$160,160) and (\$180,200). We then connect these points with a series of straight lines that result in an ogive that represents the "less than" cumulative distribution.

We also might be interested in the number of institutions that charge at least \$100 per day, \$120 per day, and so on. Table 3-1 shows that all of the institutions in our study charge \$100 per day or more, 170 charge \$120 per day or more, and so on. By connecting these points with a series of straight lines we may construct an ogive that represents the "or more" cumulative frequency distribution. Referring to Figure 3-10, we see that when an "at least"

Figure 3-8 Example of a Histogram

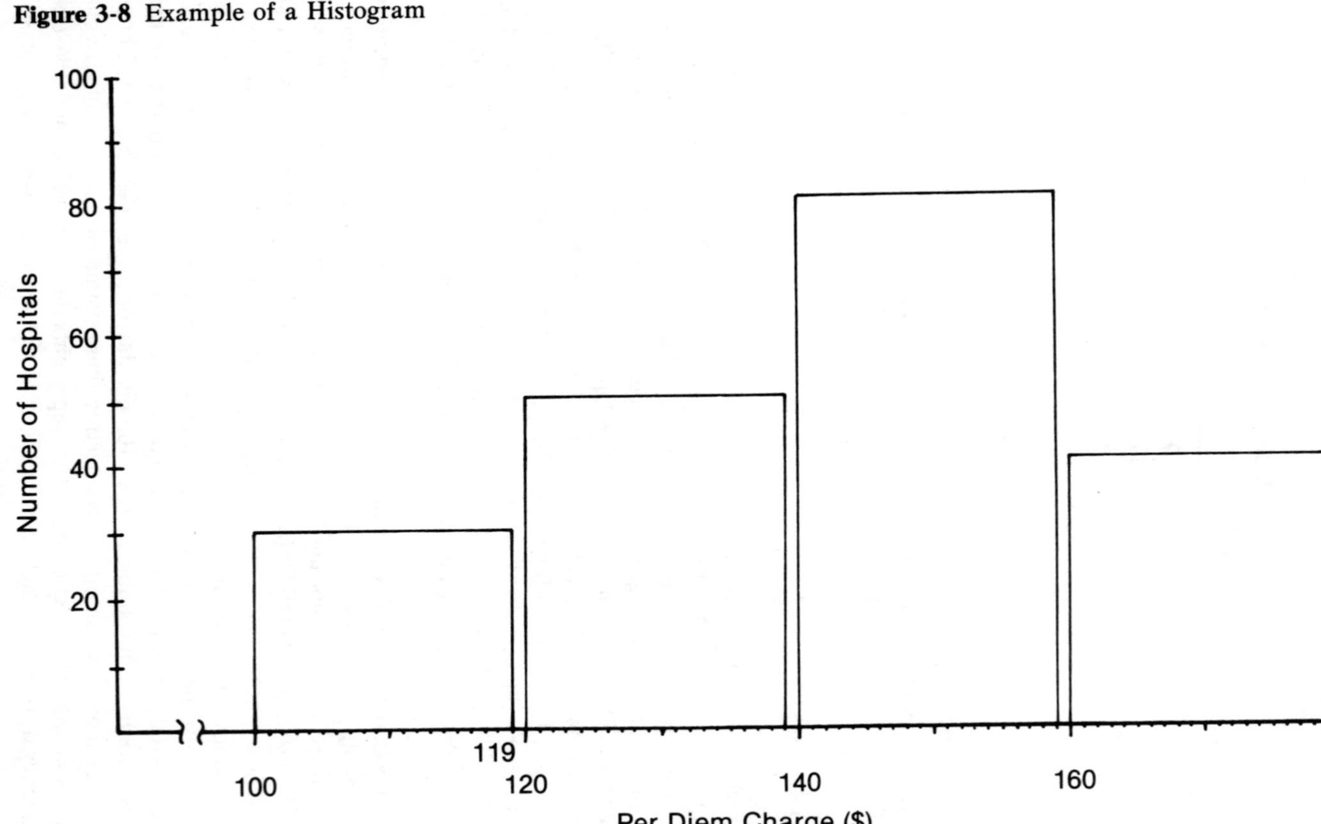

Table 3-2 Example of a Relative Frequency Distribution

Per Diem	Relative Frequency
$100–119	.15
$120–139	.25
$140–159	.40
$160–179	.20
Total	1.00

Table 3-3 Example of a Cumulative Distribution

Per Diem Less Than	Cumulative Frequency
$100	0
$120	30
$140	80
$160	160
$180	200

or an "or more" cumulative distribution is constructed, the cumulative class frequencies decline as the value of the variable recorded on the horizontal axis increases.

3.5 SCATTERGRAM(R,G,A,F)

In the previous sections, line charts and ogives were constructed by connecting successive points with a series of straight lines. In this section we consider scattergrams, which are constructed by simply plotting the paired values associated with two variables of interest. When constructing a scattergram, however, successive points are *not* connected with a series of straight lines; rather, the pattern formed by the points is used to obtain a visual image of the relation between the two variables of interest.

As an example, suppose that management is interested in examining the responsiveness of costs to changes in the rate of activity. Assume further that, after adjusting cost data to eliminate the effects of inflationary pressures,

Figure 3-9 Example of a "Less Than" Cumulative Frequency Distribution

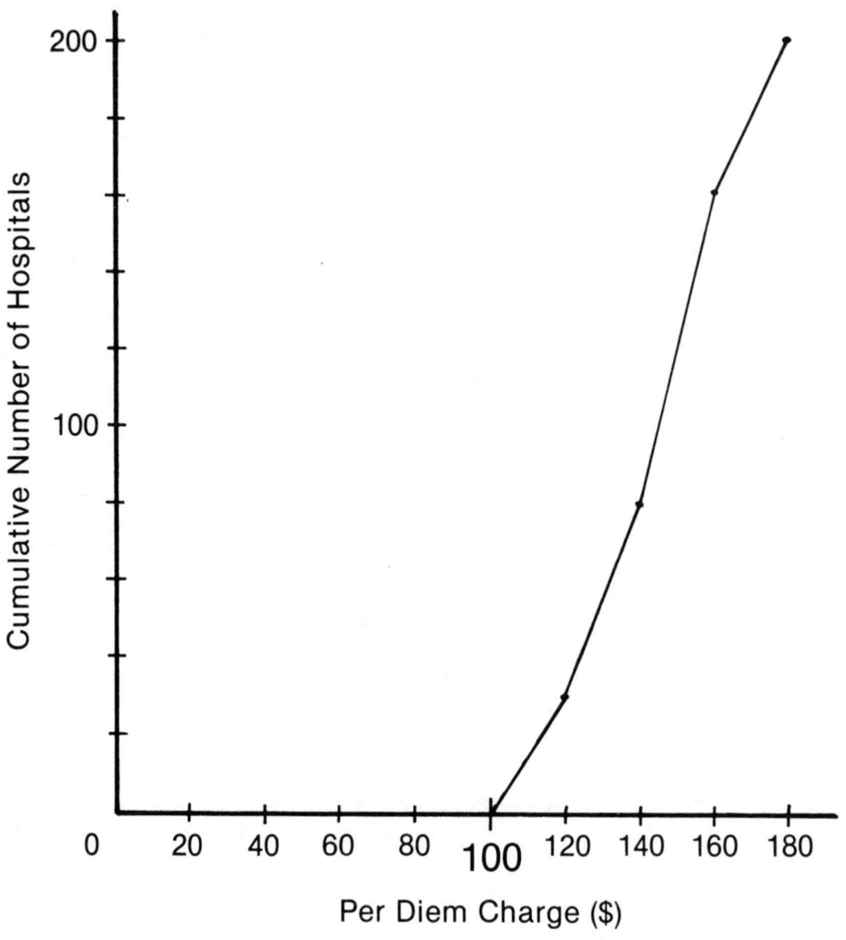

management obtains the information presented in Table 3-4. In this example, we assume that the costs of providing 50 units of care were $5,000, the costs of 70 units were $8,000, and so on. Each pair of values in the table might be represented by the general coordinate (x_i, y_i) where x_i corresponds to a given rate of activity and y_i represents the corresponding costs. Plotting all such coordinates, we obtain the pattern of the relation in Figure 3-11. Such a figure is referred to as a scattergram and may be used, in this example, to obtain a visual image of the general relation between costs and the rate of activ-

Figure 3-10 Example of an "Or More" Cumulative Frequency Distribution

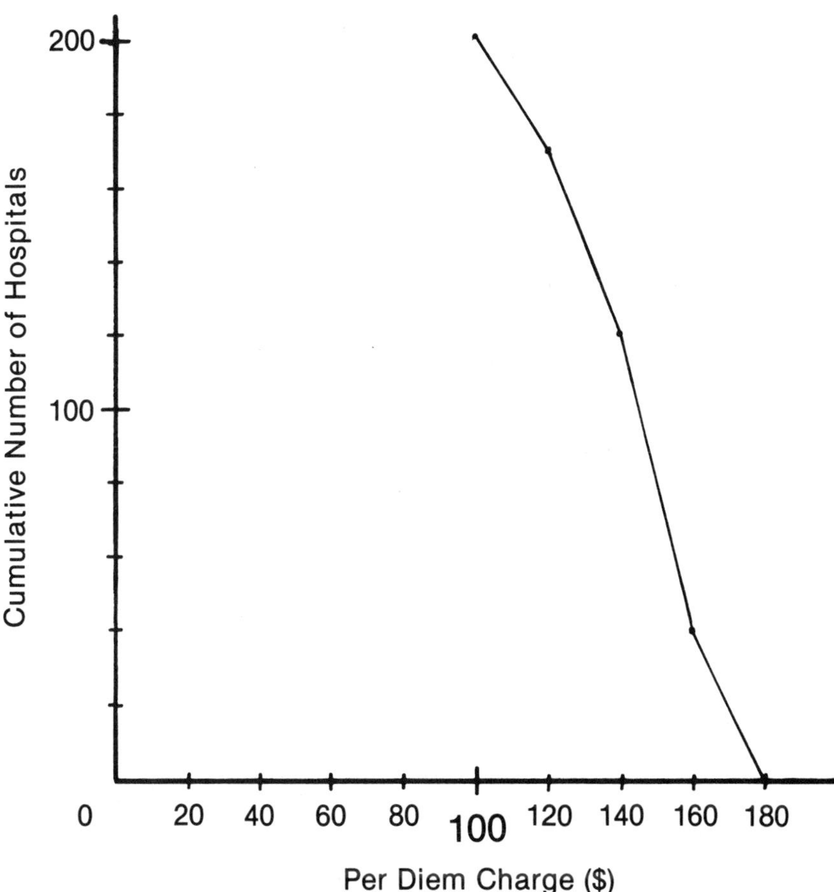

ity. Notice that the pattern of the relation indicates that costs rise as the rate of activity is increased. We also might employ a straight line to approximate this relation even though all of the paired values (x_i, y_i) would not correspond to a point on the line.

As another example, in Figure 3-12 the relation between the age of the patient and the costs of providing care has been displayed by constructing a scattergram. These data pertain to a group of patients who were treated for a single condition (Irritable Bowel syndrome) and suggest that there is a slight relationship between the age of the individual and the total cost of providing care.

Table 3-4 Costs of Sustaining Different Rates of Activity

Rate of Activity (in Units)	Costs (in Dollars)
50	$5,000
70	$8,000
90	$9,000
110	$12,000
130	$16,000
160	$20,000
220	$26,000
250	$29,000
260	$32,000
300	$38,000

Figure 3-11 Scattergram Portraying the Relation between Costs and the Rate of Activity

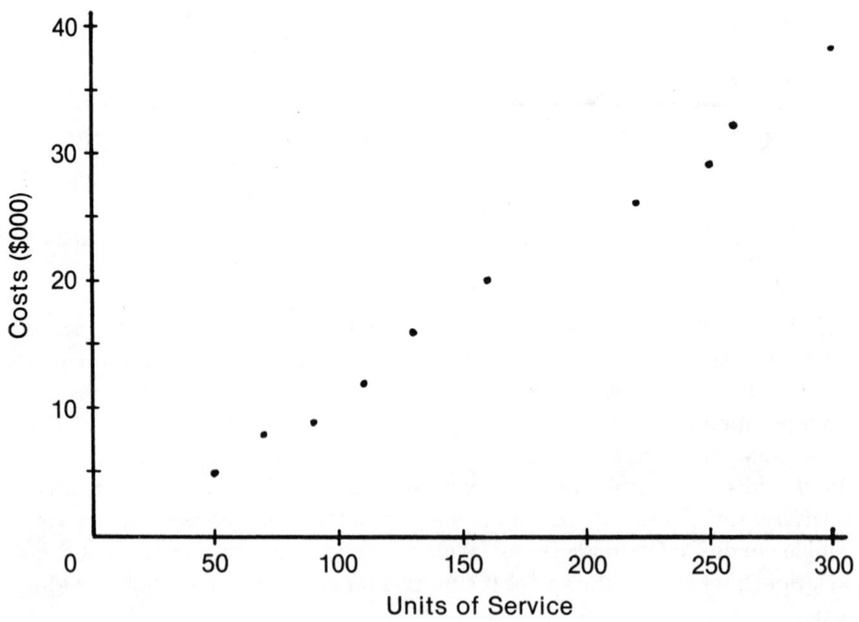

Figure 3-12 Scattergram of Cost-Age Relationship

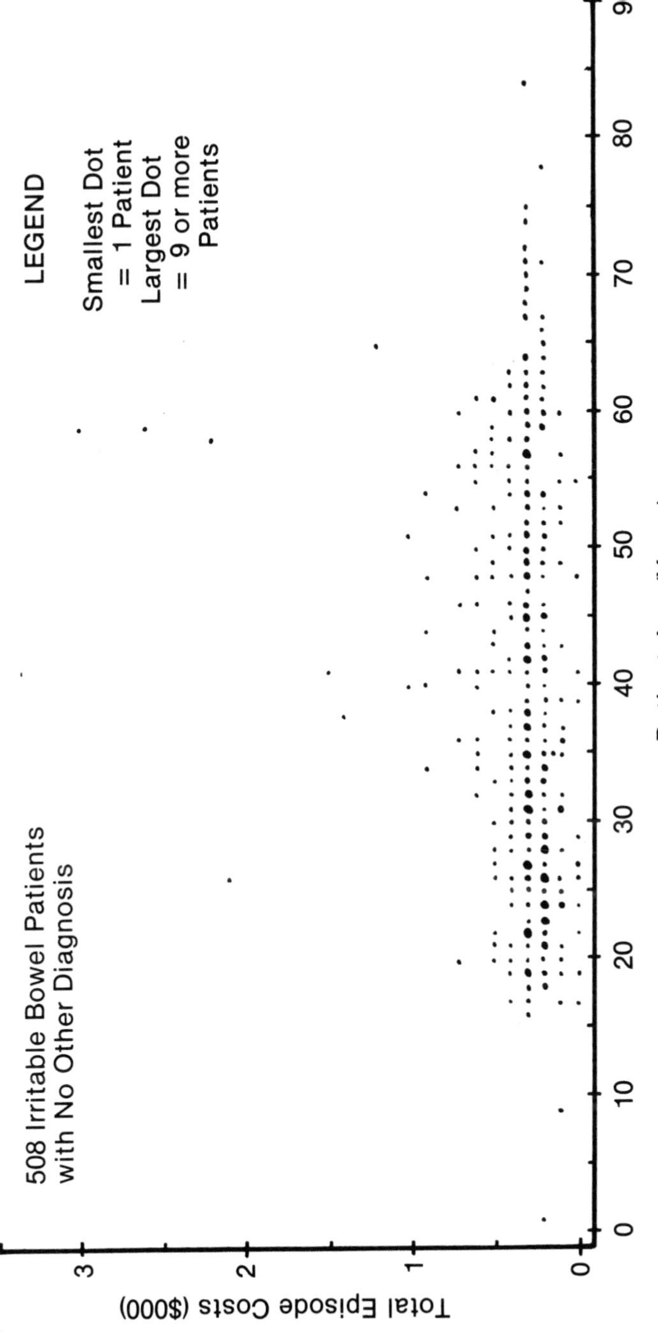

Source: Disease Costing in an Ambulatory Clinic-Disease and Physician Profiles and the Selection of Patients for Review, a Ph.D. dissertation by Colin M. Lay, © 1978.

3.6 PIE CHARTS(R,G,A,F)

This section considers the use of circle diagrams or pie charts to represent the component parts of some totality. In this case, the total area of the circle, which is given by $A = \pi r^2$, represents the totality of the phenomenon of interest. The circle is then partitioned to reflect the relative importance of each component comprising the whole. For example, if we are interested in the expenditures on all forms of health care (e.g., physician care, dental care, hospital care, etc.), the area of the curve would represent total expenditures on medical care during a specified period. Figure 3-13 shows that each component of the circle represents the relative importance of expenditures on physician care, dental care, hospital care, and so on. Also notice that the relative importance of each component usually is expressed in percentage terms and that the components sum to 100 percent. The angle for the sector representing each component is calculated by multiplying 360 degrees by the appropriate percentage.

3.7 RADIAL CHARTS(R,G,A,F)

The performance of any department, hospital, or other organizational unit may be evaluated in terms of several dimensions. Some or all of these dimensions may require quite different measurements that, when expressed in natural or epidemiological units, are not comparable. Measurements not directly comparable include length of stay, employee satisfaction, patient satisfaction, length of waiting lists, cancellations, occupancy rate, and the number of nosocomial infections. In order to evaluate the performance of the institution in several dimensions, it frequently is necessary to use a general scale that indicates the extent to which operations have resulted in acceptable outcomes. To evaluate the activity of the institution as acceptable, it often is necessary to maintain a balance among the various indicators of performance.

As an example, suppose we are interested in evaluating a nursing unit in terms of the following dimensions:

1. patient satisfaction,
2. employee morale,
3. quality of care,
4. efficiency of operations.

Notice that the four dimensions cannot be expressed in terms of a common unit of analysis and, as a result, we decide to evaluate each of these dimensions by employing the following rating scheme:

Evaluation		Score
Unacceptable	—requires immediate corrective action	1
Poor	—requires corrective action	2
Acceptable	—performance is within normal range	3
Good	—performance is above normal range	4
Outstanding	—performance is exemplary	5

Now suppose that after evaluating the unit in July and August we obtain the following information:

	Rating	
Dimension	July	August
Patient Satisfaction	3	4
Employee Morale	2	5
Quality of Care	2	4
Efficiency of Operations	3	5

We may now display these results as seen in Figures 3-14(a) and 3-14(b). The first of these figures indicates that remedial action is required to improve the quality of care as well as the morale of employees. When compared with Figure 3-14(a), the second figure indicates the rather striking improvement in the performance of the unit.

3.8 MAPS(R,G,A,F)

Health care, urban, and social service planners all benefit from the presentation of important information in the form of maps. For example, the pattern of discharges from an active treatment hospital in an urban area can help institutional managers in identifying service areas as well as in understanding the characteristics of the patients requiring care. Similarly, a set of maps depicting the pattern of discharges from all active treatment hospitals in an urban area is of obvious value to regional planners.

In this regard, health planners would be interested in the entire set of maps presented in Figure 3-15, while hospital administrators probably would be more interested in the map that depicts the service areas of their institutions. The set of maps in this figure was prepared for hospitals in the Ottawa

Figure 3-13 Pie Chart Depicting the Relative Importance of Expenditure on Various Forms of Health Care

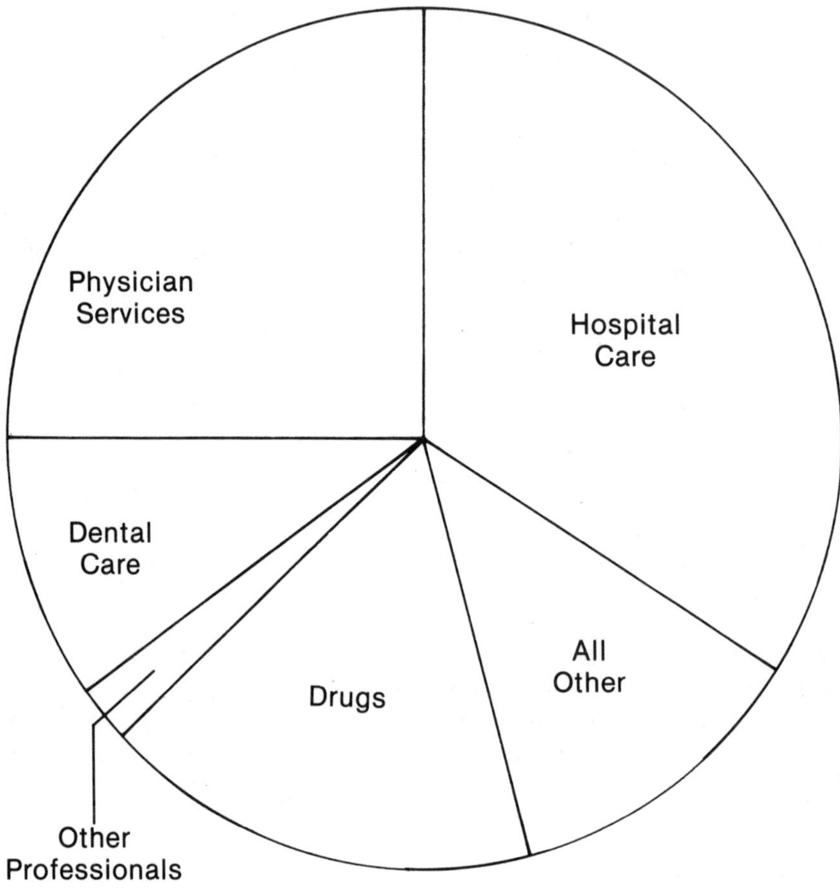

Source: B. S. Cooper and N. L. Worthington, *National Health Expenditures, 1929–1970,* Research and Statistic Note, no. 1, 1972, Social Security Administration, Office of Research and Statistics, January 1972, Table 9.

---

area, based on a sample of discharges during 1972. In this case, the patients are represented as dots on the map.[1] With an appropriate system of data collection, similar maps might be prepared either annually or at five-year intervals to illustrate changes in the service pattern of the area.

[1] The authors gratefully acknowledge the efforts of Professor D. Douglas of the University of Ottawa, who prepared these maps.

Figure 3-14(a) July Performance

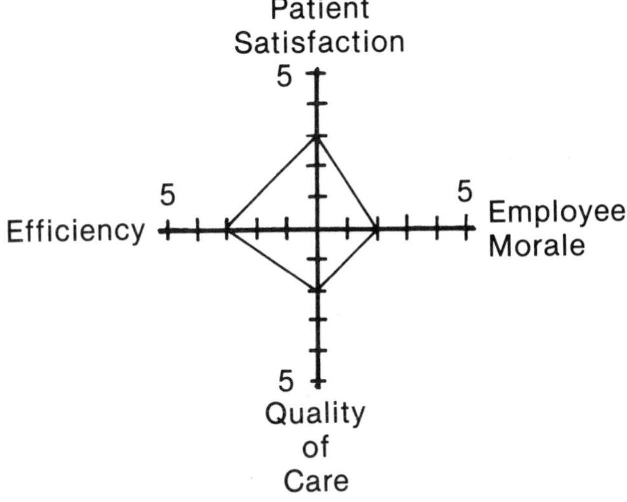

Figure 3-14(b) August Performance

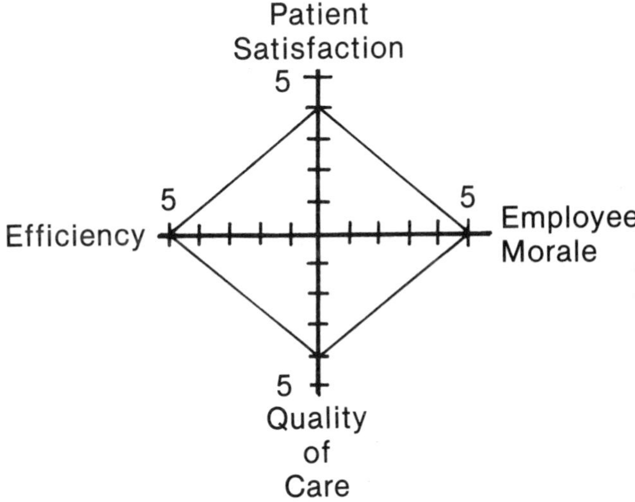

As in the discussion of pie charts, different values or characteristics of interest might be emphasized by employing different markings when preparing maps. For example, we might highlight different values of the relevance index (i.e., the percentage of patients visiting each hospital in a given planning

Figure 3-15 Patient Address Patterns, Five Ottawa Hospitals, 1972–73

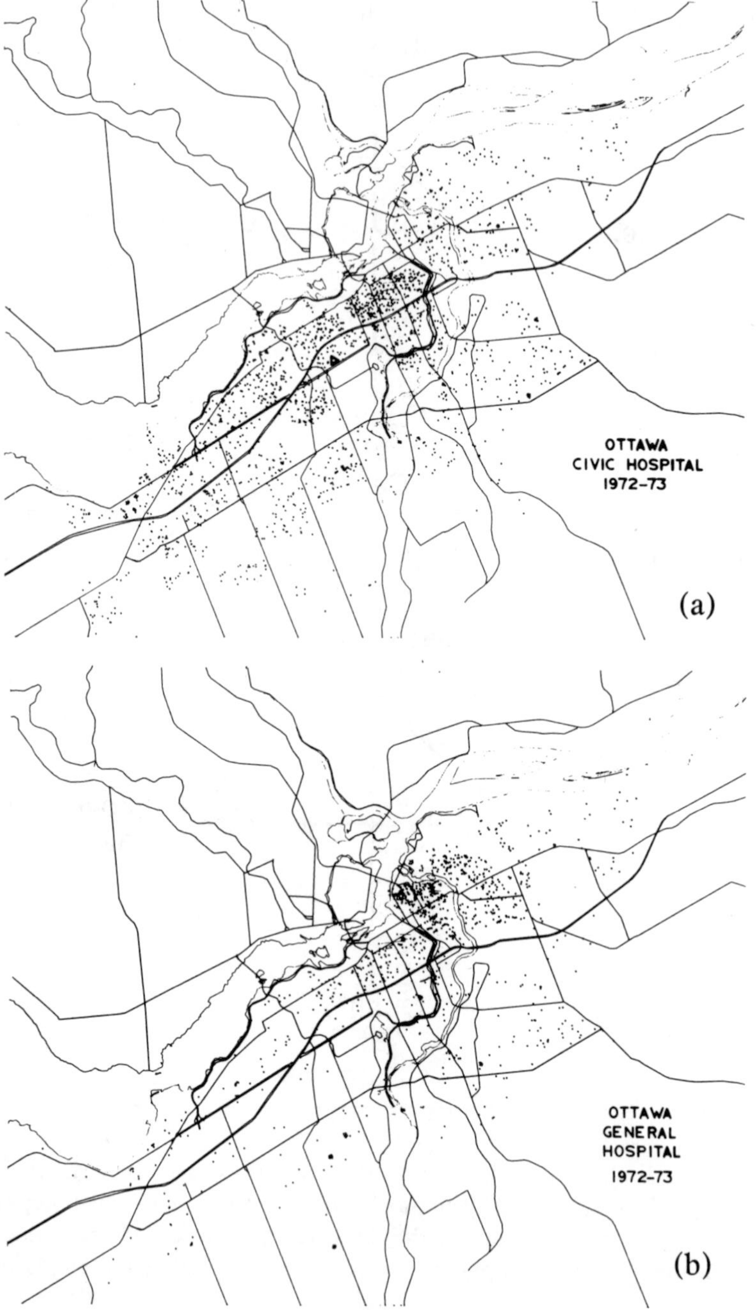

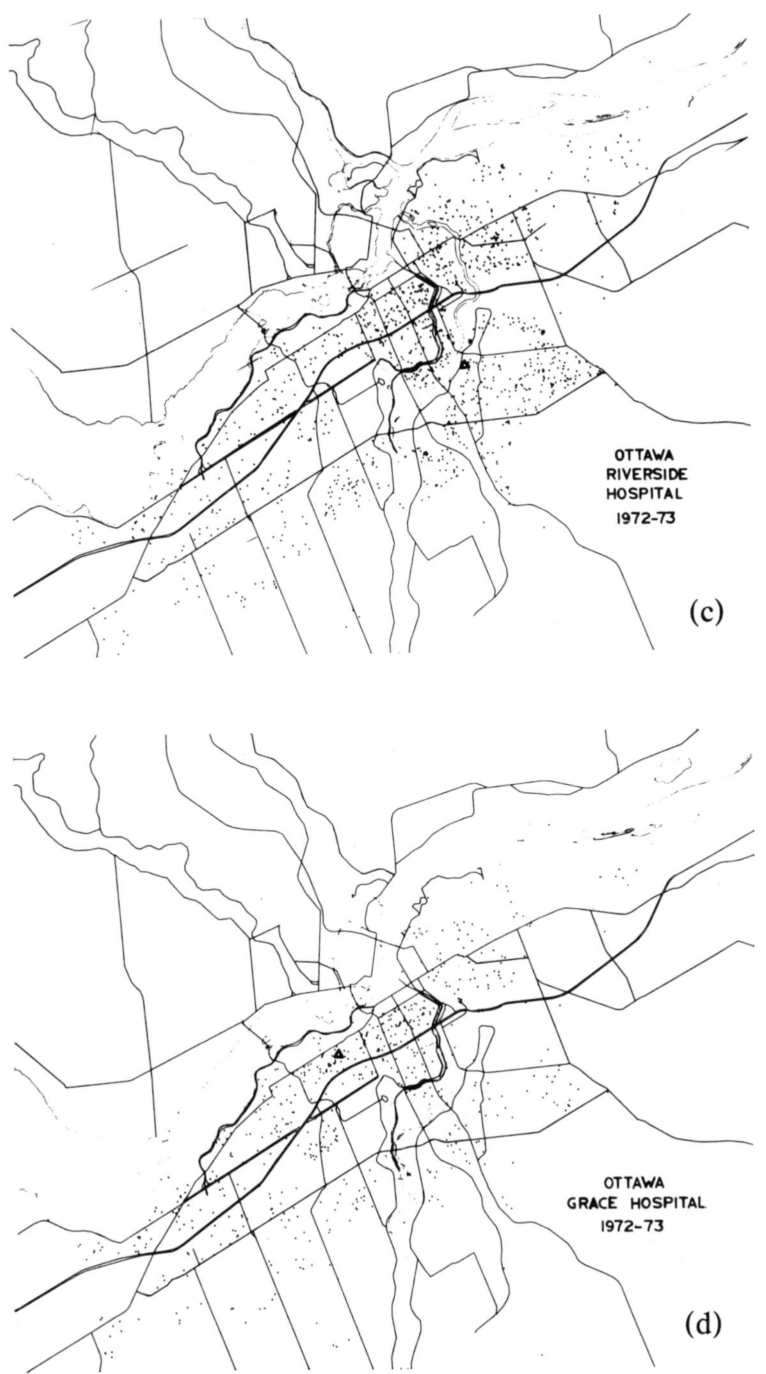
OTTAWA
RIVERSIDE
HOSPITAL
1972-73
(c)
OTTAWA
GRACE HOSPITAL
1972-73
(d)

Figure 3-15 continued

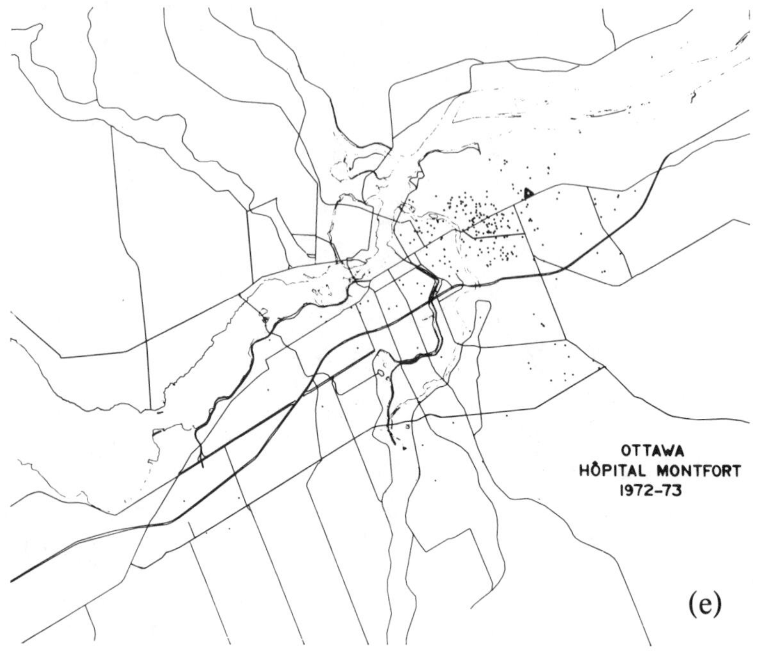

Legend: Each dot represents one patient (sample). Δ = hospital location.

---

area) as shown in Figure 3-16. A wide variety of marking patterns are available on a commercial basis, several of which are shown in Figure 3-17.

3.9 COMPUTER ASSISTANCE[O,G,F]

The manual preparation of maps is time-consuming and laborious and requires painstaking accuracy. Similarly, the preparation of tables consisting of large amounts of numeric information is a thankless, error-prone, and tiring task. Computer storage of information is now common, and systems for analyzing data are readily available to hospitals, clinics, health maintenance organizations, etc., at reasonable cost. Less common, but advancing rapidly, are systems capable of storing information, selecting subsets, displaying line charts or bar graphs on a television-type screen, and printing the corresponding graph or chart. A primary advantage of these systems is that the display screen permits management to make alterations before the line chart or bar graph is produced.

Figure 3-16 Relevance Index—Children's Hospital

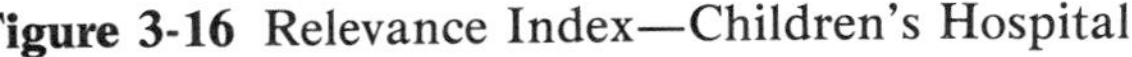

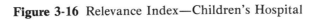

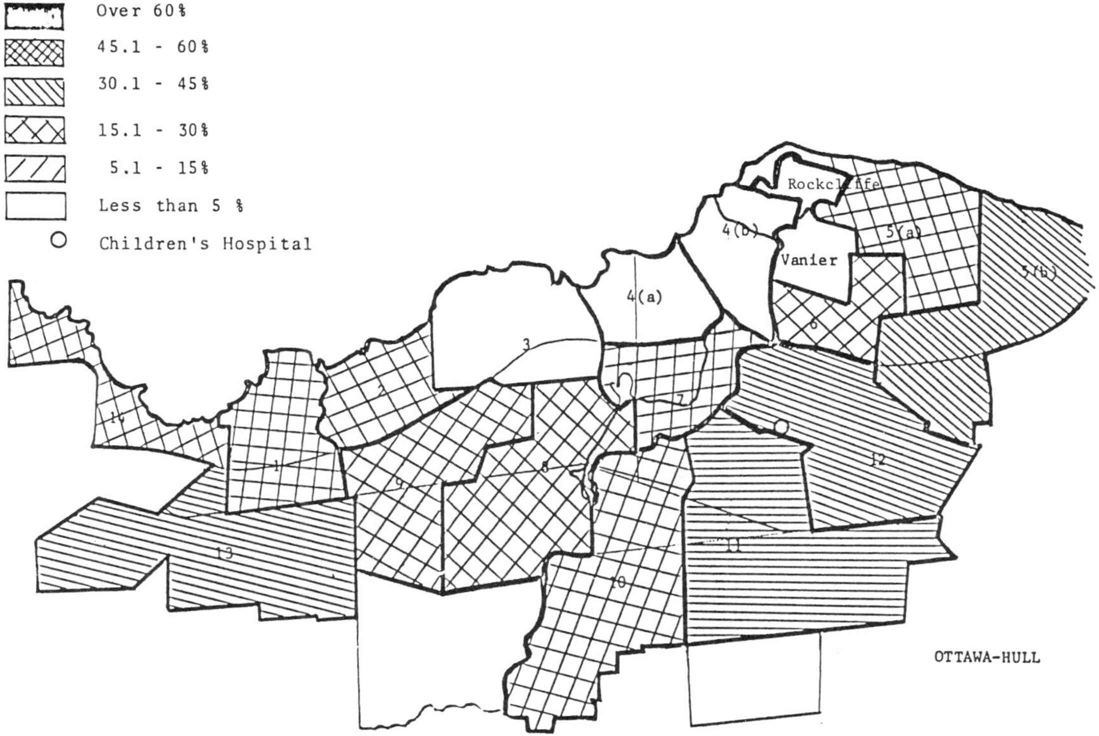

Source: H. Thornhill, "A Study of Patient Origin and Utilization Rates of Ambulatory Care Facilities at Acute Care General Hospitals, in Ottawa. Ottawa: University of Ottawa, School of Health Administration, 1977.

Figure 3-17 Samples of Patterns of Crosshatching Transfer Material

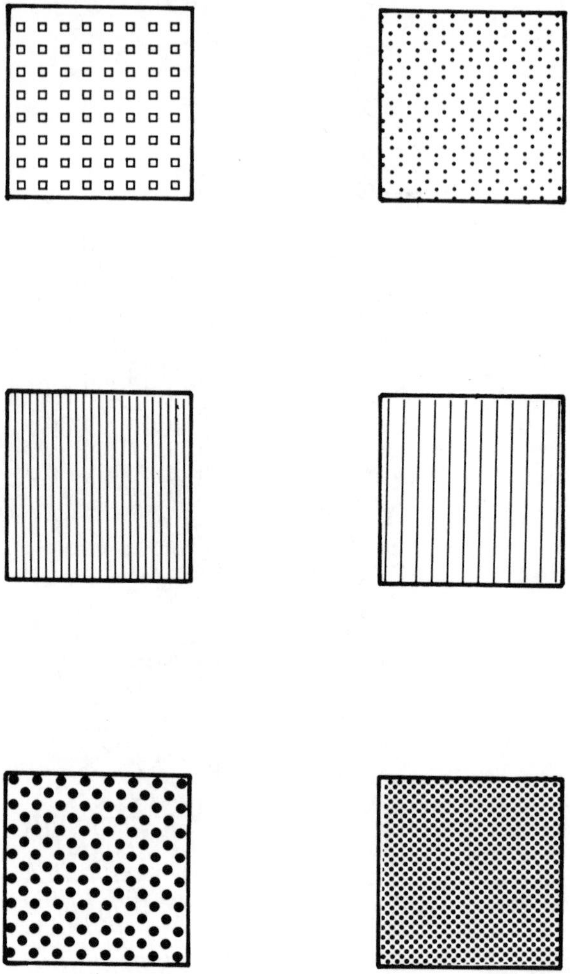

REFERENCES

R. W. Broyles and C. M. Lay, *Statistics in Health Administration: Volume I—Basic Concepts and Applications.* Germantown, Maryland: Aspen Systems Corporation, 1979.

E. S. Buffa, *Modern Production Management* (5th edition). New York: Wiley/Hamilton, 1977.

U.S. Bureau of the Census, *Bureau of the Census Manual of Tabular Presentation,* by Bruce L. Jenkinson. U.S. Government Printing Office, Washington, D.C., 1949.

V. Clover and H. Balsley, *Business Research Methods.* Columbus, Ohio: Grid, Inc., 1974.

J. R. Griffith, *Quantitative Techniques for Hospital Planning and Control.* Lexington Books, 1972.

E. G. Heidemann, "The Financial Impact of Hospital-Acquired Infection," unpublished master's research project. Ottawa: University of Ottawa, School of Health Administration, 1975.

C. M. Lay, *Disease Costing in an Ambulatory Clinic: Disease and Physician Profiles and the Selection of Patients for Review,* unpublished Ph.D. Dissertation. Cambridge, Massachusetts: Massachusetts Institute of Technology, 1978.

C. M. Lay and R. W. Broyles, *Statistics in Health Administration: Volume II—Advanced Concepts and Applications.* Germantown, Maryland: Aspen Systems Corporation, 1980.

H. Thornhill, "A Study of Patient Orgin and Utilization Rates of Ambulatory Care Facilities at Acute Care General Hospitals in Ottawa," unpublished master's research project. Ottawa: University of Ottawa, School of Health Administration, September 1977.

Problems for Solution

1. Draw a bar chart showing average length of stay from Table 3-2.
2. Draw a scatter diagram (1 point for each service or floor) showing the relationship between patients discharged and patient days from Table 3-2.
3. a. Obtain a map of your hometown (or region). Trace a copy, adding census boundaries and/or planning district boundaries. For each area determine the percentage of the population in the following age ranges: 0–14, 15–44, 45–64, 65 and over. Choose one of these age groups as the basis for cross-hatching. Differentiate betweeen 0–4%, 5–9%, 10–14%, 15–19%, 20% and over. (If these percentage ranges are not useful in your case, create your own set.)
   b. Create pie charts for several of the areas showing the variation in the share of the population for the four age groups.

Part II

Algebra

Chapter 4

Fundamentals of Algebra

Objectives

After completing this chapter, you should be able to:

1. Understand and use the basic laws of arithmetic and algebra;
2. Perform basic operations on fractional data;
3. Define algebraic expressions and identify their components;
4. Define and use powers, bases, exponents, and logarithms;
5. Expand and factor simple algebraic expressions;
6. Construct a logarithmic line chart.

Chapter Map

The sections comprising this chapter may be summarized as follows:

Section Number	Required Reading	Optional Reading	Generic Development	Application to Management	Fundamental Principles	Complex Material
	(R)	(O)	(G)	(A)	(F)	(C)
4.1	x		x		x	
4.2	x			x	x	
4.3	x		x		x	
4.4	x		x		x	
4.4.1	x		x		x	
4.4.2	x		x		x	
4.4.3	x		x		x	
4.4.4	x			x	x	
4.5	x		x		x	
4.5.1	x		x		x	
4.5.2	x		x		x	
4.5.3	x		x		x	
4.5.4		x		x	x	
4.6	x		x		x	
4.7	x		x	x	x	
4.8		x				

4.1 AN OVERVIEW(R, G, F)

In previous chapters, we have considered the system of numbers as well as the techniques by which numeric information might be presented in tabular and graphic form. This section focuses on the fundamentals of algebra and its uses in a health care managerial setting. This chapter examines the fundamental concepts required to understand the relationship of one variable to another as expressed by a function, a linear equation, or an inequality. In turn, the discussion of functions and linear equations provides the foundation for examining the process of solving a system of linear equations as presented in Chapter 6. In Chapter 7, the focus shifts from linear equations to nonlinear equations and functions. Finally, this part concludes with a discussion of the mathematics of finance, using much of the material developed earlier.

4.2 THE IMPORTANCE OF ALGEBRA(R, A, F)

The extent to which mathematics is used in discharging managerial responsibilities is perhaps not well recognized. Indeed, administrators in all branches of health care delivery employ mathematical techniques consciously or unconsciously every day and probably many times each day. For example, mathematical techniques are used when preparing the budget, evaluating current performance, estimating cash flows, predicting the use of service, forecasting resource requirements, establishing or revising the fee schedule, preparing the payroll, and developing recurrent reports. Notice that either the implicit or the explicit use of mathematics is required in discharging each of these responsibilities.

More specifically, the appropriate use of algebra may enhance the administrator's conceptual understanding of the problem or situation that must be considered. For example, we saw previously that a flexible budget might be expressed in a simplified form by employing an algebraic equation. Such an equation captures the behavior of cost in response to an alteration in the rate of activity; consequently, such a relationship is of value when addressing the problem of containing costs.

Similarly, algebra can increase the ease with which required calculations are performed. For example, assume that

$$y = 10 + 5x$$

where y corresponds to total cost and x represents the volume of care. In this case, the administrator can use a hand calculator to determine the value of y

for known or assumed values of x with considerable ease, which in turn enhances the extent to which operational performance is controlled.

Many persons approach the formal study of mathematics with anxiety and trepidation.* However, it should be recognized that mathematics is a "language" that in many respects is far simpler to understand than English. As an example of the complexity and ambiguity in English, consider the words to, too, and two. These three simply are symbols that in written form connote specific meanings. However, in isolation, the three are phonetically identical and cannot be distinguished one from the other except in the context of their use.

On the other hand, mathematics is a language that is completely unambiguous. Just as the symbol "b" represents the second letter of the alphabet, the symbol Σ is a Greek letter that indicates we are to perform the mathematical operation of addition. In turn, the operation of addition is performed according to well-defined rules. Hence, mathematics is simply a language that, unlike English, is unambiguous and is governed by well-developed rules and laws. To provide the foundation for understanding mathematics, we begin our study with a specification of the basic laws of arithmetic and algebra that probably are familiar to most.

4.3 BASIC LAWS OF ARITHMETIC AND ALGEBRA(R,G,F)

In arithmetic, numbers are represented explicitly, but algebra serves as a shorthand for expressing arithmetic ideas in which letters are used to represent numbers. A letter may represent *any* value, but only when a specific number is associated with the letter does arithmetic become possible. A letter is called a *variable* while a number is called a *constant*. Thus "a" and "b" can each represent any number, and "$a + b$" represents their sum. For example, if we let $a = 5$ and $b = 3$, their sum is represented by $a + b$, which has a specific value, 8.

There are three basic laws of arithmetic that apply to the operations of addition and multiplication but not to subtraction or division. These are the commutative law, the associative law, and the distributive law. These laws are so simple that most managers understand and use them unconsciously. However, explicit recognition of the rules is necessary in Part V of the text when we consider matrix arithmetic, to which the laws apply only partially.

The arithmetic operations of addition and multiplication are *binary* operations because they involve *two* numbers. Similarly, subtraction and division

---

*For those who feel this anxiety, we recommend a careful reading of *Overcoming Math Anxiety* by Sheila Tobias. She indicates that this anxiety is related to unfortunate learning experiences rather than a person's basic intelligence.

also are binary. The *commutative law* states that, when numbers are added or multiplied, the *order* in which *addition* or *multiplication* is performed is *unimportant.* Thus $3 + 4$ and $4 + 3$ give identical results, as do 5×9 and 9×5. Using letters, $a \times b$ is the same as $b \times a$. If a and b are separate variables, then we may express their product as $a \times b$, or $a \cdot b$, or $(a)(b)$, or ab. The commutative law permits us to reverse these expressions and obtain $b \times a$, or $b \cdot a$, or $(b)(a)$ or ba. Note, however, that $5 - 3$ is not the same as $3 - 5$, while $3 \div 5$ is not equal to $5 \div 3$. Hence, the commutative law does not apply to subtraction and division.

The *associative law* allows us to express a series of additions or multiplications without ambiguity. The numbers in the series may be associated into pairs in any manner desired and, by the commutative law, the order does not influence the result obtained when addition or multiplication is performed. Therefore, we find that

$$a + b + c = (a + b) + c = a + (b + c) = b + (c + a)$$

Also, it is easy to verify that

$$a \cdot b \cdot c = (a \cdot b) \cdot c = a \cdot (b \cdot c) = b(c \cdot a)$$

When subtraction and division are involved, the commutative and associative laws do not apply. Rather, working from left to right, we begin with the first pair of values and perform the indicated operation. The result then is paired with the next value in the series and the indicated arithmetic operation is performed. For example, we find that

$$\begin{aligned} & 7 - 3 - 2 - 4 \\ &= (7 - 3) - 2 - 4 \\ &= (4 - 2) - 4 \\ &= 2 - 4 \\ &= -2 \end{aligned}$$

The *distributive law* allows us to distribute multiplication over several additions, using parentheses, and forms the basis of factoring in algebraic expressions. Using letters, the distributive law asserts that:

$$a \times (b + c) = ab + ac^{*}$$

*Note that the parentheses are necessary to show that both b and c are multiplied by a. The expression $a \times b + c$ has an *entirely different* meaning. Parentheses can be very useful to show which operations are to be grouped.

For example, let $a = 3$, $b = 5$, and $c = 2$. After substituting appropriately we find that

$$3 \times (5 + 2) = 3 \times 7 = 21,$$

and

$$(3 \times 5) + (3 \times 2) = 15 + 6 = 21,$$

yield identical results. The reader should verify the result by substituting other specific numbers for a, b, and c.

In addition to the three basic laws (commutative, associative, and distributive) as they apply to addition and multiplication, we also consider other fundamentals of arithmetic operations: namely, identity elements, inverse operations, and operations with signed numbers.

Addition and multiplication have *identity* elements. When zero is added to "a," the sum is "a" (i.e., $a + 0 = a$). Hence, the identity element for addition is 0 (zero), and its application in the operation of addition results in the same number. On the other hand, the identity element for multiplication is 1 (one), which implies that

$$a \times 1 = a$$

When considering matrix arithmetic in a later chapter, we will consider the identity matrix **I**, which has the same function as 1.

Addition and multiplication have *inverse operations* (subtraction and division, respectively) that can be viewed as reverse or "undoing" operations. Thus, we find that

$$a + b - b = a$$

while

$$a \times b \div b = a$$

Notice that these results are valid since

$$b - b = 0$$

while

$$b \div b = 1$$

Similarly, we find that $a \times b \div b = a \times b/b = a \times b \times (1/b)$. In the last of these expressions, $1/b$ is often called "the inverse of b" and customarily is represented by b^{-1} (b inverse). Thus $b \times (1/b) = b \times b^{-1}$. (Remember that the inverse of zero is not defined.) We show in a later chapter that multiplication by the inverse of a matrix replaces the operation of division in matrix arithmetic. We also find that there are matrices for which the inverse is not defined.

Finally, positive and negative numbers have "*signs*" (+ and −) associated with them. There are rules for performing addition and multiplication as well as subtraction and division with signed numbers. Before examining these rules, we must define the *absolute value* of a number as the number to which a positive sign has been attached. The absolute value is represented by two vertical bars surrounding the number. For example, the absolute values of 5 and −6 are given by $|5| = 5$ and $|-6| = 6$, respectively. Also note that $|4|$ is less than $|-7|$ (i.e., $|4| = 4 < |-7| = 7$).

The rules for adding signed numbers are as follows:

a. If the signs are the same, add their absolute values, and assign the common sign to the result;
b. If the signs are opposite, subtract the smaller absolute value from the larger, and assign the sign of the number with the larger absolute value to the result.

The addition rule can be explained by using arrows pointing along the real number line. A positive arrow points to the right, while a negative one points to the left.

When performing the operation of *multiplication* (*division*), we simply multiply (divide) the absolute values, and

a. If the signs are the same, assign a positive sign to the result, and
b. If the signs are different, assign a negative sign to the result.

For *subtraction*, we simply change the sign of the number to be subtracted and then proceed as for addition.

4.4 OPERATIONS WITH FRACTIONS(R,G,F)

A fraction is composed of one number divided by another, referred to as the *numerator* and the *denominator*, respectively. For example, in the fraction 3/4 the numerator is 3 and the denominator is 4. Three divided by four can be represented by $(3 \div 4)$, $(3/4)$, $(\frac{3}{4})$, or .75.

4.4.1 Addition of Fractions[R, G, F]

When adding two or more fractions, each of the terms in the addition must have the same denominator. When this condition is satisfied, we simply add the numerators and retain the common denominator. For example, when performing the addition

$$3/4 + 2/4 = 5/4$$

the numerator of the result is simply the sum of the numerators 3 and 2 while the denominator of 5/4 is common to both 3/4 and 2/4. Thus,

$$\frac{3}{4} + \frac{2}{4} = \frac{3 + 2}{4} = \frac{5}{4}$$

Note that the multiplication and division operations take precedence over either addition or subtraction unless marked differently by the insertion of parentheses.

Consider next the addition of the ratios 3/4 and 5/8. In this case, the denominator of the first is 4 while the denominator of the second is 8. To perform the operation of addition, it is necessary to transform these ratios into equivalent fractions that have a common denominator. In such a situation we employ the smallest value into which each of the denominators can be divided evenly (i.e., with no remainder or decimal fraction) as the lowest common denominator. To obtain the lowest common denominator, we first find the factors of the denominators that are defined as the values that, when multiplied, yield the original number. For example, the factors of the number 8 might be expressed in the form

$$8 = 8 \times 1 = 4 \times 2 \times 1 = 2 \times 2 \times 2 \times 1$$

The next step is to identify the prime factors of the denominators. Prime numbers are those values that have only themselves and 1 as factors (i.e., 1, 2, 3, 5, 7, 11, 13, 17, 19, 23, etc.). The lowest common denominator is simply the *product* of the *smallest* set of prime factors that contains the prime factors of each of the denominators.

To illustrate this approach, suppose we want to obtain the sum of 3/4 and 5/8. The prime factors of the number 4 are 2, 2, and 1 (i.e., $4 = 2 \times 2 \times 1$) while the prime factors of 8 are 2, 2, 2, and 1 ($8 = 2 \times 2 \times 2 \times 1$). Given that the prime factors of the number 4 (2, 2, and 1) are contained in the prime factors of the number 8 (2, 2, 2, and 1), the lowest common denominator is $2 \times 2 \times 2 \times 1$, or 8. Hence, we find that

$$3/4 + 5/8 = 6/8 + 5/8 = 11/8$$

Employing a similar approach, it can be verified that

$$2/7 + 5/8 = 16/56 + 35/56 = 51/56$$

In this case, the prime factors of 7 are 7 and 1 while the prime factors of 8 are 2, 2, 2, and 1. As a result, the lowest common denominator is given by $7 \times 2 \times 2 \times 2 \times 1$ or 56.

4.4.2 Multiplication and Division of Fractions(R, G, F)

When fractions are multiplied, the lowest common denominator is not required. Rather, when multiplying one fraction by another, we simply find the products of the numerators and the denominators. For example, the product of 2/3 and 4/9 is given by

$$2/3 \times 4/9 = \frac{2 \times 4}{3 \times 9} = \frac{8}{27}$$

Similarly, the product of 2/3 and 7/9 is 14/27.

When one fraction is divided by another, the denominator is inverted and the result is multiplied by the numerator. For example, suppose we want to divide 2/3 by 3/4. We may express the division in the form

$$2/3 \div 3/4 = \frac{2/3}{3/4}$$

Following the procedure outlined above, we invert 3/4 and obtain 4/3, which then is multiplied by 2/3. Thus, we find that

$$\frac{2/3}{3/4} = \frac{2}{3} \times \frac{4}{3} = \frac{8}{9}$$

We may demonstrate the validity of these results by observing that

$$\frac{2/3}{3/4} \times 1 = \frac{2/3}{3/4} \times \frac{4/3}{4/3} = \frac{2/3}{3/4} \times \frac{4/3}{4/3} = \frac{2/3 \times 4/3}{1} = \frac{8/9}{1} = \frac{8}{9}$$

As should be verified,

$$\frac{7/8}{1/2} = \frac{7}{8} \times \frac{2}{1} = \frac{14}{8}$$

which, of course, may be simplified by applying the techniques described in the following sections.

4.4.3 Simplification and Mixed Fractions$^{(R,G,F)}$

When a fraction has common factors in the numerator and denominator, it is possible to simplify the fraction. Both numerator and denominator can be divided by the common factor, which is the equivalent of multiplying the fraction by 1. For example:

$$\frac{6}{8} \times 1 = \frac{6}{8} \times \frac{1/2}{1/2} = \frac{6/2}{8/2} = \frac{3}{4}$$

and

$$\frac{12}{56} \times 1 = \frac{12}{56} \times \frac{1/4}{1/4} = \frac{12/4}{56/4} = \frac{3}{14}$$

When a fraction has a value greater than one or some other integer, it can be changed into a mixed fraction with an integer part and a fractional part, where the fractional part must be less than one. For example:

$$\frac{11}{8} = 1\,\frac{3}{8}$$

$$\frac{17}{16} = 1\,\frac{1}{16}$$

$$\frac{38}{9} = 4\,\frac{2}{9}$$

4.4.4 Utilization Patterns Study$^{(R,A,F)}$

As part of a larger study of treatment patterns for utilization review, operational planning, and budgeting at a health care facility, the figures in Table 4-1 were obtained from a sample of 125 patients admitted with acute appendicitis.

Table 4-1 Services Provided to 125 Acute Appendicitis Patients

Type of Service	Units of Measure	Number of Patients Receiving Service	Number of Units Provided	Fraction of Patients Receiving Service	Units Per Patient Receiving Service
Laboratory	Lab units	123	1,365	123/125	1,365/123
Radiology	x-rays	75	95	75/125	95/75
Surgery	hours	70	50	70/125	50/70
Nursing:					
Level 1	days	125	614	125/125	614/125
Level 2	days	114	350	114/125	350/114
Level 3	days	40	90	40/125	90/40
Meals	meal-days	125	964	125/125	964/125

The fraction of patients receiving a given service and the number of units per patient can be used to determine "usual" levels of providing service to patients. In order to develop appropriate standards of care, these values can be judged by physicians to be just about right, too high, or too low. They also may be compared with the results of similar studies performed elsewhere. These fractions arise from simple counting operations and the desire to find ratios or averages.

As an exercise, simplify the fractions appearing in the table by expressing them as ordinary fractions, mixed fractions, and decimal fractions, rounded to three decimal places.

At this point in the analysis, the real number system described earlier as well as the commutative, associative, and descriptive laws presented in this chapter should be clearly understood. Readers who feel uncomfortable with their understanding of these areas should review the material presented earlier before proceeding to the next section, where we begin our analysis of algebraic expressions.

4.5 ALGEBRAIC EXPRESSIONS(R,G,F)

Algebraic expressions are used to simplify the presentation of general mathematical relationships. As such, algebraic expressions may represent more general collections of numbers in succinct form. To understand and use many of the concepts and mathematical techniques described in this text, it is necessary first to review the vocabulary and operations of basic algebra.

The basic building blocks of algebraic expressions are variables and constants. Mathematically, these concepts are related to each other through the basic operations of addition, subtraction, multiplication, and division. A *constant* is a number with a fixed value. Some constants, such as π and e, are represented by symbols, since the decimal fraction does not terminate. On the other hand, a *variable* may assume the value of any real number and is represented by either an uppercase or lowercase letter. When expressed using a computer language, a variable may be represented by one or more letters.

As will be seen in Chapter 17, in matrix algebra a variable may represent a row or table of values. Also, we shall encounter a special terminology when matrix algebra is considered where (1) a scalar represents a single number, (2) a vector is a row or column of numbers, and (3) a matrix represents a table of numbers. In basic algebra, however, a variable is always a scalar.

4.5.1 Simple Expressions(R, G, F)

The components of algebraic expressions may be viewed in terms of a hierarchical arrangement. Variables and constants represent the lowest level of the hierarchy. In turn, variables and constants may be combined by multiplication to form "terms" that represent the second level of the hierarchy. Finally, terms may be added to form expressions that represent the highest level of the hierarchy. An expression may have one or more terms and a term may have one or more variables or constants. Table 4-2 shows examples of expressions and terms indicating the number of components. An expression with two terms is referred to as a "binomial" expression; one with three terms, "trinomial;" and three or more, "polynomial."

Note that there are several methods of representing the multiplication of variables. For example, "*a times b*" can be represented as: ab, $(a)(b)$, $a(b)$, $a \cdot b$, or $a \times b$. The term or expression "$a \times b$" may be confusing since the

Table 4-2 Examples of Expressions, Terms, Variables, and Constants

Example	Type	Number of Components
$3a$	term	1 constant & 1 variable
	or	
	expression	1 term
$4a + 5b$	expression	2 terms (binomial)
$5x^2 + 3xy + 6y^3$	expression	3 terms (trinomial or polynomial)
$5x^2$	term	1 constant & 1 variable
$3xy$	term	1 constant & 2 variables
$6y^3$	term	1 constant & 1 variable

multiplication symbol might be interpreted as the variable x. Finally, the representation "$a \cdot b$" is usually read as "a dot b."

4.5.2 Powers$^{(R,G,F)}$

Certain terms, called *powers*, indicate that a variable is repeated as a factor two or more times. For example: the term x^2 represents $x \cdot x$; y^3 represents $y \cdot y \cdot y$; $5x^2$ represents $5 \cdot x \cdot x$; and 2^4 represents $2 \cdot 2 \cdot 2 \cdot 2$. In each of these cases, the *repeated factor* is called the *base* and the number of repetitions is called the *exponent.* In this case, the relation between the power, the exponent, and the base may be represented by

$$\text{power} = y^3 \quad (3 = \text{exponent},\ y = \text{base})$$

or by

$$\text{power} = \text{base}^{\text{exponent}}$$

Powers can be multiplied and, when the base is the same, the resulting power can be simplified. For example, the previous discussion allows us to assert that

$$a^4 = a \cdot a \cdot a \cdot a$$

$$a^3 = a \cdot a \cdot a$$

We may expand this notation and find $a^4 \cdot a^3$, which is given by

$$(a \cdot a \cdot a \cdot a)(a \cdot a \cdot a) = a \cdot a \cdot a \cdot a \cdot a \cdot a \cdot a$$

Notice that a appears four times in a^4 and three times in a^3, which implies that there are seven a's in the product. Thus,

$$a^4 \cdot a^3 = a^{4+3} = a^7$$

This example demonstrates the general result that the multiplication of powers of the same base may be performed by adding the exponents. In this regard, note that $a^4 \cdot b^3$ *cannot be simplified since the bases are different.*

The division of like powers is performed by subtracting exponents. For example:

$$a^5 \div a^2 = \frac{\not{a}\cdot\not{a}\cdot a\cdot a\cdot a}{\not{a}\cdot\not{a}} = a^3$$

In this case we might express $a^5 \div a^2$ by a^{5-2} or a^3. Also note that $y^4 \div y^6$ is given by y^{4-6} or y^{-2}. This result means that

$$\frac{y\cdot y\cdot y\cdot y}{y\cdot y\cdot y\cdot y\cdot y\cdot y} = \frac{1}{y^2} = y^{-2}$$

In particular, a *negative exponent* means the *inverse* of the power with the same positive exponent. Also, $m^3 \div m^3 = m^0 = 1$, so that any base raised to the *power zero* has the value 1. Only the base zero is excluded from this consideration, because division by zero is not defined.

A power of a power also can be obtained. For example:

$$(x^3)^2 = x^3 \cdot x^3 = x^{3+3} = x^6$$

The middle steps can be omitted by recognizing that the exponents are multiplied to obtain the correct result (i.e., $3 \times 2 = 6$). Similarly,

$$((y^4)^2)^3 = (y^8)^3 = y^{24}$$

since $4 \times 2 \times 3 = 24$. Also observe that

$$(4a)^2 = (4a)(4a) = 16a^2$$

Fractional exponents also have meaning, as is demonstrated by

$$(4^{1/2})(4^{1/2}) = 4^1 = 4$$

Therefore, the exponent ½ is equivalent to the square root, and the representations $\sqrt{4}$ and $4^{1/2}$ are equivalent to each other. The cube root of a number is indicated by the exponent ⅓, and the nth root by the exponent $1/n$. For example, consider

$$27 = 3 \times 3 \times 3$$

which implies that the cube root of 27 (i.e., $\sqrt[3]{27}$) is:

$$27^{1/3} = 3$$

Similarly,

$$32 = 2 \times 2 \times 2 \times 2 \times 2$$

which implies that the 5th root of 32 (i.e., $\sqrt[5]{32}$) is:

$$32^{1/5} = 2$$

4.5.3 Expansion and Factoring of Polynomial Expressions(R, G, F)

The distributive law of arithmetic is applicable in algebra as illustrated by the following examples.

$$4(a + b) = 4a + 4b$$
$$12(m + n + p) = 12m + 12n + 12p$$
$$a(x + y + z) = ax + ay + az$$

Consider the first of these examples. We refer to the use of $4(a + b)$ to obtain $4a + 4b$ as *expanding* the expression. Conversely, we refer to the use of $4a + 4b$ to obtain $4(a + b)$ as *factoring* the expression. Thus, the factors in the first example are 4 and the quantity $(a + b)$ while the expansion is $4a + 4b$. Notice that the terms in the expansion have 4 as the common factor.

If the first factor in each of these examples is replaced by an expression in parentheses, we form more complex patterns:

$$\begin{aligned}(3 + 4)(a + b) &= (3 + 4)a + (3 + 4)b \\ &= 3a + 4a + 3b + 4b \\ &= 7a + 7b\end{aligned}$$

which, of course, is somewhat obvious and rather trivial. However,

$$\begin{aligned}(m + n)(a + b) &= (m + n)a + (m + n)b \\ &= am + an + bm + bn\end{aligned}$$

is less trivial. A more useful form is given by

$$(a + b)(a + b) = (a + b)a + (a + b)b$$

$$= a^2 + ab + ab + b^2$$
$$= a^2 + 2ab + b^2$$

where like terms can be *collected* to simplify the final expression. Employing this approach, we obtain

$$(a + b)^2 = a^2 + 2ab + b^2$$

In a similar fashion,

$$(a - b)^2 = a^2 - 2ab + b^2$$

Notice that in these two cases the internal sign of the $(a + b)$ or $(a - b)$ term determines the sign of the term $2ab$, which is the *cross-product*. Another pattern encountered frequently is given by

$$\begin{aligned}(a + b)(a - b) &= (a + b)(a) + (a + b)(-b)\\ &= a^2 + ab - ab - b^2\\ &= a^2 - b^2\end{aligned}$$

These expansions and factorings are used often in the development of formulas in statistics.

4.5.4 Patient Treatment Cost(O, A, F)

As an example of the use of the laws of algebra in health administration, consider the budget implications of admitting n patients for a particular problem (say, acute appendicitis). From past experience we may know that each of these patients will require

a units of laboratory tests at \$1.50/unit
b x-rays at \$25.00/x-ray
c hours of surgery at \$150.00/hour
d days of level 1 nursing care at \$25.00/day
e days of level 2 nursing care at \$45.00/day
f days of level 3 nursing care at \$80.00/day

and

$d + e$ meal days at \$5.00/meal day.

The expected cost for each patient admitted with acute appendicitis is given by

$$(1.50\,a + 25\,b + 150\,c + 25\,d + 45\,e + 80f + 5\,(d + e))$$

Applying the distributive law, we may express the expected costs of treating *n* such patients in the form

$$n(1.50\,a + 25\,b + 150\,c + 30\,d + 50\,e + 80\,f)$$

In this example, the *d* term has been simplified by combining $(25\,d + 5d)$. Similarly, the *e* term was simplified by combining $(45e + 5e)$.

4.6 LOGARITHMS(R,G,F)

Before the advent of pocket calculators with impressive capabilities in arithmetic, multiplication and division by hand were tedious and time consuming. Partial relief from the tedium was obtained by using logarithms, which rely on the concept of powers presented in Section 4.5.2.

Multiplication of powers with the *same base* can be accomplished by adding their exponents:

$$a^2 \times a^4 = a^{2+4} = a^6$$

Division and powers of powers also are important, as described earlier. These properties of powers were recognized and developed into two systems of calculation known as natural and common logarithms in the early 1600s.

A logarithm is the exponent of a power with a specified base. For example, letting $a > 0$, the equation $y = a^x$ means that x is the logarithm of y to the base a. More specifically, let $y = 100$ and $a = 10$. In this case, the equation

$$100 = 10^2$$

implies that 2 is the logarithm of 100 to the base 10. Similarly, the equation

$$1{,}000 = 10^3$$

means that 3 is the logarithm of 1,000 to the base 10. We may also express these two results by

$$\log_{10} 100 = 2$$

and by

$$\log_{10} 1{,}000 = 3$$

respectively. A common characteristic of these examples is the use of the base 10 to determine the corresponding logarithm. Such logarithms to the base 10 are called *common logarithms.*

On the other hand, *natural logarithms* employ the base e, the value of which is given by the sum of the infinite series

$$\begin{aligned} e &= 1 + \frac{1}{1} + \frac{1}{2 \times 1} + \frac{1}{3 \times 2 \times 1} + \frac{1}{4 \times 3 \times 2 \times 1} + \cdots \\ &\cong 2.71828 \end{aligned}$$

Natural logarithms with the base e are of theoretical importance and are used extensively in the study of calculus and, to a lesser extent, statistics.

The "logarithm of base e" usually is abbreviated "ln" while that with base 10 is "log." Because 10 and e are constants, there is a simple relationship between common and natural logarithms. Pocket calculators with scientific capabilities include one or more functions for $\log x$, $\ln x$, 10^x, e^x, or x^y. All of these allow the user to work directly with logarithms.

It is easiest to explain the properties of logarithms using common logarithms, but the same properties apply for base e or any other base that might be chosen.

In common logarithms we have seen that $\log 100 = 2$ and $\log 1{,}000 = 3$, and we can easily see that adding logarithms gives us the logarithm of the product. For example, since $100 = 10^2$ while $1{,}000 = 10^3$, we find that

$$\begin{aligned} 10^2 \times 10^3 &= 10^{2+3} \\ &= 10^5 \end{aligned}$$

Similarly, employing the notation developed earlier, we find that the product of 100 and 1,000 might be expressed in the form

$$\begin{aligned} \log(100 \times 1{,}000) &= \log 100 + \log 1{,}000 \\ &= 2 + 3 \\ &= 5 \end{aligned}$$

which, of course, agrees with the results obtained previously.

We frequently require the logarithms for values other than those that may be expressed as an integer power of 10. For example, we might require the logarithm of numbers such as 3, 30, and 300. To understand the logarithm of such a value, we observe that

$$10^0 = 1; \quad 10^1 = 10; \quad 10^2 = 100; \quad 10^3 = 1{,}000$$

which allows us to assert that

$$\log 1 = 0; \quad \log 10 = 1; \quad \log 100 = 2; \quad \text{and} \quad \log 1{,}000 = 3$$

respectively. Referring to the values 3, 30, and 300 we may assert that

$$\log 1 < \log 3 < \log 10$$

while

$$\log 10 < \log 30 < \log 100$$

and

$$\log 100 < \log 300 < \log 1{,}000$$

To obtain the precise value of log 3, log 30, and log 300, we observe that the logarithm of any number is composed of two components. The first is an integer value and is called the *characteristic.* The second is a fractional decimal and is called the *mantissa.* For example, it can be verified that

(1) $3 = 3 \times 10^0$ and $\log 3 = 0.4771$;

(2) $30 = 30 \times 10^1$ and $\log 30 = 1.4771$; and

(3) $300 = 3 \times 10^2$ and $\log 300 = 2.4771$.

At this point, note that the mantissas (i.e., .4771) in these three cases are identical. On the other hand, the characteristic (i.e., the integer values 0, 1, and 2) indicates the power to which the base 10 must be raised in order to locate the decimal point. In a similar fashion, the reader should verify that

(1) $2.5 = 2.5 \times 10^0$ and $\log 2.5 = 0.3979$;

(2) $25 = 2.5 \times 10^1$ and $\log 25 = 1.3979$; and

(3) $250 = 2.5 \times 10^3$ and $\log 250 = 2.3979$.

As before, the mantissa (.3979) is used to determine the digits of the original number while the characteristics (i.e., the integer values 0, 1, and 2) locate the decimal place.

Historically, logarithms have been used to perform calculations such as

$$y = 375 \times 521 \times 75.1$$

As can be verified easily, the logarithms of 375.0, 521.0, and 75.1 are given by

Number	Logarithm
375.0	2.5740
521.0	2.7168
75.1	1.8756

The desired product is obtained by first calculating the sum

$$\log 375 + \log 521 + \log 75.1 = 7.1664$$

The antilogarithm, which is simply the number that has 7.1664 as its logarithm, is then found as follows. As seen in any table of logarithms, .1644 = log 1.46 and 7 is the exponent of 10 of the number we desire. Hence, the antilogarithm is given by

$$\begin{aligned} 10^{7.1664} &= 10^7 \times 10^{.1664} \\ &= 10^7 \times 1.4669 \\ &\cong 14{,}669{,}000 \end{aligned}$$

Similarly, the operation of division may be performed using logarithms. For example, dividing 375 by 75.1 we obtain

$$\begin{aligned} \log (375 \div 75.1) &= 2.5740 - 1.8756 \\ &= .6984 \end{aligned}$$

Employing this approach, we find that the antilogarithm of .6984 (i.e., $10^{.6984}$) is approximately 4.993.

Logarithms also may be used when dealing with fractional exponents. For example, suppose we wish to find $(75)^{1/4}$. In this case, we find that

$$\begin{aligned} (75)^{1/4} &= \frac{1}{4} \log 75.1 \\ &= .4689 \end{aligned}$$

The antilogarithm of .4689 is approximately 2.944. Obviously, greater precision may be achieved by using a larger number of significant digits in our calculations.

4.7 LINE CHARTS USING LOGARITHMS(R, G, A, F)

The previous discussion provides the basis for an examination of the construction of line charts based on logarithms. When logarithms are employed in constructing line charts, we normally use *semilogarithmic paper* (see Figure 7-7 in Chapter 7). In this case, the horizontal axis is divided into normal units while the vertical axis is divided into ranges representing powers of 10. Thus, the scale represented on the y axis is expressed in natural units but is measured in logarithms of the natural units. Each range of 10 on the scale is referred to as a cycle, and paper consisting of 1, 2, 3, 4, or 5 cycles is commonly available. The number of cycles should be selected so as to accommodate the nature and the range of data that pertain to a particular problem. Semilogarithmic paper is commonly used to graph processes that are characterized by a more or less constant percentage rate of increase (growth rate). When plotted on arithmetic paper, such processes quickly exceed the limits of the arithmetic scale. However, when plotted on semilogarithmic paper, these processes usually are portrayed by a straight line, the slope of which represents the growth rate. As an example, consider Figures 4-1 and 4-2, which portray the populations of the United States and Canada. Figure 4-1 would lead you to believe that the two countries have had different growth patterns, but Figure 4-2, where these data have been recorded on semilogarithmic paper, shows that the two populations have increased at similar rates during the period 1850–1971. This is so because the lines are approximately parallel, which implies that the slopes of the lines, and hence the growth rates, are similar.

4.8 CONCLUSIONS(O)

In this chapter we have reviewed the basic laws of arithmetic as well as the basic concepts of algebraic expressions and their manipulation. These topics are fundamental to a further discussion of algebra as well as vital tools in mathematics for health care managers. Further, we described areas in which basic mathematical concepts may be used by health care professionals such as financial managers, regional planners, directors of laboratories, nursing units, and so on. We believe that mathematical skills are a very useful supplement to the other basic skills that health professionals require in order to discharge their responsibilities.

Figure 4-1 The Populations of the United States and Canada from 1850/51 to 1970/71

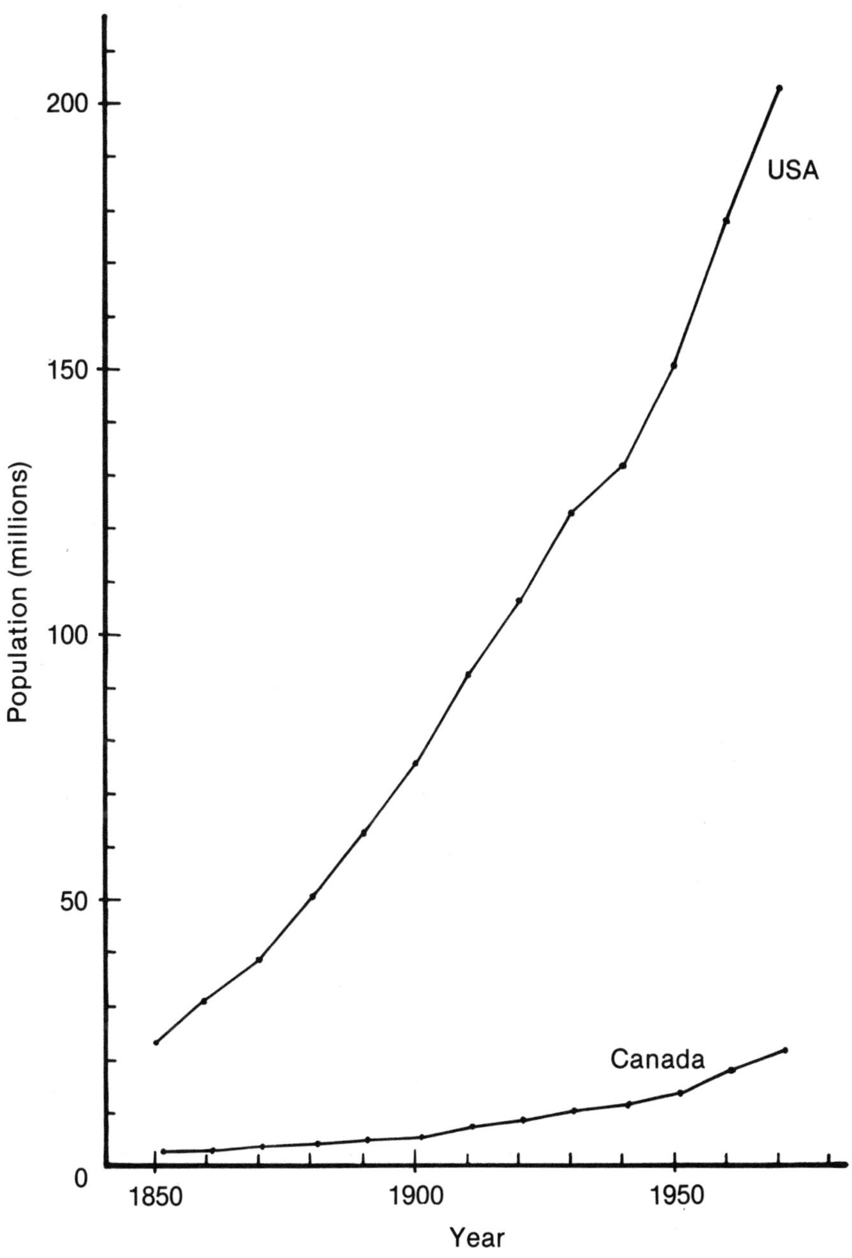

Figure 4-2 Populations of the United States and Canada from 1850/51 to 1970/71 (Semilogarithmic Plot)

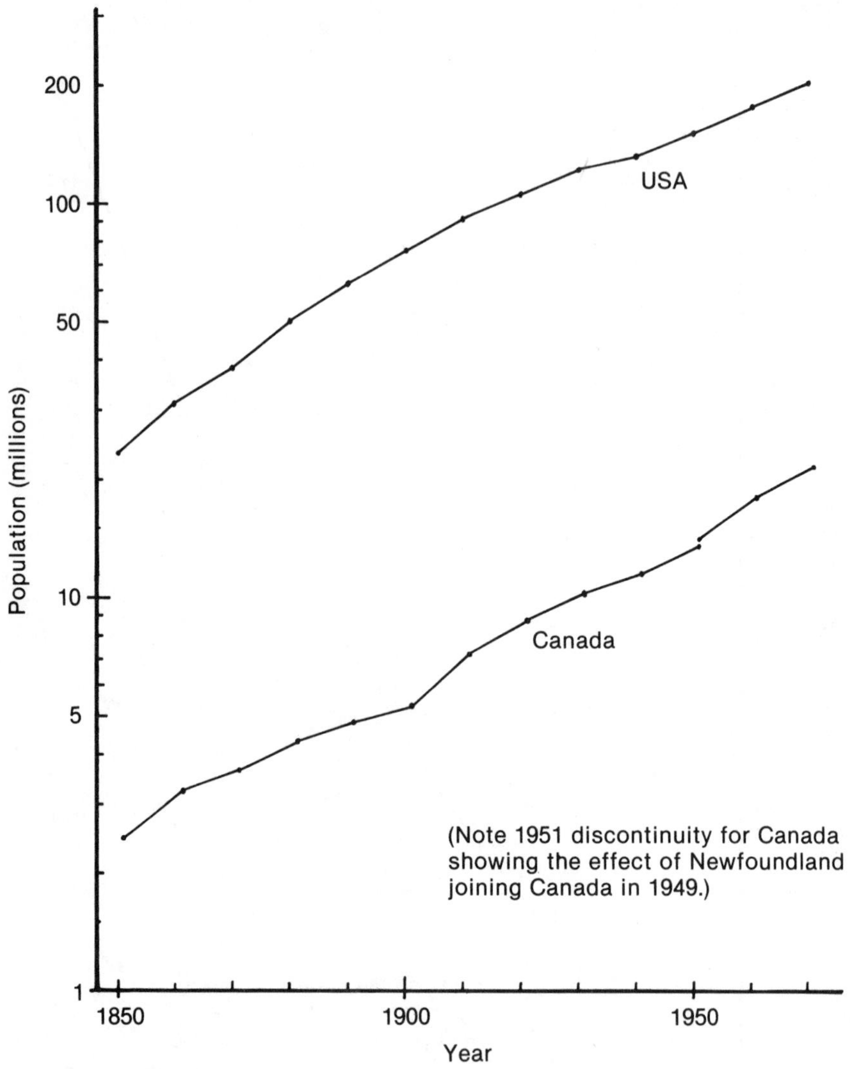

Problems for Solution

1. List the first five:
   a. positive integers,
   b. integers less than 12,
   c. integers greater than -7,
   d. prime numbers greater than 6,
   e. fractions that are integer multiples of 3/8.

2. a. Simplify the fractions in 1 (e).
   b. Express each of the above as a mixed fraction.
   c. Express each of the above as a decimal.
   d. Round the decimal fractions to 2 places.

3. Demonstrate the associative and commutative laws for the following:

   a. $3 + 7 + 5 + 9$
   b. $3 \times 7 \times 5 \times 9$

4. Demonstrate that the commutative and associative laws do not hold for:

   a. $3 - 7 - 5 - 9$
   b. $3 \div 7 \div 5 \div 9$

5. Demonstrate the distributive law for:

$$(m + n)(a + b)$$

   for the following values

   a. $m = 3$, $n = 4$, $a = 7$ and $b = 10$
   b. $m = 6$, $n = -2$, $a = 5$ and $b = -8$

6. Perform the following, and simplify:

   a. $\frac{3}{4} + \frac{2}{4} + \frac{6}{4}$

b. $\frac{3}{16} + \frac{7}{16}$

c. $\frac{5}{9} + \frac{3}{4}$

d. $\frac{5}{6} + \frac{3}{5} + \frac{6}{7}$

7. Perform the following, and simplify:

a. $\frac{1}{2} \times \frac{3}{4}$

b. $\frac{5}{6} \times \frac{3}{11}$

c. $\frac{4}{3} \times \frac{3}{4}$

d. $\frac{7}{5} \times \frac{14}{15}$

e. $\frac{1}{2} \div \frac{3}{4}$

f. $\frac{5}{6} \div \frac{3}{11}$

g. $\frac{4}{3} \div \frac{3}{4}$

h. $\frac{7}{5} \div \frac{14}{15}$

8. Identify each of the following, and list each component, giving its type:

a. x
b. 17
c. $13x$
d. $4xyz$
e. $15 + 12mn$
f. $3a^2 + 4ab + 5b^2$

9. Perform each of the following, and simplify where possible:

a. $z^2 \times z^3$
b. $mn^2 \times m^2$
c. $a^3b^4 \times ab$
d. $c^3 \div c^4$
e. $y^5 \div xy$
f. $4a^2b \div 2ab$
g. $(5a^2)^3$
h. $64^{1/3}$

10. a. Simplify the fractions in Table 4-1, expressing them as ordinary or mixed fractions.
    b. Express them as decimal fractions.

Chapter 5

Relationships, Functions, Linear Equations, and Inequalities

Objectives

After completing this chapter you should be able to:

1. Define a relationship and a function using words;
2. Graph a set of ordered pairs to show a binary relationship;
3. Determine whether or not a relationship is a function;
4. Define the inverse of a relationship;
5. Determine whether the inverse of a given function is also a function;
6. Use functional notation in the general (or nonspecific) sense and with specified expressions;
7. Differentiate between linear and nonlinear functions;
8. Set up the explicit form of a linear equation using either general functional notation or a specified expression;
9. Graph a linear equation;
10. Define and calculate the slope and intercept of a linear equation;
11. Graph a linear inequality in the explicit form;
12. Starting from an implicit (conditional) form of an equation and using the appropriate difference reduction operators (axioms), solve the equation by transforming it to the explicit form.

Chapter Map

The sections comprising this chapter may be summarized as follows:

Section Number	Required Reading	Optional Reading	Generic Development	Application to Management	Fundamental Principles	Complex Material
	(R)	(O)	(G)	(A)	(F)	(C)
5.1	x			x	x	
5.2	x		x	x	x	
5.3	x		x		x	
5.3.1	x		x		x	
5.3.2	x			x	x	
5.4		x	x		x	
5.4.1		x	x		x	
5.4.2		x	x		x	
5.4.3		x	x		x	
5.4.4		x	x		x	
5.4.5		x	x		x	
5.4.6		x	x		x	
5.4.7		x	x		x	

5.1 INTRODUCTION[R,A,F]

In problem solving in health care management it often is helpful to understand relationships among characteristics of the problem situation. Problems may arise because relationships are not recognized or are poorly understood. In this case, part of the solution lies in achieving a better understanding of the relationship so that the probable impact of proposed actions can be better assessed. This chapter considers the use of relationships to capture the various dimensions of a given problem or situation mathematically. However, such an approach assists only in improving understanding and does not lead automatically to a solution. Judgment and intuition still are absolutely necessary.

A problem in scheduling the operating room may emanate from an inadequate understanding of the relationship between various types of surgery and the amount of required operating time. The presence of such a problem may be indicated by the frequent need to reschedule surgery, inadequate use of the operating room, and excessive costs. On the other hand, a number of different methods based on the relationship between various types of procedures and the required amount of operating time have been used successfully in many hospitals to avoid these problems.

The satisfaction of patients or the morale of employees may be related to various personal characteristics or to situational characteristics. Sometimes important characteristics of the situation are easily identified, as when nurses are assigned tasks that they feel are inappropriate and should be performed by other employees. The nurses are likely to express their dissatisfaction and the reason for it. The relationship of interest, then, is between type of task and satisfaction.

In other cases the important relationship may not be very obvious. In monitoring patient satisfaction, an excessive number of patients may be dissatisfied and their unhappiness may be related to several aspects of their situation. Often studies designed to examine patient satisfaction are performed with no real idea of what relationship should be investigated, and the study may be undertaken for the sole purpose of obtaining accreditation. On the other hand, formalizing the study of relevant relationships may assist substantially in understanding and solving the problem.

The study of relationships has been formalized and usually is referred to as *causal analysis*. Epidemiologists frequently employ causal analysis in statistical studies that are designed to determine the causes of specific diseases or factors contributing to diseases. These concepts are well expounded in the book *Causal Thinking in the Health Sciences* by Merwyn Susser. Because of the necessity for action, the major disadvantage health care administrators face is the lack of time to study a relationship adequately. However, the concepts of causal analysis are central to problem solving. Understanding the concepts of relationships, and expressing relationships in the form of linear equations and inequalities as presented in this chapter, will help managers who want to improve their problem-solving skills or researchers who wish to explore and use causal analysis in formalized research settings.

5.2 RELATIONSHIPS AND FUNCTIONS (R,G,A,F)

The study of a relationship deals with the association between two or more characteristics of interest. The characteristics usually are referred to as *variables* and the basic elements of the study (e.g., patients, objects, or situations) are called *cases* or *observations.* To study a relationship, two or more variables must be measured for each case or observation. For example, in studying the relationship between height and weight, these two variables must be *observed* or *measured* for each person in the study. The simplest relationships are *binary* in that only two variables are studied. However, more complex relationships often are of interest, as in the differential study of height and weight for males and females. It is possible to study the relationship between patient status and amount of nursing care required; such studies have resulted in systems for determining areas in which additional nurses are most required during any given shift.

When investigating a binary relationship, one of the variables is regarded as the dependent variable and the other as the independent variable. The exact nature of the relationship can be seen by arranging the information into a *set of ordered pairs* of values of the two variables. For each observation, two numbers (or other appropriate symbols such as M, F for sex) are presented within parentheses, with the independent variable followed by the dependent. These often are referred to as the *domain* and the *range* or the x and y variables, respectively. The relationship, then, consists of the paired values.

$$\boxed{\{(x_1,y_1), (x_2,y_2), (x_3,y_3), \cdots, (x_n,y_n)\}}$$

Here, the subscript refers to the case number and indicates that each pair of values contains the data for one and only one case.

In more complex relationships, there often are several independent variables and one dependent variable. If two independent variables are incorporated in the study, the relationship might be portrayed by

$$(y_1, x_1, w_1), (y_2, x_2, w_2), \cdots (y_n, x_n, w_n)$$

where y is the dependent variable while x and w are the independent variables. Often the data in more complex relationships are presented with the dependent variable either first or last in the list, and the person conducting the study must specify the order of the variables and their meaning. Thus, if the independent variables are p, q, and r, and the dependent variable is y, the relationship could be represented as:

$$\{(p_1, q_1, r_1, y_1), (p_2, q_2, r_2, y_2), (p_3, q_3, r_3, y_3), \cdots\}$$

or as

$$\{(y_1, p_1, q_1, r_1), (y_2, p_2, q_2, r_2), (y_3, p_3, q_3, r_3), \cdots\}$$

A binary relationship consists of a subset of all of the possible (x, y) pairs. A more complex relationship is a subset of all the possible combinations of the dependent and independent variables. For example, it would be quite usual to find a 6-foot male weighing 176 pounds, but not 3 pounds, and not 500 pounds. Thus, only some of the possible combinations form the relationship.

The relationship also can be displayed in tabular form where the columns represent variables and the rows represent cases. Thus Table 5-1 might represent the relationship between nurses, their training, their years of experience, and their salary, for a given floor of a hospital.

From this set of data, three binary relations of interest to the personnel director can be extracted: (*Training, Experience*), (*Training, Salary*), and (*Experience, Salary*). The triple relationship of (*Training, Experience, Salary*) is probably the most meaningful, but it also is the most difficult to portray.

Relationships can be portrayed graphically, and the visual presentation often is extremely effective for conveying information. The independent (domain) variable is measured on the x axis and the dependent (range) variable on the y axis. When the variables have natural scales or orders, the process of plotting data poses no problems. However, when there is no natural order, the decision concerning the scale that will be used is arbitrary. In the case of the (Training, Salary) relationship, the training variable is categorical, but there is a natural order from aide to registered nurse. In such a graph, each observation is represented by a point and, as seen earlier, the relationship is given by the pattern formed by the points. The (Training, Salary) relationship is shown in Figure 5-1, which depicts vividly the nature of the relationship in which the salary paid by the institution increases as the level of training rises.

Table 5-1 Nurses, Training, Experience

Number	Initials	Training	Years of Experience	Annual Salary
1	J.M.R.	Dip	3	12,000
2	M.K.L.	R.N.	5	15,000
3	D.E.S.	Dip	2	10,000
4	A.M.F.	Aide	1	5,000
5	B.L.C.	Aide	3	7,000
6	M.J.L.	R.N.	1	12,000
7	S.K.	R.N.	2	13,000
8	J.T.	Dip	5	13,000
9	M.J.M.	Dip	2	11,000
10	C.J.	Aide	2	5,500
11	N.K.	Aide	3	6,000
12	K.A.W.	Aide	4	7,000

Figure 5-1 Training and Salary of Nursing Personnel

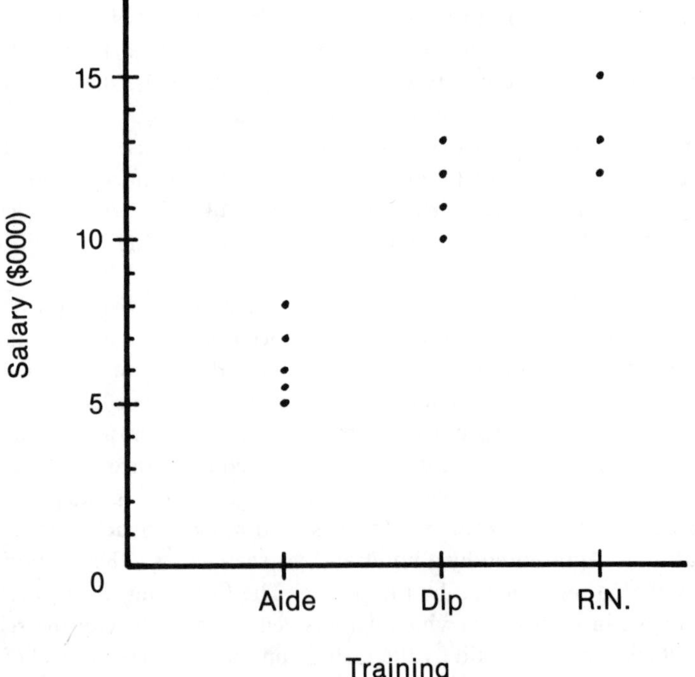

A relationship is called a *function* if each value of the domain (x) variable has only *one* value for the range (y) variable. Obviously, the relationship of training and salary is not a function, because at each level the nurses receive different salaries. In terms of our example, Figure 5-2 reveals that the relationship between the base salary and each level of training is a function. Similarly, the relationship represented by (Initials, Salary) is also a function, but of course such a relationship is less useful to study than any of the others.

For a binary relationship, the order of the variables can be reversed, which reverses their roles, and the result is called the *inverse* relationship. We often are interested in knowing whether the inverse of a function is still a function. For example, the relationship

$$\{(12, \text{JMR}), (15, \text{MKL}), (10, \text{DES}), \cdots\}$$

is not a function because there are two 12s, two 13s, and two 7s with different sets of initials. Any such pair of values eliminates the possibility that the inverse is a function.

Figure 5-2 Training and Base Salary of Nursing Personnel

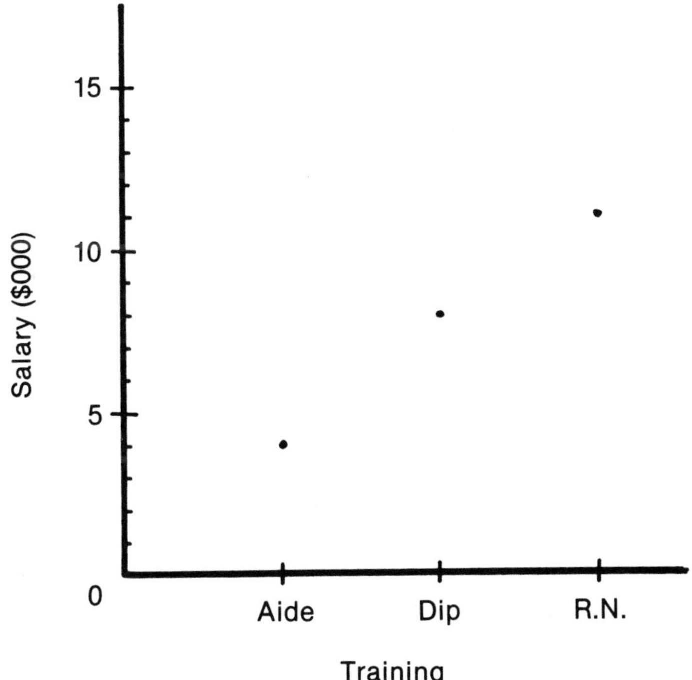

It should be obvious that when dealing with data collected in managerial studies or in formal research there will always be relationships, but only a few of them will be "strong" enough to qualify as functions. In this text, however, we examine algebraic functions in which the practicalities concerning variability in real data can be ignored.

5.3 ALGEBRAIC FUNCTIONS AND EQUATIONS (R,G,F)

In algebra, we assume that there is no statistical uncertainty in our data, which implies that we may employ a functional relationship, as expressed by an equation, when dealing with a given set of data. An equation is a mathematical statement in which the left- and right-hand sides are equal or can be made equal if the variables of interest assume appropriate values. The relationship is represented by an equation for a curved or straight line. For example, the equation

$$y = 3 + 4x$$

represents a straight line while

$$y = 7 + 2x + 3x^2$$

represents a curved line known as a parabola. Similarly, the equation

$$y = 4e^{-2x}$$

represents an exponential curve. When we want to avoid being specific about the exact form of an equation, we can use functional notation. Thus, $y = f(x)$, which is read as "y equals f of x," can represent a straight line or a curve. The algebraic expression represented by $f(x)$ can include:

1. a constant,
2. the variable x,
3. any power of x (e.g., x^2, x^3, etc.), or
4. a special function of x such as an exponential (e^x) or a trigonometric function (e.g., $\sin(x)$, $\cos(x)$, etc.).

Consequently valid expressions represented by $f(x)$ and the values they assume when substituting specific values of x could include:

$$f(x) = x, \quad f(1) = 1, \quad f(2) = 2, \quad \text{etc.}$$

$$f(x) = 5x, \quad f(1) = 5, \quad f(2) = 10, \quad \text{etc.}$$

$$f(x) = 2 - 3x, \quad f(1) = -1, \quad f(2) = -4, \quad \text{etc.}$$

More complex expressions could include:

$$f(x) = -1 + 2x - .5x^2 \qquad f(x) = \cos(3x)$$

$$f(x) = 2.5e^{.125x} \qquad f(x) = \sin(-x)$$

$$f(x) = a + bx^2 + cx^3$$

In many problems, it is necessary to represent several different functions that are of interest. In such a situation, a different letter is substituted for "f" in $f(x)$ for each function. For example, $g(x)$, $h(x)$, and $k(x)$ might represent three functions of x that differ from $f(x)$. The letters f, g, and h are ones that are commonly employed to represent functional relationships.

5.3.1 Linear Functions and Equations (R,G,F)

A linear function is one in which the independent variable is either absent (leaving only a constant) or appears as the first power only. (The "line" part of "linear" means "straight.") If x is the independent variable and y the dependent, the general form of the linear function is

$$y = a + bx \qquad \textbf{(5.1)}$$

where a and b are constants. If b is 0 (zero), then the value of y is fixed. If b is other than 0, then the value of y depends on the value of x. To obtain the value of y, we simply substitute the values of a, b and x into the expression and perform the required multiplication and addition. For example, consider the equation $y = a + bx$, where $a = 3$, $b = 2$, and $x = 5$. In this case we find that

$$y = 3 + (2 \times 5) = 3 + 10 = 13$$

For the values of a and b given above, the values of y corresponding to specified values of x are easily calculated, and the results can be displayed in tabular form as in Table 5-2. The center column of the table shows the substitution of values for the independent variable into the expression and should be required only by those who are unfamiliar with the process of substitution. Those who are familiar with this process may ignore this column.

In Section 3.2 the plotting of line charts was discussed. For plotting of straight line equations the previous discussion usually must be extended to allow negative x and y values. Also, since all points are expected to fall on a straight line, it is simplest to join the two most extreme points ((−10, −17) and (10, 23)) by a straight line. If any points do not fall on that line it is necessary to check the accuracy of the plotted points (using judgment to decide which one(s) might be in error).

The (x, y) pairs from Table 5-2 are plotted in Figure 5-3, and all points fall on the straight line. Referring to our earlier discussion, it will be observed that the values of x and y are ordered pairs, and the line in the figure represents the relationship between x and y. The relationship is a function (in terms of Section 5.2) because each value of x has one, and only one, value of y.

In the preceding paragraphs we have used the *explicit* form of the equation $y = a + bx$. Such an expression is called explicit because the value of y (left-hand side) can be found directly by substituting the value of x into the right-hand side of the equation and no further manipulation is necessary. We shall deal with implicit forms later.

The explicit form, $y = a + bx$, is also called the slope-intercept form because the constants a and b play specific roles. Here, the term "a" is called the y intercept and gives the value of y when $x = 0$. Table 5-2 shows that the y

Table 5-2 Calculation of the Values of the Dependent Variable for a Specified Linear Function, $y = 3 + 2x$

x	Substituted Values	y
−10	$3 + 2(-10) = 3 - 20$	−17
−5	$3 + 2(-5) = 3 - 10$	−7
−1	$3 + 2(-1) = 3 - 2$	1
0	$3 + 2(0) = 3$	3
1	$3 + 2(1) = 3 + 2$	5
2	$3 + 2(2) = 3 + 4$	7
5	$3 + 2(5) = 3 + 10$	13
10	$3 + 2(10) = 3 + 20$	23

intercept is 3. Similarly, the "x intercept" is the value of x when $y = 0$. When the equation $y = a + bx$ is plotted, the value of y (if one exists) that corresponds to the point at which the graph intersects the y axis is identified as the y intercept. Similarly, the x intercept is the value of x that corresponds to the point (if one exists) at which the function intersects the x axis.

The constant "b" represents the *slope* of the line that is defined as the change in the value of y that results when x changes by one unit. Returning to Table 5-2 and focusing on the range $-1 \leq x \leq 2$, we see that an increase of one unit in the value of x is accompanied by a two unit increase in the value of y. As a consequence, the slope of the line is 2. The change in y is represented as Δy (Δ is the Greek letter "delta") while the change in x is represented as Δx. The slope of a straight line is the ratio $\Delta y / \Delta x$ and a line is straight if the slope is the same at all points as well as for any size of interval, Δx. The special problem of $\Delta x = 0$ is considered in Chapter 9, in which we introduce differential calculus.

Returning to the example in which

$$y = 3 + 2x$$

we find that the term a in $y = a + bx$ assumes the value 3, which of course is the y intercept. Similarly, Figure 5-3 reveals that when the function $y = 3 + 2x$ is portrayed graphically, the resulting graph intersects the vertical axis at the point (0,3) that also indicates that the y intercept is 3. Since the slope of the line is given by $\Delta y/\Delta x$, we may verify that the term b in $y = a + bx$ is the slope by

$$\text{Slope} = \frac{\Delta y}{\Delta x} = \frac{y_2 - y_1}{x_2 - x_1} \tag{5.2}$$

where (x_1, y_1) and (x_2, y_2) are two points. Referring to the expression $y = 3 + 2x$ and letting $(x_1, y_1) = (0,3)$ while $(x_2, y_2) = (10, 23)$, the slope of the line between these two points is given by

$$\frac{y_2 - y_1}{x_2 - x_1} = \frac{23 - 3}{10 - 0} = \frac{20}{10} = 2 = b$$

When employing Equation 5-2, it is necessary to maintain the relative positions of the members of the ordered pairs (x_1, y_1) and (x_2, y_2). We can just as easily calculate the slope between any two points on a straight line, say (5, 13) and $(-10, -17)$, as follows

$$\text{slope} = \frac{\Delta y}{\Delta x} = \frac{y_2 - y_1}{x_2 - x_1} = \frac{(-17) - 13}{(-10) - 5} = \frac{-30}{-15} = 2$$

In the explicit form, the *b* value in the equation for a *straight* line always gives the slope. Thus, in terms of our example, the slope of the line is given

Figure 5-3 Plotting the Equation $y = 3 + 2x$

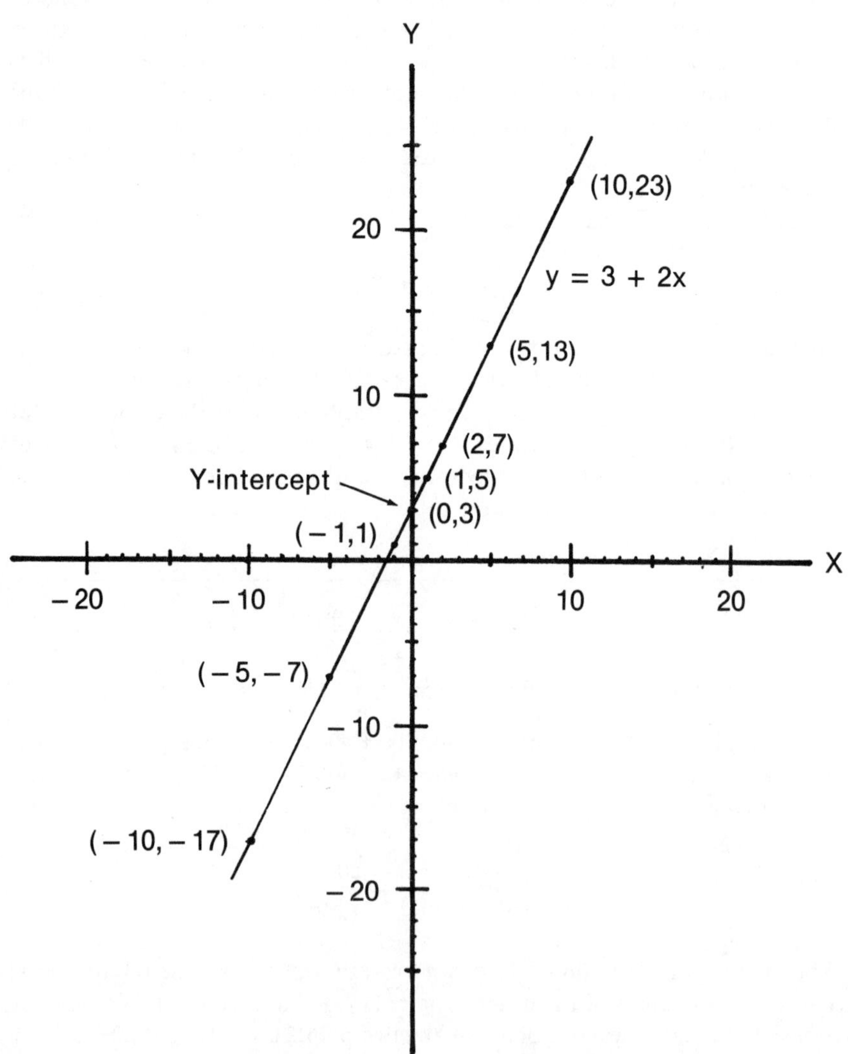

by $b = 2$. In Chapter 9 we consider the problem of determining the slope of a curved line.

5.3.2 Linear Inequalities (R,A,F)

There are many situations requiring techniques that are simple extensions of the material discussed previously. We begin with a simplified example. Assume that the hospital frequently operates at an occupancy rate less than 95 percent. If we let y represent the occupancy rate, the statement $y \leq .95$ means that y can assume any value less than or equal to .95. Assuming that the occupancy rate can never fall below zero, the statement $y \leq .95$ means that y can range from 0 percent to 95 percent inclusively. This situation can be expressed in the form

$$0.00 \leq y \leq .95 \tag{5.3}$$

Sometimes the relationship between two variables assumes a similar form. Suppose we know that the optimum number of labor hours (y) a laboratory requires to produce a given number of hundreds of tests (x) is expressed by the equation

$$y = 25 + 3x \tag{5.4.1}$$

However, there are conditions under which the lab might be less than perfectly efficient and more labor than indicated by the equation might be required. Such a situation could be represented by:

$$y \geq 25 + 3x \tag{5.4.2}$$

which is called an inequality and is linear in form. This relationship can also be portrayed graphically, as in Figure 5-4. This figure reveals that negative values of x are not possible. Consequently, the inequality

$$x \geq 0 \tag{5.5}$$

implies that the level of activity can never be reduced below zero. Such an inequality often is referred to as a "nonnegativity constraint." The area on or above the line $y = 25 + 3x$ and on or to the right of the y axis is where we would expect to find the relationship of labor hours and hundreds of tests. The area outside has been shaded to indicate that it is excluded from the relationship. Mathematicians frequently speak of inequalities by defining *feasible* and *infeasible* regions. Sometimes the shading convention is used to indicate the feasible region and sometimes, as in this example, the infeasible region.

5.4 SOLUTION OF LINEAR EQUATIONS (O,G,F)

Thus far, the discussion of relationships has been limited to those that have been expressed in an explicit form similar to

$$y = a \tag{5.6}$$

$$y = a + bx \tag{5.7}$$

Figure 5-4 Possible Relationship between Laboratory Labor Hours and Hundreds of Tests Performed

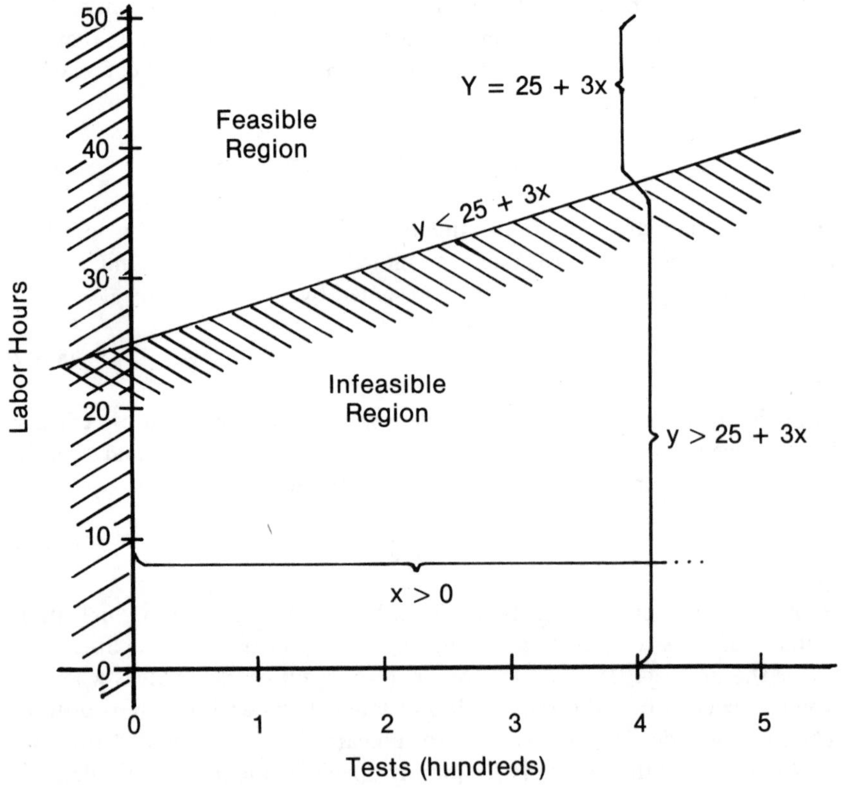

$$y = a + bx + cx^2 \tag{5.8}$$

$$y = ae^{bx} \tag{5.9}$$

where y is the dependent variable, x the independent variable, and the a, b, and c are constants. Using functional notation, the *explicit* form of an equation with two variables is

$$y = f(\text{constant and } x) \tag{5.10.1}$$

or

$$\text{dependent variable} = f(\text{constant and independent variable}) \tag{5.10.2}$$

Occasionally, however, the relationship between two different functions of the dependent variable, the independent variable, and the constants is known or given. Suppose that

$$f_1(x, y, \text{constant}) = f_2(x, y, \text{constant}) \tag{5.11}$$

In this case, Equation 5.11 is an example of a functional relation that has been expressed in implicit form. The process of transforming a relationship from an implicit form into an explicit form is called solving the equation for the dependent variable. Thus, we begin with an implicit form such as Equation 5.11 and employ a process of solution that results in the explicit form of Equation 5.10.1.

The characteristics of an equation that is expressed in an explicit form are as follows:

a. On the left-hand side of the equation is an expression that contains only a single term that consists of the dependent variable, multiplied by one; and
b. On the right-hand side of the equation is an expression consisting of one or more terms in which only constants and independent variables are represented.

By way of contrast, the dependent variable is *not* isolated on the left-hand side of an equation that is expressed in implicit form. When dealing with relationships expressed in implicit form, the dependent variable, the independent variable, and constants may be found in expressions appearing in both the right- and left-hand sides of the equation. Each term may contain: (1) either the dependent or the independent variable(s), (2) both the dependent and the independent variable(s), or (3) both the dependent and the independent variables as well as the constants that define the functional relation.

In explaining the process of solution, we employ the notion of differences between the implicit and explicit forms as well as the concept of difference reduction operators. The basic approach introduced in this section relies heavily on the concepts developed by Allen Newell and Herbert Simon in describing *Human Problem Solving*. The difference reduction operators represent steps that are employed to transform an equation that is expressed initially in implicit form into an equation that more closely resembles the explicit form described earlier. Thus, for any equation, our task is to select and apply an appropriate operator so as obtain an equation that more closely resembles the desired explicit form. Hence, the process of solution involves the identification of the sequence of operators that must be applied in a given situation so as to obtain the explicit form.

In the following discussion, we examine seven differences in which the use of each operator requires an understanding of all preceding operators. In each example, the current difference is presented first and its reduction is illustrated next. If any of the remaining differences require additional reductions, the required calculations involve the techniques developed previously and the process advances in incremental steps so that the derivation of the solution is seen easily.

In summary, the method of solving an equation consists of: (1) identifying the most complex difference, (2) applying the appropriate operator, and (3) transforming the equation into one that is less complex. This process is repeated until the desired explicit form is obtained.

5.4.1 Dependent Variable Multiplied by Another Factor[O,G,F]

Consider first a situation in which only the dependent variable appears in a single term on the left-hand side of the equation and it has been multiplied by one or more factors other than one. In this case, the difference between the implicit and explicit forms consists of the nonunitary factors. Consequently, we may eliminate the difference by multiplying both sides of the equation by the inverse of the nonunitary factors. As an example, let

$$4y = 3x + 2$$

Multiplying both sides of the equation by the inverse of 4 (i.e., $1/4$) we obtain

$$\frac{1}{4}(4y) = \frac{1}{4}(3x + 2)$$

$$\boxed{y = \frac{3}{4}x + \frac{1}{2}}$$

Similarly, if

$$3y = 5$$

we find that

$$\frac{1}{3}(3y) = \frac{1}{3}5$$

$$\boxed{y = \frac{5}{3}}$$

A similar approach is employed in each of the following examples.

Example 1

Let:

$$5xy = 20$$

Solution:

$$\frac{1}{5x}(5xy) = \frac{1}{5x}(20)$$

$$\boxed{y = \frac{4}{x}}$$

Example 2

Let:

$$\frac{7y}{3x} = 21$$

Solution:

$$\frac{3x}{7}\left(\frac{7y}{3x}\right) = \frac{3x}{7}(21)$$

$$\boxed{y = 9x}$$

Example 3

Let:

$$(a + bx)y = 25$$

where a and b are any nonzero constants
Solution:

$$\frac{1}{a + bx}(a + bx)y = \frac{1}{a + bx}(25)$$

$$\boxed{y = \frac{25}{a + bx}}$$

This section covers the most elementary, but also the most important, principle for the solution of linear equations. We have chosen to orient this discussion around the multiplication by the inverse, rather than around the (sometimes simpler) division by the coefficient of y. Our reason is that the inverse *must* be used when we extend our analysis of one equation to the solution of a system of several linear equations using matrix notation, as developed in Chapter 19. In matrix algebra there is no operation of division, but there is an inverse operation and its use allows us to simplify the solution of simultaneous linear equations that is the topic of Chapter 6.

5.4.2 Additional Terms of Left-Hand Side(O,G,F)

Consider next a slightly more complex situation represented by

$$21y + 14x + 3 = 35x + 24$$

On the left-hand side of the equation are not only the dependent variable but also an expression $(14x + 3)$ that contains the independent variable and a constant. In this case, we simply subtract the expression $(14x + 3)$ from both sides of the equation and obtain

$$21y + (14x + 3) - (14x + 3) = 35x + 24 - (14x + 3)$$
$$21y = 35x - 14x + 24 - 3$$
$$21y = 21x + 21$$

We may now solve for y using the technique described earlier. Thus, multiplying both sides of the equation by the inverse of 21, we obtain

$$\frac{1}{21}(21)y = \frac{1}{21}(21x + 21)$$
$$\boxed{y = x + 1}$$

5.4.3 Dependent Variable on Both Sides$^{(O,G,F)}$

When the dependent variable appears in a simple term on both sides of the equation, our objective is to rearrange the equation so that only the dependent variable appears on the left-hand side. To achieve this objective, we identify the term on the right-hand side that contains the dependent variable and subtract the term from both sides of the equation. We then add those factors that are multiplied by the dependent variable and solve as before. As an example, let

$$23y + 14x + 3 = 2y + 35x + 4$$

In this case, we simply subtract $2y$ from both sides of the equation and obtain

$$23y - 2y + 14x + 3 = 2y - 2y + 35x + 4$$
$$21y + (14x + 3) = 35x + 4$$

Notice that the resulting equation assumes a form that may be solved using the technique described above. Subtracting the expression $(14x + 3)$ from both sides of the equation and solving for y, we find that

$$y = x + 1/21$$

5.4.4 Dependent Variable As a Denominator$^{(O,G,F)}$

Suppose the dependent variable appears in the denominator of a fraction that is the only term on the left-hand side of the equation. The difference between the implicit and explicit forms is reduced by finding the inverse of each side of the equation. Notice that if two expressions are equal, the corresponding inverses also are equal. However, when terms appearing on both sides of the equation are inverted, the dependent variable appears in the numerator of the fraction. To illustrate this approach, suppose that

$$\frac{a + bx}{cy} = d + ex$$

where:

(1) a, b, c, d, and e are constants and
(2) c as well as the expressions $a + bx$ and $d + ex$ are not equal to zero.

In this case, we find the inverse of each side of the equation, which yields

$$\frac{cy}{a + bx} = \frac{1}{d + ex}$$

We can obtain the same results by multiplying both sides of the original equation by the expression

$$\left(\frac{cy}{a + bx}\right)\left(\frac{1}{d + ex}\right)$$

Employing this procedure, we find that

$$\left(\frac{a + bx}{cy}\right)\left(\frac{cy}{a + bx}\right)\left(\frac{1}{d + ex}\right) = (d + ex)\left(\frac{1}{d + ex}\right)\left(\frac{cy}{a + bx}\right)$$

Simplifying this result yields

$$\frac{1}{d + ex} = \frac{cy}{a + bx}$$

which is equivalent to

$$\frac{cy}{a + bx} = \frac{1}{d + ex}$$

We may now use the techniques already developed and proceed as follows:

$$\frac{a + bx}{c}\left(\frac{cy}{a + bx}\right) = \frac{a + bc}{c}\left(\frac{1}{d + ex}\right)$$

$$\boxed{y = \frac{a + bx}{c(d + ex)}}$$

As an additional illustration, let

$$\frac{2 + 3x}{4y} = 5 + 6x$$

Inverting both sides and solving for y, we find that

$$y = \frac{2 + 3x}{4}\left(\frac{1}{5 + 6x}\right)$$

$$\boxed{y = \frac{2 + 3x}{20 + 24x}}$$

Similarly, if

$$\frac{7x}{3y} = 14$$

we find that

$$\frac{3y}{7x} = \frac{1}{14}$$

and

$$\boxed{y = x/6}$$

5.4.5 Dependent Variable in Several Denominators(O,G,F)

When the dependent variable appears in the denominator of several terms, a common denominator must be found in order to derive a single term. For example, suppose that

$$\frac{2x}{3y} + \frac{7x}{4y} = 26$$

In this case, the common denominator of the terms appearing on the left-hand side of the equation is $12y$. We then proceed as follows:

$$\frac{4(2x) + 3(7x)}{12y} = 26$$

$$\frac{8x + 21x}{12y} = 26$$

$$\frac{29x}{12y} = 26$$

$$y = \frac{29x}{12(26)}$$

$$\boxed{y = \frac{29x}{312}}$$

In the examples thus far, the dependent variable has appeared in (1) *single term expressions* (e.g., $4y$); (2) terms that also include the independent variable (e.g., $5xy$); (3) terms in which the multiplication of the dependent variable is distributed over a function of x (e.g., $(a + bx)y$); (4) the denominator of a fraction [e.g., $(2 + 3x)/4y$]; and (5) *several term expressions* that are the sum of simple terms containing the dependent and independent variables (e.g., $21y + 14x + 3$). The single or several term expressions introduced above frequently are regarded as simple additive forms. In the following discussion, we illustrate the techniques that may be used to change other forms into simple additive forms.

5.4.6 Conversion to a Simple Additive Form(O,G,F)

When the multiplication of the independent variable is distributed over an expression containing the dependent variable, the transformation of the expression into a simple additive term is accomplished by (1) performing the required multiplication and (2) removing the dependent variable as a common term from the expression. For example, let

$$3x(3y + 5) = 20x$$

After performing the required multiplication we find that

$$9xy + 15x = 20x$$

Subtracting $15x$ from both sides of the equation yields

$$9xy = 20x - 15x$$

We may now remove $9x$ from the left side, as follows:

$$\frac{1}{9x}(9xy) = \frac{1}{9x}(5x)$$

When both sides of the equation are multiplied by $1/(9x)$, we obtain

$$\boxed{y = \frac{5}{9}}$$

As an additional illustration, suppose that

$$3x(4x + 3y + 5) = 20x$$

After performing the required multiplication, we find that

$$12x^2 + 9xy + 15x = 20x$$

Rearranging slightly, the expression becomes

$$9xy + (12x^2 + 15x) = 20x$$

Subtracting $(12x^2 + 15x)$ from both sides yields

$$9xy = 20x - (12x^2 + 15x)$$

We may now multiply both sides of the resulting equation by $1/(9x)$ and obtain

$$\frac{1}{9x}(9xy) = \frac{20x}{9x} - \frac{1}{9x}(-12x^2 + 5x)$$

$$y = \frac{-12x}{9} + \frac{5}{9}$$

After simplifying the equation, we find that

$$y = -\frac{4x}{3} + \frac{5}{9}$$

5.4.7 Dependent Variable As an Exponent[O,G,F]

In the final situation in this section, suppose the dependent variable and perhaps the independent variable as well as a constant appear as the exponent of some base. In such a case, we may reduce the difference between the implicit and explicit forms by employing the logarithms of all terms on both sides of the equation. At this point in the analysis, recall that

$$\log_{10}(10^{\text{exponent}}) = \text{exponent}$$

$$\log_{10}(\text{constant}^{\text{exponent}}) = (\log_{10}\text{constant}) \times \text{exponent}$$

$$\ln(e^{\text{exponent}}) = \text{exponent}$$

$$\ln(\text{constant}^{\text{exponent}}) = \ln(\text{constant}) \times \text{exponent}$$

The first two definitions use logarithms of the base 10 ($\log_{10}$) that, as mentioned earlier, are called "common logarithms." The last two definitions use logarithms of the base "e" that are called natural logarithms (ln). The two systems of logarithms are equivalent when each is multiplied by an appropriate constant. In this case, we find that

$$\ln(10^{\text{exponent}}) = \ln(10) \times \text{exponent}$$

and

$$\log_{10}(e^{\text{exponent}}) = \log_{10}(e) \times \text{exponent}$$

where $\ln(10) \cong 2.30259$ and $\log_{10}(e) \cong .43429$.

On the basis of these definitions, we see that the difference between

$$e^{f(x,y,\text{constant})} = f(x, \text{constant})$$

and the corresponding explicit form may be reduced by the use of logarithms. In this case, we find that using logarithms on both sides of the equation yields

$$f(x,y,\text{constant}) = \ln(f(x, \text{constant}))$$

More specifically, suppose that

$$e^{3y + 4x} = 25$$

Employing the approach outlined above, we find that

$$3y + 4x = \ln(25)$$

where $\ln(25) \cong 3.2188$. After substituting appropriately, and rearranging slightly, we obtain

$$3y = 3.2188 - 4x$$

$$y = \frac{1}{3}(3.2188 - 4x)$$

$$\boxed{y = 1.0729 - \frac{4x}{3}}$$

REFERENCES

Merwyn Susser, *Causal Thinking in the Health Sciences.* Oxford, England: Oxford University Press, 1973.

Allen Newell and Herbert Simon, *Human Problem Solving.* Englewood Cliffs, N.J.: Prentice Hall, 1972.

Problems for Solution

1. Draw a graph similar to Figure 5-1 showing the relationship between training and years of experience for the nurses in Table 5-1. Is the relationship functional? Why?
2. Draw a scattergram of the relationship between years of experience and salary for each level of training in Table 5-1.
3. Determine the values of y for values of x in the set $\{-5, -2, -1, 0, 1, 2$ and $5\}$ for each of the following and plot the equations on graph paper. What is the value of the y intercept and the slope for each?

   a. $y = 3 + 4x$
   b. $y = -2 + 5x$
   c. $y = 4 - 2x$

4. Cross-hatch the regions $\leq$, $\geq$, and $\leq$ for the equations in 3a, b, and c, respectively.
5. Solve the following equations for y:

   a. $3y = 4x + 6$
   b. $-2y = 6x - 5$
   c. $6xy = 33$
   d. $\frac{6y}{x} = 14$
   e. $(3 + 4x)y = 5$

6. Solve the following equations for y:

   a. $3x + 4y = 17$
   b. $3x + 4xy = 21$

7. Solve the following equations for y:

   a. $3y = -4y + 7x + 9$
   b. $4x - 3y = -7y + 8 + 2x$

8. Solve the following equations for y:

   a. $\dfrac{3x}{4y} = 5$

   b. $\dfrac{7x + 3}{6y} = 17$

   c. $\dfrac{3 - 2x}{7y} = 4 + 2x$

9. Solve the following equations for y:

   a. $\dfrac{5x}{6y} + \dfrac{9x}{7y} = 14$

   b. $\dfrac{4x}{3y} + \dfrac{2x}{5y} = -7$

   c. $\dfrac{3x}{5y} - \dfrac{7x}{10y} = 12$

10. Solve the following equations for y:

    a. $5x(6y + 4) = 15x$

    b. $\dfrac{1}{3x}(7y + 9) = 14$

    c. $-2x(-3x + 4y - 7) = 25x$

11. Solve the following equations for y:

    a. $10^{5x + 3y} = 17$
    b. $e^{3y + 5} = 21$

Chapter 6

Simultaneous Linear Equations

Objectives

After completing this chapter you should be able to:

1. Solve a system of two (or three) simultaneous linear equations in two (or three) variables by the method of elimination and substitution;
2. Solve a system of two simultaneous linear equations in two variables by means of a graph;
3. Describe the process of solving a system of simultaneous linear equations by the elimination and substitution method;
4. Describe the conditions under which there is no single or unique solution for a system of equations. Use a two-equation example to demonstrate the conditions and the solution process.

Chapter Map

The sections comprising this chapter may be summarized as follows:

Section Number	Required Reading	Optional Reading	Generic Development	Application to Management	Fundamental Principles	Complex Material
	(R)	(O)	(G)	(A)	(F)	(C)
6.1	x		x		x	
6.2	x		x	x	x	
6.2.1	x		x	x	x	
6.2.2	x			x	x	
6.3	x		x		x	
6.3.1	x		x		x	
6.3.2	x		x		x	
6.3.3		x	x		x	
6.3.4		x	x			x
6.4		x	x		x	
6.4.1		x	x		x	
6.4.2		x	x		x	
6.5		x		x		

6.1 INTRODUCTION(R,G,F)

The solution of single equations considered in the preceding chapter is important both in its own right and as a building block for this and succeeding chapters. In this chapter we examine the solution of systems of simultaneous linear equations in which the number of variables is equal to the number of equations. Here we deal with algebraic and graphic methods of solution. In Chapter 19 we return to the algebraic method and use matrix algebra rather than ordinary algebra to solve a system of linear equations simultaneously.

Linear equations are used to represent relations among two or more variables, and we assume that these relations are linear in form and additive. This means that the variables appear in the equation only in the first power (i.e., x or y, not x^2, x^3, y^4, e^x, etc.) and that different variables are separate terms that are added rather than multiplied (i.e., $3x + 4y$, not $3xy$). Frequently, the assumption of linearity is overly restrictive and must be relaxed in order to accommodate nonlinear relationships. Single nonlinear equations and systems of a quadratic and a linear equation are covered in the next chapter.

A system of simultaneous linear equations consists of 2, 3, 4, or more linear equations, with the same number of variables as equations. The objective of solving the system of equations is to find a set of values that satisfies all of the equations in the system simultaneously. In many cases there is one set of values that satisfies all of the equations simultaneously; this set is called the *solution*. However, sometimes there is no solution, and sometimes there is an infinite number of solutions. These are special cases in which two or more equations can be expressed as linear combinations of each other.

The process of solving a system of linear equations is an important component of the operations research tool known as linear programming. Operations research evolved from the need to apply mathematical tools to the problems of management that emerged before World War II, were made urgent by military requirements during the war, and were facilitated by the development of computers toward the end of the war. In the 1950s, linear programming emerged as one of the most important tools for finding the best (*optimal*) allocation of resources among competing uses in problems of production. Although the field of health care management has lagged 10 to 15 years in the adoption of linear programming and other operations research techniques, these methods are very clearly applicable to many health care situations.

6.2 SYSTEMS OF LINEAR EQUATIONS IN HEALTH CARE MANAGEMENT(R,G,A,F)

6.2.1 Supply and Demand Equations(R,G,A,F)

For purposes of illustration, let us consider the traditional interaction between supply and demand as an example in which a system of equations is useful. More specifically, consider, as an extremely simplified situation, the supply and demand for aspirin. If aspirin are cheap, consumers will tend to buy many, while if they are expensive people will buy fewer. Such a situation may be represented by a *demand equation*, which is assumed to have a *negative slope* under normal conditions. If the quantity demanded is represented by y and the price by x, then a demand equation might assume the form

$$y = 3 - \frac{1}{4}x \qquad \text{(Demand)}$$

Conversely, drug firms are not likely to produce a large volume of aspirin during periods in which the price is low. However, we would expect the production of aspirin to increase in response to a rising price. As a result, *supply equations* are characterized by a *positive slope* and might assume the form

$$y = 1 + \frac{1}{3}x \qquad \text{(Supply)}$$

These two equations are in a *slope-intercept* form that may be solved explicitly for y. Together they form a system of equations when we assume that a market exists in which buyers and sellers interact so as to determine the quantity and price that will satisfy both sets of expectations at once. Thus, the system of equations representing *market equilibrium* is

$$y = 3 - \frac{1}{4}x \qquad \text{(Demand)}$$

$$y = 1 + \frac{1}{3}x \qquad \text{(Supply)}$$

These equations are portrayed graphically in Figure 6-1, where the point of equilibrium is represented by the intersection of the two lines (i.e., the point at which supply and demand are equal).

Usually a system of linear equations is not presented in the slope-intercept form, because neither the y variable nor the x variable is clearly the dependent or the independent variable. The equations usually are rearranged so that the terms containing the variables are on the left-hand side (L.H.S.), while the constants are on the right-hand side (R.H.S.). Thus, the market equilibrium for aspirin may be represented by

$$\frac{1}{4}x + y = 3 \qquad \text{(Demand)}$$

$$\frac{1}{3}x - y = -1 \qquad \text{(Supply)}$$

Notice that the slope of each of these lines is not nearly as discernible as it was previously. In the real world, the slopes of the two equations are even less discernible because the coefficients associated with x and y usually are not fractional. To approximate such a situation, multiply the demand equation by 8 and the supply equation by 9. After performing these operations we obtain the system of equations

$$2x + 8y = 24 \qquad \text{(Demand)}$$

$$3x - 9y = -9 \qquad \text{(Supply)}$$

where the coefficients are now integer values.

The process of solution results in values of x and of y that satisfy both of the equations simultaneously. Referring to our example, we find that, when $x = 3\ 3/7$ and $y = 2\ 1/7$, the demand and supply equations are satisfied simultaneously. This assertion can be verified by substituting these values in any of the forms of the set of equations. For example, the original slope-intercept form of the demand equation was given by

$$y = 3 - \frac{1}{4}x$$

Substituting 3 3/7 for x in this equation, we find that

$$\begin{aligned} y &= 3 - 1/4\,(3\ 3/7) \\ &= 3 - 24/28 \\ &= 3 - 6/7 \\ &= 2\ 1/7 \end{aligned}$$

Figure 6-1 Hypothetical Demand and Supply Equations for Aspirin

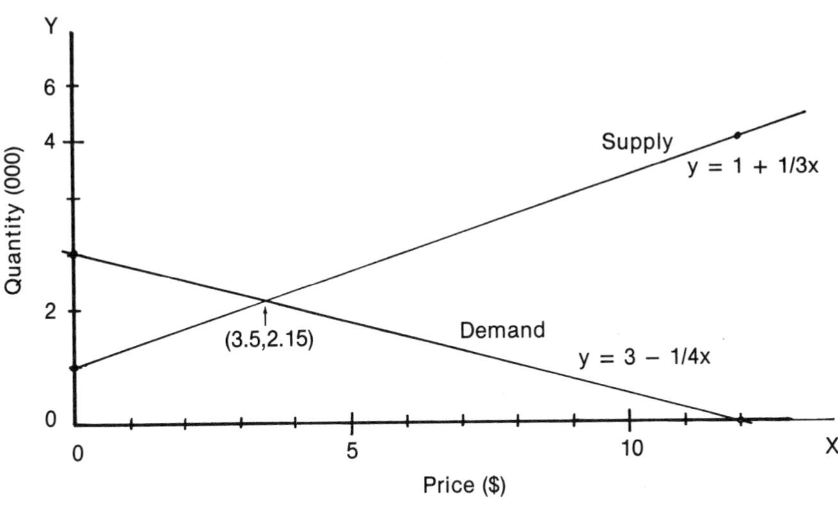

which, of course, is the required value of y. Similarly, if we substitute 3 3/7 for x in the supply equation

$$y = 1 + (1/3)x$$

we obtain

$$\begin{aligned} y &= 1 + 1/3\,(3\ 3/7) \\ &= 1 + 24/21 \\ &= 1 + 8/7 \\ &= 2\ 1/7 \end{aligned}$$

Notice that when $x = 3\ 3/7$, the value of y in both the supply and demand equations is 2 1/7. Thus, the values $x = 3\ 3/7$ and $y = 2\ 1/7$ represent the simultaneous solution to the system of equations.

In Figure 6-1, the intersection of the lines represents the point at which supply and demand are in equilibrium. In this case the point (3.429, 2.143) is the solution of the system of supply and demand equations. Thus, if we measure quantity in 1,000 units, we find that when a price of \$3.43 prevails in the market, 2,143 units will be supplied by producers and demanded by consumers.

6.2.2 Capacity Constraints(R,A,F)

In another simplified example, assume that the operational activity of our hospital is limited to providing care to either a medical or a surgical patient. The average medical patient requires 1 x-ray and 4 lab tests, while the average surgical patient requires 3 x-rays and 2 lab tests. The laboratory can produce 5,000 lab tests per month and the radiology department can provide 3,000 x-rays per month. Our objective in this exercise is to determine the mix of patients that results in the use of the maximum capacity of both departments simultaneously.

Let the number of medical patients be represented by x_1 and the number of surgical patients by x_2. The total number of lab tests will be $4x_1 + 2x_2$ and the maximum capacity is 5,000. The utilization of maximum capacity is represented by the equation:

$$4x_1 + 2x_2 = 5{,}000 \qquad \text{(Lab)}$$

Similarly, the utilization of maximum capacity in radiology is represented by

$$1x_1 + 3x_2 = 3{,}000 \qquad \text{(x-ray)}$$

The simultaneous solution of these two equations suggests that 900 medical patients and 700 surgical patients in one month will exhaust the capacity of both the laboratory and the radiology department. The reader should verify that these two values satisfy both equations simultaneously. The equations are plotted in Figure 6-2 and the graph shows the solution as well.

6.3 METHODS OF SOLVING SYSTEMS OF EQUATIONS(R,G,F)

The purpose of this section is to describe several methods of obtaining a solution to a system of simultaneous equations involving

a. two linear equations in two unknowns, and
b. three equations in three unknowns.

In addition, we examine situations in which no solution exists as well as situations in which an infinite number of solutions exists.

Figure 6-2 Hypothetical Laboratory and Radiology Capacity Equations

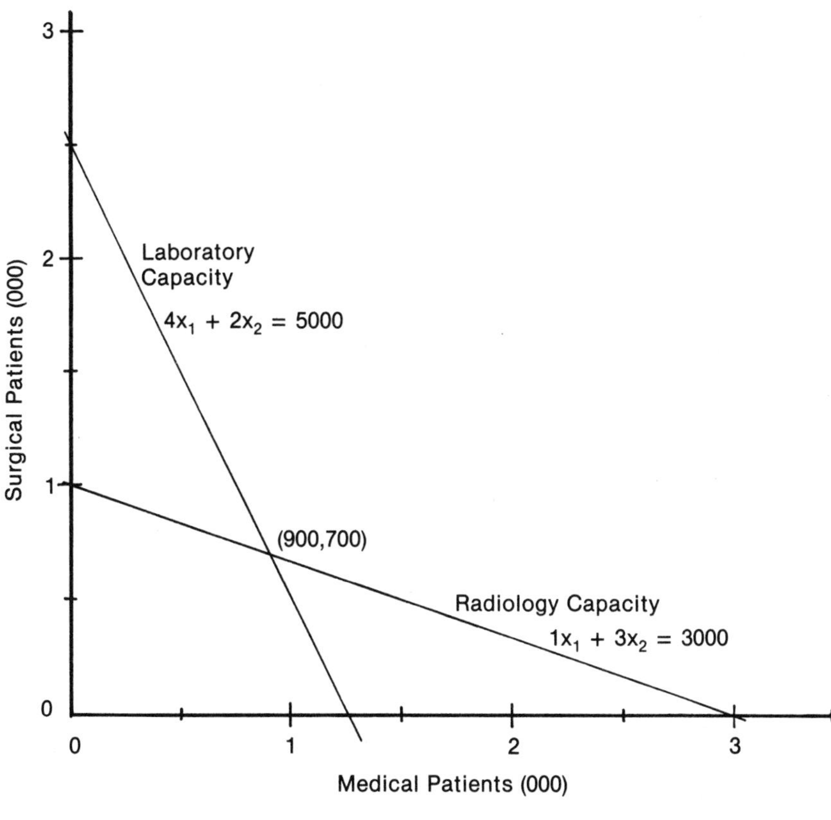

6.3.1 Solving 2 Equations in 2 Unknowns: The Method of Substitution(R,G,F)

Suppose we were given the system of linear equations represented by

$$3x - y = 4 \tag{6.1}$$

$$2x + y = 6 \tag{6.2}$$

When the substitution method is employed to derive the solution to the system of equations, we first solve one of the equations explicitly for y. Returning to our example, we may solve Equation 6.1 for y and obtain

$$y = 3x - 4$$

We may now substitute $3x - 4$ for y in Equation 6.2 and obtain

$$2x + (3x - 4) = 6 \tag{6.3}$$

in which x is the only unknown. Solving Equation 6.3 for x we obtain

$$5x = 6 + 4$$

$$\boxed{x = 2}$$

We may now substitute the value 2 for x in Equation 6.1 and solve for y as follows

$$3(2) - y = 4$$

$$-y = 4 - 6$$

$$\boxed{y = 2}$$

As can be verified, Equations 6.1 and 6.2 are satisfied simultaneously when $x = 2$ and the resulting value of y is 2.

6.3.2 Solving 2 Equations in 2 Unknowns: The Method of Elimination(R,G,F)

The example in the previous section also may be solved using the method of elimination. In this case, we need only add Equations 6.1 and 6.2, eliminating y:

$$3x - y = 4 \tag{6.1}$$

$$\underline{2x + y = 6} \tag{6.2}$$

$$5x \quad = 10$$

These results indicate that, as before, $x = 2$. Substituting the value 2 for x in one of the original equations, we find that y is equal to 2 which, of course, agrees with our earlier results.

Consider next a slightly more complex example. In this case we assume that

$$5x - 2y = 40 \qquad \textbf{(6.3)}$$

$$10x + 6y = 60 \qquad \textbf{(6.4)}$$

represents the system of equations of interest. By multiplying Equation 6.3 by 3, we obtain

$$15x - 6y = 120 \qquad \textbf{(6.3.1)}$$

We may now add Equations 6.3.1 and 6.4 so as to eliminate y and obtain

$$\begin{array}{rcl} 15x - 6y & = & 120 \\ 10x + 6y & = & 60 \\ \hline 25x \quad\quad & = & 180 \end{array}$$

which implies that x is equal to 7.2. Substituting 7.2 for x in Equation 6.4 we obtain

$$\begin{aligned} 10(7.2) + 6y &= 60 \\ 6y &= -12 \\ y &= -2 \end{aligned}$$

Hence, we find that the simultaneous solution to this system of equations is given by (7.2, −2).

6.3.3 Solving 2 Equations in 2 Unknowns Graphically(O,G,F)

A rough idea of the point at which the two equations have the same value can be gained by plotting them as linear equations on graph paper. To plot the equations we need two points on each line, and usually the ones we would like to choose are $(x, 0)$ and $(0, y)$. The first point is the x intercept and the second is the y intercept. For each line, the x and y intercepts are found, as shown previously. First, let $y = 0$ and solve (explicitly) for the x intercept. Next, let $x = 0$ and solve for the y intercept. Referring to our earlier example, we may solve Equation 6.1 for the x intercept and obtain

$$3x - 0 = 4$$

$$x = \frac{4}{3}$$

$$\boxed{x = 1\frac{1}{3}}$$

We then determine the y intercept as follows. When $x = 0$,

$$3(0) - y = 4$$

which implies that

$$\boxed{y = -4}$$

Solving Equation 6.2 for the x and y intercepts, we find that when $y = 0$,

$$2x + 0 = 6$$

$$\boxed{x = 3}$$

while, when $x = 0$,

$$2(0) + y = 6$$

$$\boxed{y = 6}$$

Also, solving Equation 6.1 explicitly for y in order to find the slope of the line, we find:

$$\boxed{y = -4 + 3x}$$

This indicates that Equation 6.1 has a positive slope with value 3. Similarly the slope of Equation 6.2 is negative and assumes the value -2.

By considering the y intercepts and the slopes we observe that Equations 6.1 and 6.2 must intersect to the right of the y axis. Considering next the x intercepts, we see that Equation 6.1 crosses the x axis at the point $(\frac{4}{3}, 0)$ with a positive slope, while Equation 6.2 crosses the x axis at the point (3, 0) with a negative slope. This indicates that the x coordinate of the point of intersection must lie between 1 1/3 and 3, while the y coordinate must be in excess of zero (i.e., it must be positive). This suggests that we must plot the two equations for values of x ranging from 0 to 3 and y values from -4 to $+6$, in order to identify the point of intersection accurately. These lines are plotted in Figure 6-3 using the considerations of this paragraph.

Depending on the care with which the lines are plotted and the range of values required, the intersection of the two lines can be determined visually from the graph. Figure 6-3 reveals that the intersection of the two lines corresponds approximately to the point (2, 2). That this is correct can be determined by substitution into the original equations. In spite of the closeness to the solution in this particular case, the graphic solution usually is only approximate because slight errors in positioning each line may lead to large errors in determining the point of intersection.

6.3.4 Algebraic Solution of 3 Linear Equations in 3 Unknowns(O,G,C)

Often a physical system will be described by a system of three or more linear equations. In such a situation, we may generalize the procedures outlined in the previous sections to accommodate three or more equations. For a system of three equations, the procedure is to solve explicitly for one of the variables in one of the equations in terms of the other variables, and to substitute the expression in the other equations, reducing the system to one in which only two equations and two unknowns appear. The procedure then is repeated, solving for one variable in terms of the other in one equation, and substituting the result in the remaining equation. This process results in a value for one variable that can be substituted in a two-variable equation to obtain the solution value for the second variable. The values of these two variables can then be substituted in a three-variable equation to obtain the solution value for the third variable.

This process can be represented schematically (Figure 6-4). In this case, we begin at the upper left-hand side of the figure. As we move down the figure, the solution for the system requires the elimination of equations and variables until only a single variable equation remains. On the right-hand side of the figure, the remaining equation is solved for that variable, and

Figure 6-3 Graphic Solution for a System of 2 Equations with 2 Unknowns

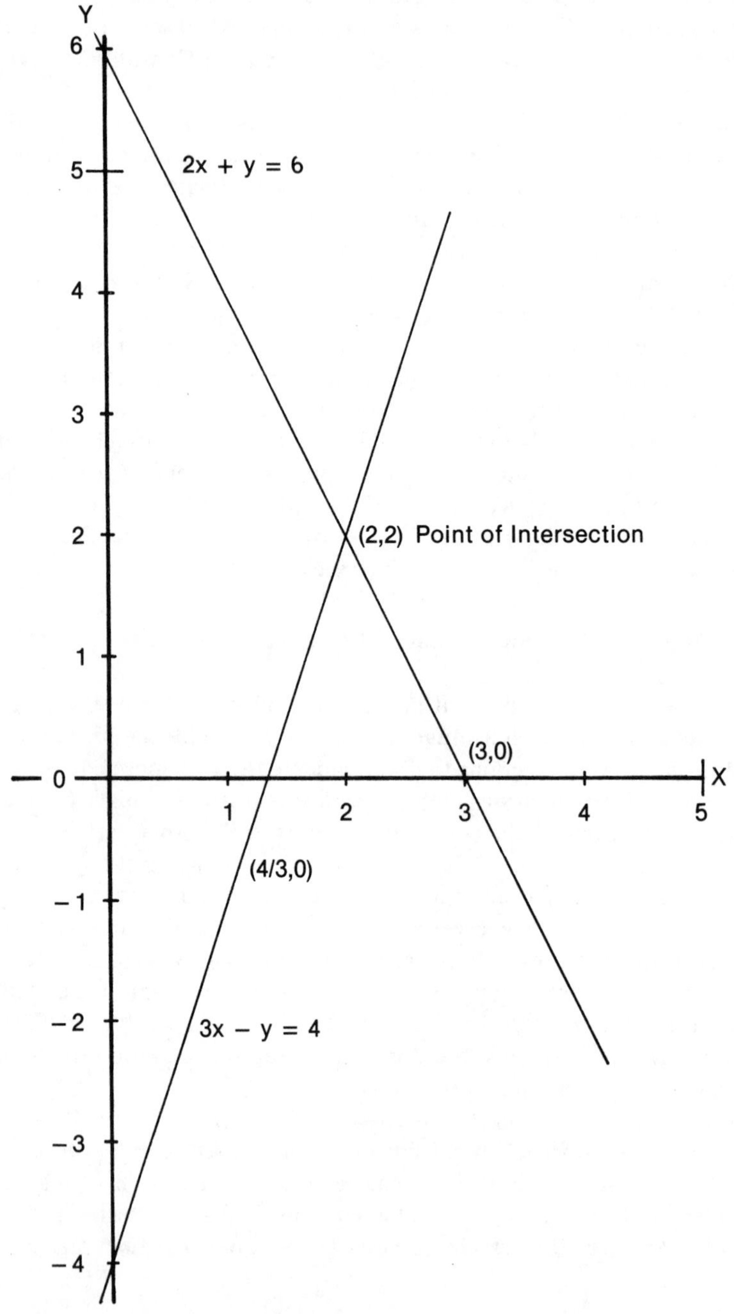

then successive substitutions are used to find the values of the second and third variables. Notice that the figure may be expanded so as to accommodate more equations and more variables. Also notice that, in order to derive a simultaneous solution, the number of variables must equal the number of equations. The equations also must be linearly independent (i.e., the slope of the lines must differ). Two situations of linear dependence are discussed in Section 6.4.

To illustrate this approach, assume that we are interested in the following system of equations

$$3x_1 + 7x_2 + 5x_3 = 9 \qquad \textbf{(6.5)}$$

$$2x_1 + 11x_2 + 3x_3 = 13 \qquad \textbf{(6.6)}$$

$$5x_1 + 3x_2 + 7x_3 = 17 \qquad \textbf{(6.7)}$$

Let us solve Equation 6.7 explicitly for x_3. We then may eliminate x_3 in Equations 6.5 and 6.6 as follows

$$x_3 = \frac{17}{7} - \frac{5}{7}x_1 - \frac{3}{7}x_2 \qquad \textbf{(6.7.1)}$$

$$x_3 = 2.429 - .714x_1 - .429x_2 \qquad \textbf{(6.7.2)}$$

Substituting these results in Equations 6.5 and 6.6, we find that

$$3x_1 + 7x_2 + 5(2.429 - .714x_1 - .429x_2) = 9 \qquad \textbf{(6.5.1)}$$

$$(3 - 3.57)x_1 + (7 - 2.145)x_2 = 9 - 12.145$$

Figure 6-4 The Solution of 3 Linear Equations in 3 Variables by the Process of Elimination and Substitution

Elimination	Substitution
Three Equations	Three Variables
↓	↑
Two Equations	Two Variables
↓	↑
One Equation ⟶	One Variable

$$-.57x_1 + 4.855x_2 = -3.145 \qquad \textbf{(6.5.2)}$$

$$2x_1 + 11x_2 + 3(2.429 - .714x_1 - .429x_2) = 13 \qquad \textbf{(6.6.1)}$$

$$(2 - 2.142)x_1 + (11 - 1.287)x_2 = 13 - 7.287$$

$$-.142x_1 + 9.713x_2 = 5.713 \qquad \textbf{(6.6.2)}$$

Next, we solve Equations 6.5.2 and 6.6.2 by eliminating x_2 as follows. Employing Equation 6.6.2, we find that

$$x_2 = \frac{5.713}{9.713} + \frac{.142}{9.713}x_1$$

$$x_2 = .588 + .0146x_1 \qquad \textbf{(6.6.3)}$$

Substituting this result in Equation 6.5.2 we obtain the value for x_1

$$-.57x_1 + 4.855(.588 + .0146x_1) = -3.145$$

$$(-.57 + .0709)x_1 = -3.145 - 2.855$$

$$-.4991x_1 = -6$$

$$x_1 = 12.022 \qquad \textbf{(6.5.3)}$$

The value of x_1 can then be substituted into Equation 6.6.3, yielding

$$x_2 = .588 + .0146(12.022)$$

$$x_2 = .764$$

which in turn can be substituted into Equation 6.7.2

$$x_3 = 2.429 - .714(12.022) - .429(.764)$$

$$x_3 = 2.429 - 8.584 - .328$$

$$\boxed{x_3 = -6.483}$$

The point defined by $(x_1, x_2, x_3) = (12.022, .764, -6.483)$ represents the solution of the system of Equations 6.5, 6.6, and 6.7. Such a point represents the values of x_1, x_2, and x_3 that satisfy the three equations simultaneously. The values can be substituted into each of the equations to verify that they are the correct values. Employing Equation 6.5, we find that

$$\text{L.H.S.} = 3(12.022) + 7(.764) + 5(-6.483)$$

$$= 8.999 \cong 9$$

$$\text{R.H.S.} = 9$$

Similarly, when considering Equation 6.6, we obtain

$$\text{L.H.S.} = 2(12.022) + 11(.764) + 3(-6.483)$$

$$= 12.999 \cong 13$$

$$\text{R.H.S.} = 13$$

Finally, referring to Equation 6.7, observe that

$$\text{L.H.S.} = 5(12.022) + 3(.764) + 7(-6.483)$$

$$= 17.021 \cong 17$$

$$\text{R.H.S.} = 17$$

The slight inaccuracies found in verifying these values are caused by rounding the calculations to three or four decimal places at each stage of the solution process. If eight digits of accuracy had been retained throughout the calculations, the values that would have been obtained are given by the point (12.000001, .7647048, −6.4705883).

6.4 CONDITIONS FOR NO SOLUTION TO A SYSTEM OF EQUATIONS(O,G,F)

The solution process just described will fail to yield a simultaneous solution under two related conditions. The first involves one or more equations that may be expressed as the sum of appropriate multiples of the other equations. In the two-variable/two-equation situation, this condition implies that the two equations represent a single line. The second condition involves a system of equations in which the left-hand side of one or more of the equations may be expressed as the sum of appropriate multiples of the corresponding sides of the other equations while the right-hand side may not be so expressed. Such a situation results in a system of parallel lines for which no point of intersection occurs. Both of these conditions are known as *linear dependence*.

6.4.1 Systems with No Solutions(O,G,F)

A system of two equations representing parallel lines has no solution, because by definition the lines never intersect. Such a system is given by

$$3x_1 - x_2 = -4 \tag{6.8}$$

$$6x_1 - 2x_2 = -12 \tag{6.9}$$

Solving Equation 6.8 explicitly for x_2 we obtain

$$x_2 = 3x_1 + 4 \tag{6.8.1}$$

Substituting the results expressed by Equation 6.8.1 in Equation 6.9 we obtain

$$6x_1 - 2(3x_1 + 4) = -12$$
$$6x_1 - 6x_1 - 8 = -12$$
$$-8 = -12$$

The x_1 term disappears, which results in an impossible situation ($-8 \neq -12$). This implies that there is no set of values (x_1, x_2) that simultaneously satisfies Equations 6.8 and 6.9. This is in accord with the fact that the lines are parallel. The reader should verify that both lines have a slope of 3, but have different intercepts. In addition, the reader should construct a graph of these lines to demonstrate that they are parallel.

6.4.2 Systems with an Infinite Number of Solutions[O,G,F]

Another situation in which the solution process leads to an impasse occurs when the two equations describe the same line. For example, consider a slight modification of the previous system so that

$$3x_1 - x_2 = -4 \tag{6.10}$$

$$6x_1 - 2x_2 = -8 \tag{6.11}$$

As before, from Equation 6.10 we obtain

$$x_2 = 3x_1 + 4 \tag{6.10.1}$$

Substituting this result in Equation 6.11 yields

$$-8 = -8$$

The x_1 term has disappeared but in this case the result is an "identity" (i.e., it is always true). Choosing *any* value of x_1, Equation 6.10 can be used to find the value of x_2 that satisfies the equation and the same pair (x_1, x_2) also satisfies Equation 6.11. This indicates that there is an infinite number of solutions, and that there is no unique solution.

6.5 CONCLUSIONS[O,A]

Although the methods presented in this chapter are useful only for small problems, they form an essential basis for solving larger, more complex problems. In Part V we present a matrix algebra approach to problems similar to those in this chapter. The use of matrix algebra allows us to express the simultaneous solution to a system of linear equations in concise form. In two-equation systems, the matrix algebra approach probably is easier (after practice) than the method of elimination and substitution. Three-equation systems are complex for solution by hand regardless of which method is chosen. Systems involving four equations are prohibitive when solved manually. However, once the system has been expressed in the notation and operations of matrix algebra, programs may be developed that permit the use of high-speed electronic computers to perform the required calculations. As an example, in the financial management of health care institutions, cost-finding techniques may be translated into systems of linear equations that can be solved using matrix algebra and a computer to perform required calculations.

Basic descriptions of overhead cost allocation using the direct and step-down methods are presented in textbooks on managerial accounting such as those by Anthony and Horngren. An approach using simultaneous linear equations solved by matrix algebra is given in the book by Livingstone.

REFERENCES

R. N. Anthony, *Management Accounting: Text and Cases,* 4th ed. Homewood, Illinois: Richard D. Irwin, Inc., 1970.

C. T. Horngren, *Accounting for Management Control: An Introduction,* 2nd ed. Englewood Cliffs, New Jersey: Prentice-Hall, Inc., 1970.

J. L. Livingstone, ed., *Management Planning and Control: Mathematical Models.* New York: McGraw-Hill Book Company, 1970.

Problems for Solution

1. Solve for x_1 and x_2 graphically and algebraically. Solve each explicitly for x_2. Plot x_1 horizontally and x_2 vertically.

   a. $4x_1 + 5x_2 = 5$
   $3x_1 + 7x_2 = 6$

   b. $7x_1 - 3x_2 = 4$
   $5x_1 + 2x_2 = -5$

   c. $2x_1 + 3x_2 = 1$
   $-3x_1 + 2x_2 = 5$

   d. $5x_1 + 7x_2 = 15$
   $3x_1 + 9x_2 = 18$

   e. $6x_1 + 3x_2 = 24$
   $4x_1 + 5x_2 = 30$

2. Solve for x_1, x_2, and x_3 algebraically.

   a. $2x_1 + 3x_2 - 5x_3 = 5$
   $x_1 - 3x_2 + 4x_3 = 3$
   $3x_1 - 2x_2 + 3x_3 = 7$

   b. $4x_1 - 2x_2 + 3x_3 = 3$
   $5x_1 + 3x_2 - 2x_3 = 2$
   $-2x_1 + 4x_2 + 5x_3 = 5$

Chapter 7

Nonlinear Functions and Equations

Objectives

After completing this chapter you should be able to:

1. Name the forms of nonlinear functions used in this chapter (quadratics, trinomials, polynomials, exponentials, and trigonometric), and describe their general shapes;
2. Calculate values for these functions and graph them on arithmetic graph paper;
3. Factor a quadratic expression using difference of squares, perfect squares, or the general formula, whichever may be the most appropriate;
4. Find the intersection point(s) of a linear and a quadratic equation, if there is an intersection, and plot the equations to demonstrate the intersections.

Chapter Map

The sections comprising this chapter may be summarized as follows:

Section Number	Required Reading	Optional Reading	Generic Development	Application to Management	Fundamental Principles	Complex Material
	(R)	(O)	(G)	(A)	(F)	(C)
7.1	x		x		x	
7.2	x			x	x	
7.3	x		x			x
7.4	x			x	x	
7.4.1	x			x	x	
7.4.2	x			x	x	
7.5	x		x		x	
7.5.1	x		x		x	
7.5.2	x		x		x	
7.5.3		x	x			x
7.6	x		x	x	x	
7.6.1	x		x		x	
7.6.2	x		x		x	
7.6.3	x		x	x	x	
7.6.4	x		x	x	x	
7.7		x		x	x	
7.8		x	x		x	

7.1 INTRODUCTION(R,G,F)

The linear equations and systems of equations in Chapters 5 and 6 have an important property that is both an advantage and a limitation for managers of health care and nonhealth care organizations. The basic limitation of the material in those two chapters involves the assumption of linearity. The primary advantage of linear equations and systems of linear equations is that they are easy to understand and to manipulate mathematically. However, their primary disadvantage is that many processes found in the "real world" are represented more accurately by nonlinear equations. Unfortunately, equations for nonlinear curves are more difficult to manipulate mathematically.

7.2 APPLICATIONS IN HEALTH CARE MANAGEMENT(R,A,F)

It is important for a health care manager to be aware of the differences between straight and curved line functions, the specific types of nonlinear functions, under what circumstances either type may be most appropriate, and how to deal with each.

For example, consider the variable and fixed portions of the cost of operating a radiology department. The *variable costs* include direct labor and supplies, while the *fixed costs* include the salaries of employees who are not directly involved in providing service to patients. First we examine the direct labor and then the supplies component of variable costs. A radiology department has a certain number of technicians, and perhaps salaried radiologists as well, whose primary duties are to perform radiological examinations. These employees position patients, load film into the machines, expose and develop the films, and label and file the charts. Salaried radiologists (i.e., not paid on a fee for service basis) might perform some of these functions as well as examine exposed films to determine whether the patient's bones and other internal organs are normal or abnormal. All of these activities have a common feature in that they are associated *directly* with providing a radiological service to the patient.

Depending on the type of examination, each of these activities can be expected to take a given number of minutes for a particular kind of technician at a particular rate of pay. These are referred to as "direct labor" costs and their amount varies linearly with the volume of x-rays. Similarly, the cost of films and chemicals is the variable cost component that refers to the use of consumable supplies. For each kind of x-ray it is possible to determine a normal or standard cost for direct labor and supplies and a linear equation adequately expresses the variable portion of total cost.

For example, suppose that the radiology department offers three types of services and that the quantity of each provided during the month is represented by x_1, x_2, and x_3. Suppose further that the direct labor costs of these three services are represented by a_1, a_2, and a_3 while b_1, b_2, and b_3 represent the direct supply costs. Consequently, the total variable cost (TVC) for the department may be represented by

$$TVC = (a_1 + b_1)x_1 + (a_2 + b_2)x_2 + (a_3 + b_3)x_3$$

Regardless of the volume of service provided, we find that $(a_1 + b_1)$, $(a_2 + b_2)$, and $(a_3 + b_3)$ correspond to the average variable costs of service 1, 2, and 3, respectively.

In addition to the variable costs of providing care, several components—called *fixed costs*—are invariant with respect to the volume of service provided. For example, once management has determined the salary of the department head, the number of secretaries that will be employed, and the amount of floor space allocated to the unit, the costs associated with these decisions remain constant, at least in the short run. In this case, since fixed costs remain constant, average fixed cost or the fixed cost per unit of service declines as the volume of care provided is increased. Letting "F" correspond to the fixed cost component, total cost (TC) may be expressed in the form

$$TC = F + (a_1 + b_1)x_1 + (a_2 + b_2)x_2 + (a_3 + b_3)x_3$$

In turn, we find that the average total cost (ATC) of providing service 1, 2, and 3 is given by

$$ATC_1 = \frac{F}{x_1 + x_2 + x_3} + (a_1 + b_1);$$

$$ATC_2 = \frac{F}{x_1 + x_2 + x_3} + (a_2 + b_2); \text{ and}$$

$$ATC_3 = \frac{F}{x_1 + x_2 + x_3} + (a_3 + b_3)$$

respectively. In this case, the average fixed cost (AFC) per x-ray is given by

$$AFC = \frac{F}{x_1 + x_2 + x_3}$$

As suggested earlier, this expression implies that when the sum $x_1 + x_2 + x_3$ is small, *AFC* is large. However, as the volume of care is increased, average fixed costs decline in nonlinear fashion.

Table 7-1 presents the average fixed costs associated with different volumes of service. In this example, we assume that the total fixed cost of the unit is \$5,000 per month. When these data are portrayed graphically, as in Figure 7-1, the nonlinearity of the resulting curve is obvious. Consequently, a linear function does not represent the curve adequately.

7.3 TYPES OF NONLINEAR FUNCTIONS AND EQUATIONS (R,G,C)

A nonlinear equation is one that, when plotted, produces a curved line. This usually results when a power other than one is assigned to the independent variable (i.e., a polynomial), or when an exponential or trigonometric function is used. The general (explicit) form of a polynomial equation is:

$$y = a + bx + cx^2 + dx^3 + \cdots$$

Although the powers have been shown with positive integer exponents, the exponents can be *any* real number and there may be one or more terms containing x.

The general explicit form of an exponential function is

$$y = ae^{bx}$$

where a and b can be any real constants.

Table 7-1 Average Fixed Cost per x-ray at Various Levels of Volume

Total Volume	Average per x-ray
500	\$10.00
1000	5.00
1500	3.33
2000	2.50
2500	2.00
3000	1.67
4000	1.25

Figure 7-1 Average Fixed Cost per x-ray at Various Levels of Volume

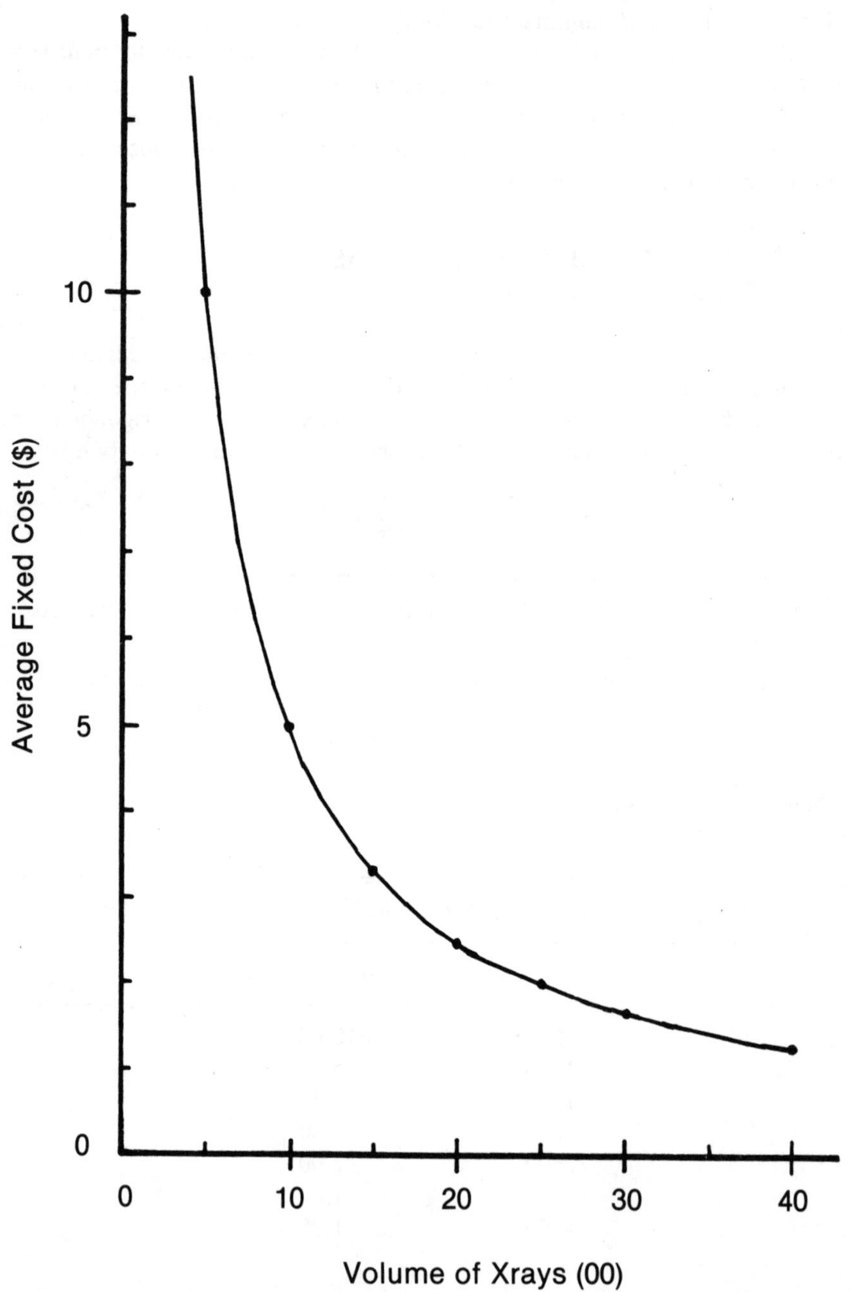

The three primary trigonometric functions are the sine, cosine, and tangent. They are defined as functions of angles (θ) of triangles and are useful for describing regular fluctuations (e.g., business cycles, seasonal changes in accident rates, etc.). The explicit forms are:

$$y = \sin(\theta); \qquad y = \cos(\theta); \qquad y = \tan(\theta)$$

These functions are described in detail later.

7.4 GRAPHING NONLINEAR FUNCTIONS (R,A,F)

When dealing with nonlinear functions, the value of the dependent variable is obtained by substituting known or assumed values for the independent variable(s) into the equation. This approach is similar to the method used when linear equations were discussed. In this section, we employ two examples to demonstrate this method.

7.4.1 The Cost of Service (R,A,F)

Let us assume that the cost per service provided by the laboratory department depends on the number of examinations performed per month. Assume further that when x^* units of service are provided, average total cost is minimized and equal to ATC^*. However, if either more or fewer than x^* units of service are provided, average total cost exceeds ATC^*. Such a relationship might be represented by a parabola that assumes the form

$$y = a + bx + cx^2$$

where

y represents average total cost
x represents the monthly volume of care (in 100 units of service)
a, b, and c are constants.

In practice, the constants a, b, and c might be determined by applying multiple regression (statistical) analysis to a sample of cost-volume data.

Employing this approach, suppose a recent study of the cost-volume relationship in the laboratory revealed that

$$a = 4.556$$

$$b = -.3556$$

$$c = .0089$$

Recalling that the volume of care is measured by 100 units of service, we may now determine the unit costs associated with different volumes of service as seen in Table 7-2. When $x = 10$ (i.e., 1,000 units of service), the corresponding value of y (unit costs) is \$1.889. As before, columns (2) and (3) contain the detailed calculations that are required to obtain the values appearing in column (4). In the table, the calculations are rounded to 3-decimal digits. Also note the changing impact of the x and x^2 terms in the function. The x^2 term is relatively small for low volumes of output but increases rapidly as volume increases. At first, the unit cost (y) decreases from 1.889 to 1.000, then begins to rise as the monthly volume of tests increases. The minimum unit cost occurs at a volume of 2,000 tests per month. The cost-volume relationship defined by $y = 4.556 - .3556x + .00889x^2$ is displayed graphically in Figure 7-2.

7.4.2 Population Growth(R,A,F)

As a second example of a nonlinear function, we might examine the increase in population size over a period of years. In planning for the delivery of health care services it is necessary to project the size of the population for a period of a few years and base the projection on past population figures. A very simple model of population growth assumes both a fixed birth rate and a fixed death rate per thousand population. Thus, we assume that the difference between the birth rate and the death rate represents the rate of natural increase. If the birth rate is b and the death rate d, then the natural rate of increase is $(b - d)$, which we can call n. Let us assume a 5 percent birth rate and a 1 percent death rate. Thus, the natural rate of increase is $(5 - 1)\% = 4\%$.

Table 7-2 Calculations for Cost/Volume Relationship for Laboratory Tests

x (hundreds) (1)	Substituted Values $f(x) = 4.556 - .3556(x) + .00889(x^2)$ (2)	Partial Calculations (3)	y (4)
10	$4.556 - .3556(10) + .00889(100)$	$= 4.556 - 3.556 + .889$	1.889
15	$4.556 - .3556(15) + .00889(225)$	$= 4.556 - 5.334 + 2.000$	1.222
20	$4.556 - .3556(20) + .00889(400)$	$= 4.556 - 7.112 + 3.556$	1.000
25	$4.556 - .3556(25) + .00889(625)$	$= 4.556 - 8.890 + 5.556$	1.222
30	$4.556 - .3556(30) + .00889(900)$	$= 4.556 - 10.668 + 8.001$	1.889
35	$4.556 - .3556(35) + .00889(1225)$	$= 4.556 - 12.446 + 10.890$	3.000

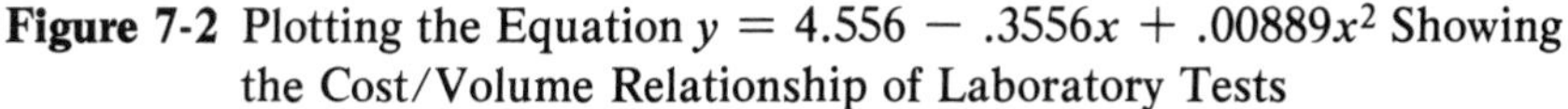

Figure 7-2 Plotting the Equation $y = 4.556 - .3556x + .00889x^2$ Showing the Cost/Volume Relationship of Laboratory Tests

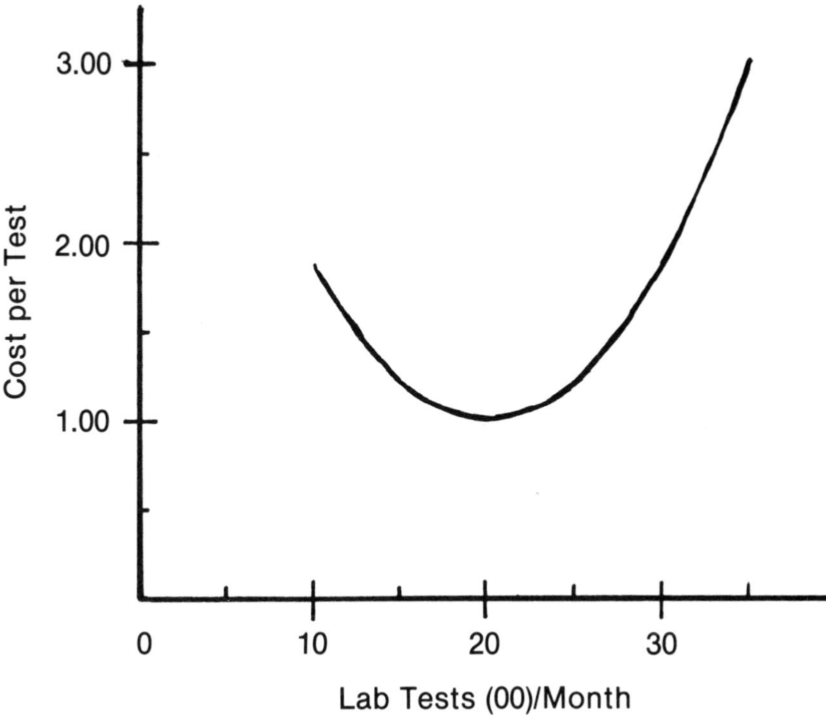

A continuously growing population is represented by an exponential curve that might be represented in the form

$$y = Pe^{nt}$$

where y is the predicted population in year t, P is the actual population in year 0, and n is the natural rate of increase as expressed by a decimal fraction (e.g., 4% = .04). Notice that time, t, rather than x, is now the independent variable. As before, the values of the dependent variable can be calculated and the results displayed (as in Table 7-3). We have assumed an initial population of 200,000 in year 1970 ($t = 0$ in 1970). Many calculators have an e^x function available, and we have used that function in this table; otherwise we would have to obtain values of e^x in a set of math tables. The corresponding values are plotted in Figure 7-3.

Several aspects of the table and the corresponding figure should be noted. The first is that the value of e^0 is one. Second, the values of nt appear in column (3) of Table 7-3 while the entries in column (4) represent the term e^{nt} and illustrate the nonlinear effects that are caused by the process of compounding. A comparison of the decimal portion of e^{nt} with the value of nt (e.g., 1.5527 vs. .44) reveals that the discrepancy widens as the value assumed by t increases. This result is clearly evident in Figure 7-3. And finally, the larger the value of nt (e.g., .04, .12, .24, .44), the greater the discrepancy; and the larger the value of n, the quicker the discrepancy becomes greater.

7.5 GENERAL EXAMPLES OF NONLINEAR FUNCTIONS (R,G,F)

The nonlinear functions used as examples in Section 7.3 are supplemented in this section by some basic nonlinear functions. The basic purpose is to illustrate the basic shapes and relationships among these types of curves.

7.5.1 Quadratic and Cubic Curves (R,G,F)

Figure 7-4 shows the values of the functions

$$y = x^2$$

and

$$y = x^3$$

plotted for values in the range $-9 \leq x \leq 9$. Note that the value of $x^2 \geq 0$ for all values of x, while

$$x^3 < 0 \text{ for } x < 0$$

$$x^3 = 0 \text{ for } x = 0$$

and

$$x^3 > 0 \text{ for } x > 0$$

Both curves pass through the origin ($x = 0$, $y = 0$) and $y = x^3$ increases much more rapidly than $y = x^2$ for values of $x > 0$. Also note that the curve $y = x^2$ opens upward but that if we had plotted $y = -x^2$ it would have opened downward. We return to this observation in Chapter 9 on the introduction to differential calculus where we note that $y = x^2$ has a positive second derivative, which means that the corresponding curve passes through a

Table 7-3 Calculation of the Values of Population Using the Exponential Equation $y = 200e^{.04t}$

Year (1)	t (2)	nt ($n = .04$) (3)	Substituted Values $200e^{nt}$ (4)	Population y (5)
1970	0	0	$200e^{0} = 200 \times 1.0000$	200
1971	1	.04	$200e^{.04} = 200 \times 1.0408$	208.16
1972	2	.08	$200e^{.08} = 200 \times 1.0832$	216.64
1973	3	.12	$200e^{.12} = 200 \times 1.1274$	225.48
1974	4	.16	$200e^{.16} = 200 \times 1.1735$	234.70
1975	5	.20	$200e^{.20} = 200 \times 1.2214$	244.28
1976	6	.24	$200e^{.24} = 200 \times 1.2712$	254.24
1977	7	.28	$200e^{.28} = 200 \times 1.3231$	264.62
1978	8	.32	$200e^{.32} = 200 \times 1.3771$	275.42
1979	9	.36	$200e^{.36} = 200 \times 1.4333$	286.66
1980	10	.40	$200e^{.40} = 200 \times 1.4918$	298.36
1981	11	.44	$200e^{.44} = 200 \times 1.5527$	310.54

minimum value. However $y = -x^2$ has a negative second derivative, which implies that the corresponding curve passes through a maximum point.

7.5.2 Exponential Curves (R,G,F)

Figures 7-5 and 7-6 compare the behavior of

$$y = 10^x$$

and

$$y = e^x$$

where $-5 < x < 5$ and $-1.8 < x < 1.0$, for Figures 7-5 and 7-6, respectively. Note that the scale of Figure 7-6 has been expanded to show the relative behavior of these two curves in the region $x < 0$. Figure 7-5 shows that $y = 10^x$ increases much more rapidly than $y = e^x$. The same general tendency can be seen in Figure 7-6 for values of $x > -.64$. Both of these curves are exponentials and are related to each other in that

$$10^x \cong e^{2.30259x}$$

Compound interest and population growth are examples of situations in which exponential curves are involved. When a wide range of x values is used to plot these functions, the resulting curve quickly exceeds the bounds of arithmetic paper. However, in semilogarithmic paper, the vertical scale is compressed automatically by using the logarithm of the y value that, in turn, permits the display of a wide range of x values. This transforms any curve involving exponentials into a straight line on the paper. Figure 7-7 presents an example of two-cycle semilog paper, where the number of cycles gives the number of powers of 10 represented on the vertical scale. Papers with 1, 2, 3, 4, and 5 cycles are commonly available.

Figure 7-3 Growth in Population with a Growth Rate of 4%

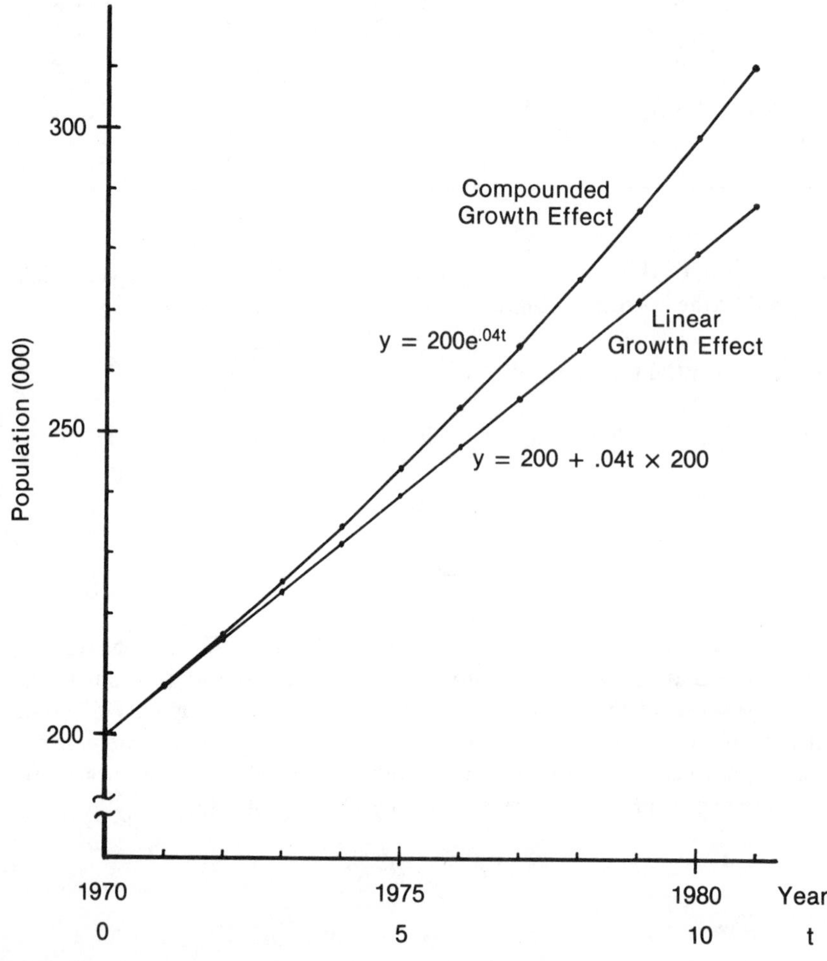

Figure 7-4 Comparison of the Behavior of $y = x^2$ and $y = x^3$ Over the Range $-9 \leq x \leq 9$

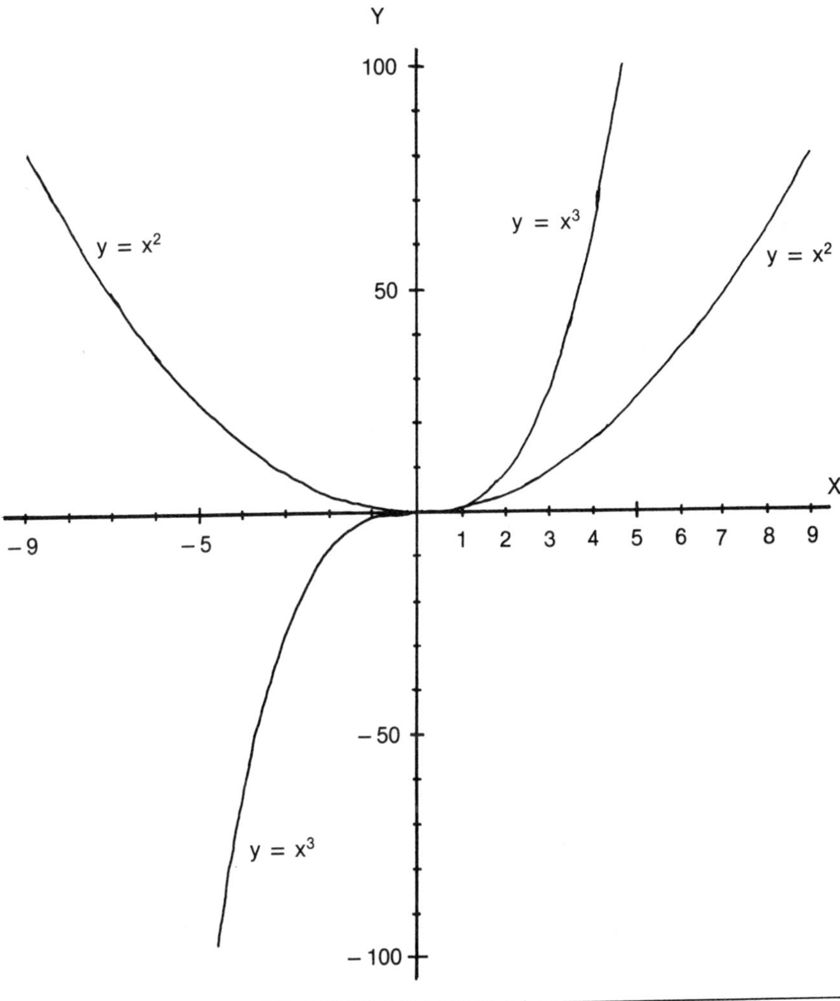

7.5.3 Trigonometric Functions (O,G,C)

Examples of some trigonometric functions are:

$$y = \sin(\theta) \qquad y = \arcsin(x)$$

$$y = \cos(\theta) \qquad y = \arccos(x)$$

$$y = \tan(\theta) \qquad y = \arctan(x)$$

Figure 7-5 Comparison of the Behavior of $y = 10^x$ and $y = e^x$ Over the Range $-2 \leq x \leq 5$

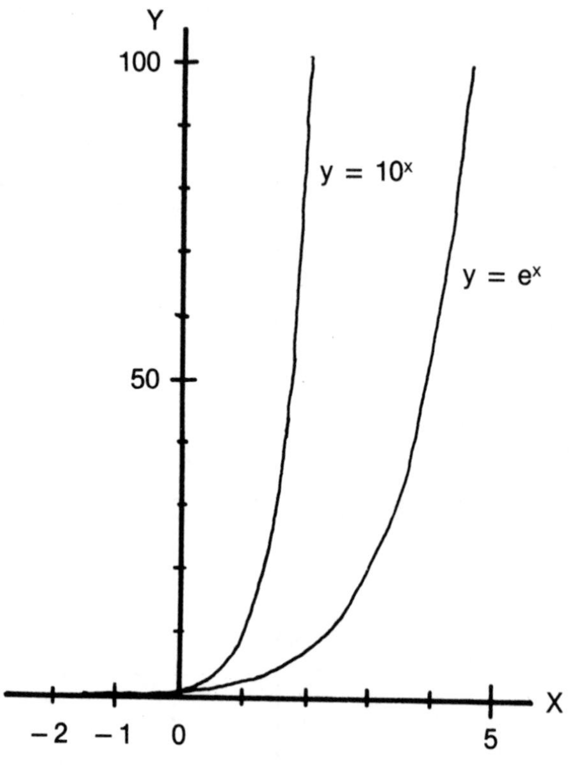

The functions sine (sin), cosine (cos), and tangent (tan) convert an angle (θ), which is measured in radians or degrees, into a numeric value. These functions usually are defined in terms of the ratios of the sides of a right-angled triangle. The functions arcsine (arcsin), arccosine (arccos), and arctangent (arctan) are the inverse of the functions sine, cosine, and tangent, respectively. These functions convert a numeric value into an angle. A radian is an angle formed by a sector of a circle bounded by two radius lines (from the center to the circumference) and a portion of the circumference that is equal in length to a radius. Consequently, a radian is slightly smaller than the 60° angle of an equilateral triangle. A circle contains 2π radians that are equivalent to 360 degrees. The following conversion equations are useful. Letting

$D =$ the number of degrees and $R =$ the number of radians,

we find that

$$D = \frac{360}{2\pi} \times R \qquad \text{while } R = \frac{2\pi}{360} \times D$$

The right triangle relationships that define the trigonometric functions are shown in Figure 7-8. Line c in each triangle is the radius of the circle, and c can be thought of as rotating counterclockwise from the initial zero position of b. The triangle in Figure 7-8(a) has a very small angle θ, while the one in Figure 7-8(b) has a larger angle θ. As the angle θ changes, the position, length, and direction of the line a change, while the length and direction of b change so that a and b are at right angles to each other. It should also be noted that the value of c is always assumed to be positive but the value of a and b may be positive or negative.

The definitions of sine, cosine, and tangent also are shown in Figure 7-8. The definition of the tangent function is the same as the slope of a straight

Figure 7-6 Comparison of the Behavior of $y = 10^x$ and $y = e^x$ Over the Range $-1.8 \leq x \leq .8$

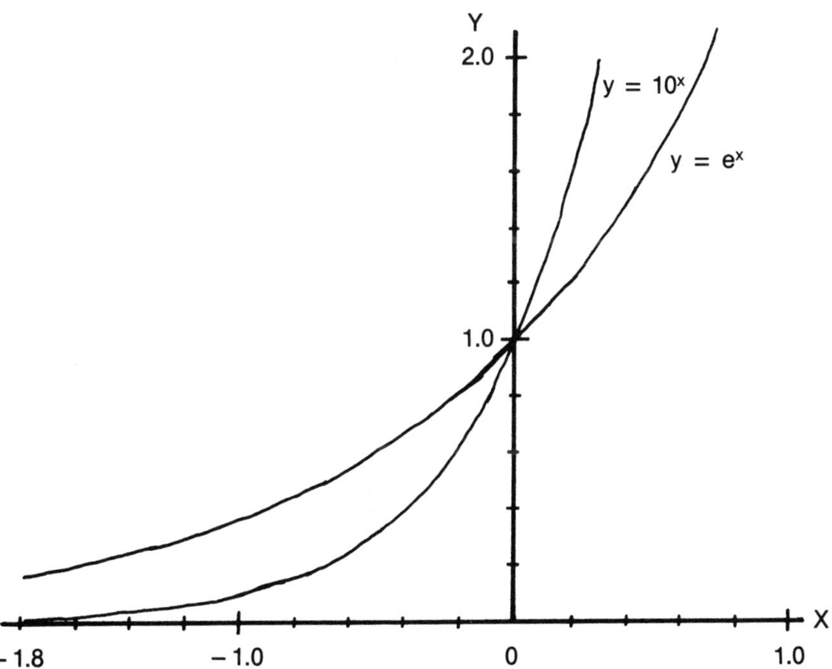

Figure 7-7 Two-Cycle Semilog Paper

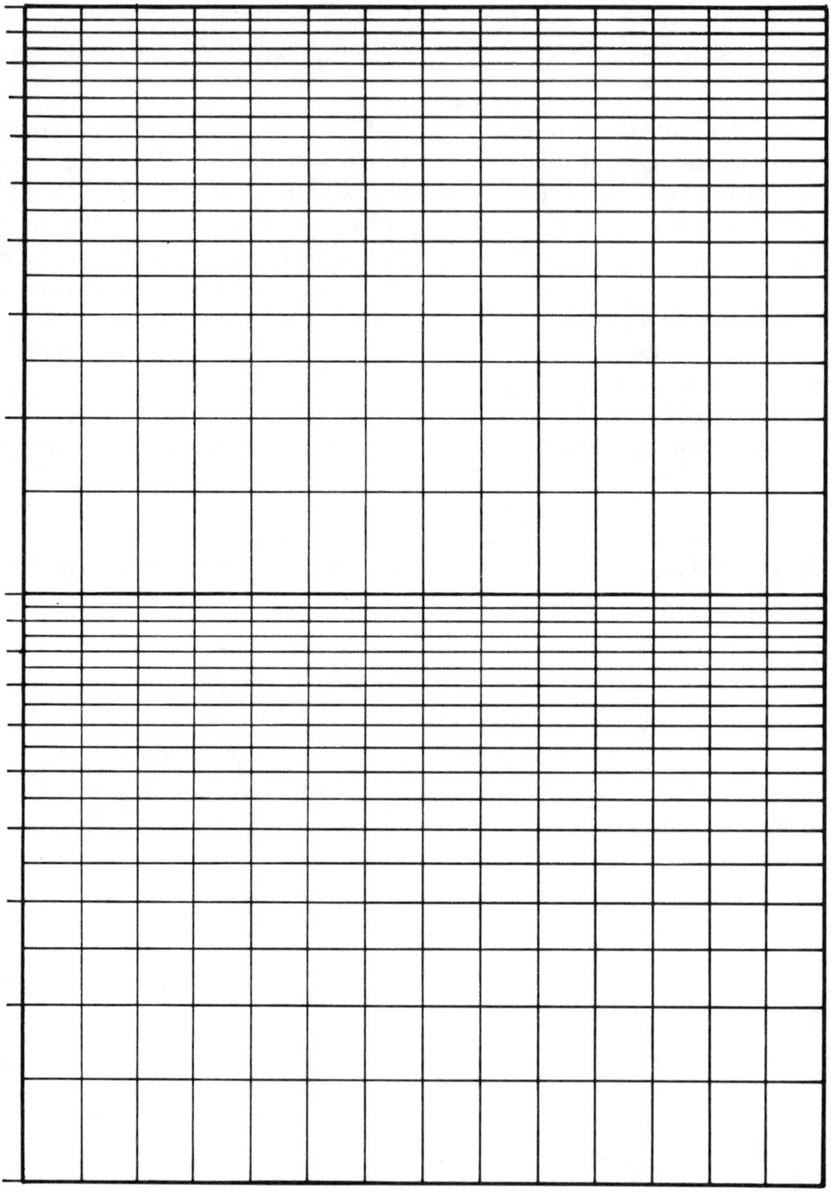

Figure 7-8 A Right-Angled Triangle and the Definition of Trigonometric Functions

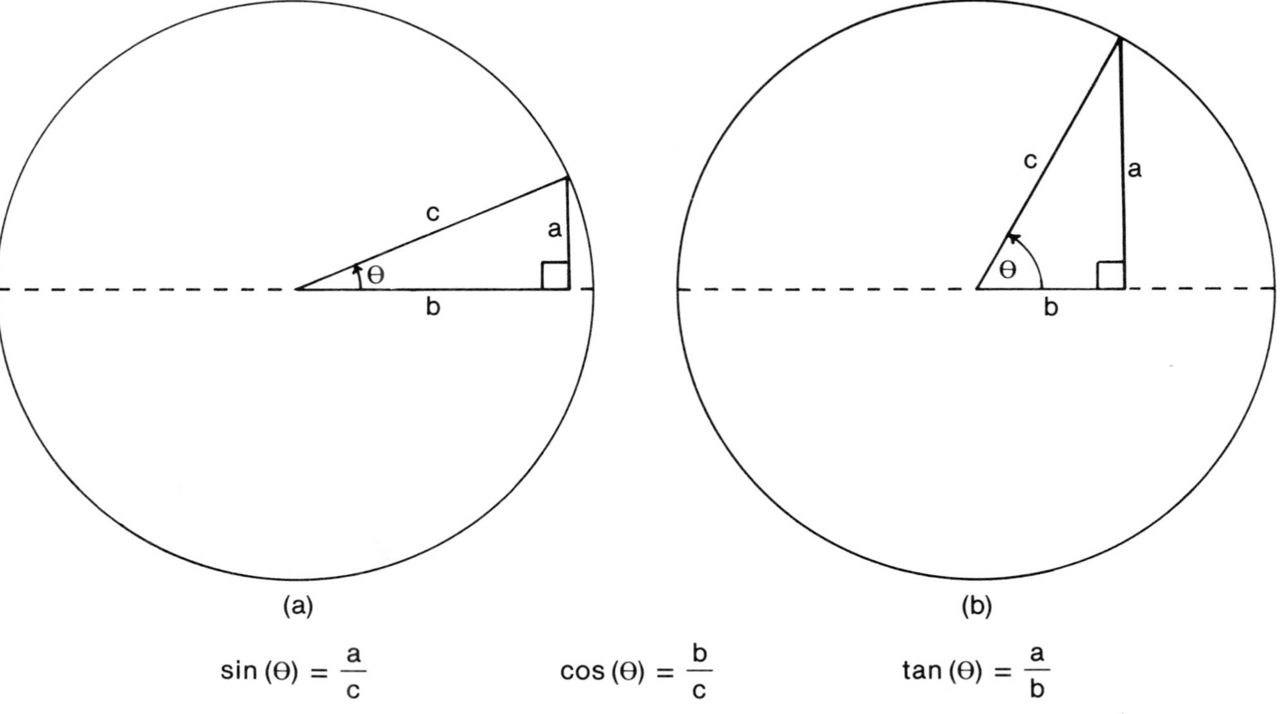

$\sin(\Theta) = \frac{a}{c}$ $\qquad$ $\cos(\Theta) = \frac{b}{c}$ $\qquad$ $\tan(\Theta) = \frac{a}{b}$

line. When θ is close to zero, the values of the tangent and of the sine are close to 0, while the value of the cosine is close to 1. As θ increases and approaches a right angle (90°), the values of the sine and tangent increase, with the sine approaching a value of 1 and the tangent increasing indefinitely. At the same time the cosine function is decreasing and approaches 0.

As the angle θ passes 90 degrees, the base b passes to the left and becomes negative. Thus, the cosine and tangent also become negative. The cosine becomes more negative (larger absolute value), while the absolute value of the tangent decreases (i.e., it approaches zero from negative infinity). At the same time, the sine starts to fall from 1.0 and approaches zero.

The values of the sine, cosine, and tangent functions are shown in Figure 7-9 for values of θ from 0° to 360°. This represents one complete revolution of c. As c continues to rotate, the patterns shown repeat themselves indefinitely. The sine and cosine have identical patterns, except that they are 90° *out of phase* with each other. This type of pattern is called a *sine wave* pattern.

The study of any regular variation frequently requires the use of sine and cosine waves of different frequencies so as to represent any pattern of variation, which may have a very complex shape and which may not even resemble a sine wave. Seasonal and cyclical patterns are explicitly incorporated in a number of forecasting methods that, in the future, will find increasing use in estimating patterns of demand for health care. Different kinds of illnesses and accidents appear to be seasonal, and good forecasts will become more necessary in order to avoid substantial costs incurred by being understaffed during periods of peak demand and overstaffed during periods of declining or low demand for service.

7.6 FACTORING (R,G,A,F)

In this section we introduce methods of determining factors that have resulted in a quadratic expression. This process is known as *factoring* and, as will be seen later, produces not only the determination of those factors that result in a quadratic expression but also the solutions to a polynomial equation.

In general, factoring simply reverses the operation of multiplication. The product of two linear factors results in a quadratic expression. For example, we find that

$$\begin{aligned}(x+4)(x+3) &= x(x+3) + 4(x+3)\\ &= x^2 + 3x + 4x + 12\\ &= x^2 + 7x + 12\end{aligned}$$

The product of the linear factors $(x + 4)$ and $(x + 3)$ is the quadratic expression $x^2 + 7x + 12$. On the other hand, given the expression

$$x^2 + 7x + 12$$

the process of factoring allows us to assert that $(x + 3)$ and $(x + 4)$ are the factors of the expression.

7.6.1 Basic Patterns (R,G,F)

Let us consider two factors $(x + a)$ and $(x + b)$ where the a and b are constants. The product of these factors is found as follows:

$$\begin{aligned}(x + a)(x + b) &= x(x + b) + a(x + b)\\ &= x^2 + bx + ax + ab\\ &= x^2 + (b + a)x + ab\end{aligned}$$

Substituting particular numeric values for a and b such as 5 for a and 3 for b we obtain

$$\begin{aligned}(x + 5)(x + 3) &= x(x + 3) + 5(x + 3)\\ &= x^2 + 3x + 5x + 15\\ &= x^2 + 8x + 15\end{aligned}$$

If we let $b = a$, we obtain a result that is called a *perfect square*. In such a situation, we find that

$$\begin{aligned}(x + a)(x + a) &= x(x + a) + a(x + a)\\ &= x^2 + ax + ax + a^2\\ &= x^2 + 2ax + a^2\end{aligned}$$

Now, if we let $a = 3$, we obtain

$$\begin{aligned}(x + 3)(x + 3) &= x(x + 3) + 3(x + 3)\\ &= x^2 + 3x + 3x + 9\\ &= x^2 + 6x + 9\end{aligned}$$

Employing a similar approach, notice that

$$\begin{aligned}(x - a)(x - a) &= x(x - a) - a(x - a)\\ &= x^2 - ax - ax + a^2\\ &= x^2 - 2ax + a^2\end{aligned}$$

Finally, if we substitute the value 3 for a in this expression, we obtain

$$\begin{aligned}(x - 3)(x - 3) &= x(x - 3) - 3(x - 3)\\ &= x^2 - 3x - 3x + 9\\ &= x^2 - 6x + 9\end{aligned}$$

These results allow us to define a perfect square by

$$(x + a)^2 = x^2 + 2ax + a^2$$

and by

$$(x - a)^2 = x^2 - 2ax + a^2$$

If we let $b = -a$, we obtain a result that is called the *difference of squares*. In this case, we find that

$$\begin{aligned}(x + a)(x - a) &= x(x - a) + a(x - a)\\ &= x^2 - ax + ax - a^2\\ &= x^2 - a^2\end{aligned}$$

If we substitute the value 6 for the a's in this expression, we obtain

$$\begin{aligned}(x + 6)(x - 6) &= x(x - 6) + 6(x - 6)\\ &= x^2 - 6x + 6x - 36\\ &= x^2 - 36\end{aligned}$$

Thus, we define a difference of squares as

$$(x + a)(x - a) = x^2 - a^2$$

Note that this specifically does *not* create a pattern such as $x^2 + a^2$.

Figure 7-9 Values of Trigonometric Functions

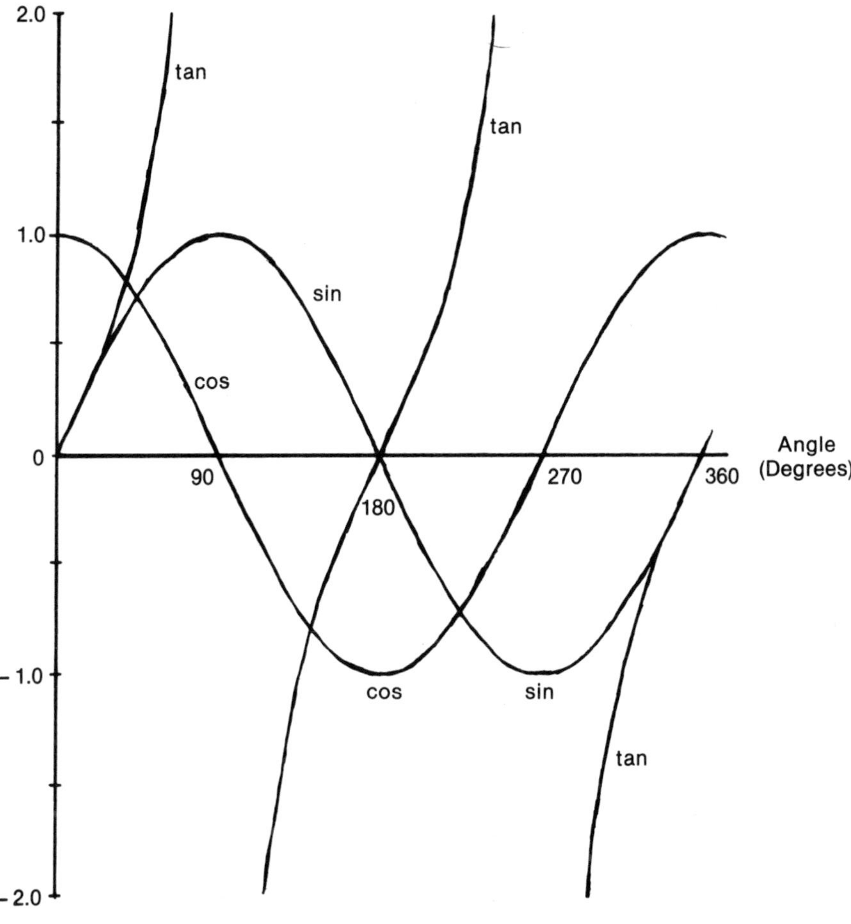

In summary, the three basic patterns described above may be used to determine the factors of a polynomial expression. The general case is represented by

$$(x + a)(x + b) = x^2 + (a + b)x + ab$$

Similarly, a perfect square may assume the form

$$(x + a)^2 = x^2 + 2ax + a^2$$

or the form

$$(x - a)^2 = x^2 - 2ax + a^2$$

Finally, the difference of squares may be represented by

$$(x + a)(x - a) = x^2 - a^2$$

7.6.2 Using the Basic Patterns (R,G,F)

In this section, the basic patterns are used to determine the factors of a quadratic expression. Consider first the quadratic expression

$$x^2 - 25$$

that corresponds to the difference of squares pattern. The factors of the expression are $(x - 5)$ and $(x + 5)$. Similarly, the factors of

$$x^2 - 41$$

are $(x + \sqrt{41})$ and $(x - \sqrt{41})$ where $\sqrt{41}$ is not an integer. In general, the factors of a quadratic expression of the form

$$x^2 - m$$

are given by $(x + \sqrt{m})$ and $(x - \sqrt{m})$.

Consider next the quadratic expression

$$x^2 - 10x + 25$$

for which the factors are $(x - 5)$ and $(x - 5)$. Similarly, we find that the factors of the expression

$$x^2 + 12x + 36$$

are $(x + 6)$ and $(x + 6)$ or $(x + 6)^2$. In general, then, the factors of the expression

$$x^2 + 2mx + m^2$$

are $(x + m)$ and $(x + m)$, or $(x + m)^2$. Finally, the factors of

$$x^2 + 15x + 54$$

are $(x + 9)$ and $(x + 6)$. Such an expression might assume the general form

$$x^2 + (a + b)x + ab$$

To find the factors of such an expression, we must find two values that have the sum $(a + b)$ and the product (ab).

7.6.3 The Roots of a Quadratic Expression(R,G,A,F)

In this section we apply the results of the factoring process to solve a quadratic expression. As seen earlier, the factors of the quadratic expression

$$x^2 - 10x + 24$$

are $(x - 6)$ and $(x - 4)$. When solving the quadratic equation, we require that

$$x^2 - 10x + 24 = 0$$

which represents the x intercept or the point at which the curve defined by the expression touches the x axis. Given that the expression $x^2 - 10x + 24$ may be expressed in the form

$$(x - 6)(x - 4)$$

we find that the solution to the quadratic equation requires that

$$(x - 6)(x - 4) = 0$$

This condition is satisfied when the factor

$$(x - 6) = 0$$

or $x = 6$, and when

$$(x - 4) = 0$$

which implies that $x = 4$. In this case, the factors $(x - 4)$ and $(x - 6)$ are used to find the solution (i.e., $x = 4$ and $x = 6$) to the quadratic expression. It should be noted that the values of x that satisfy the condition

$$y = f(x) = 0$$

where $f(x)$ is a quadratic expression are called the *roots* of the equation.

As an example of a situation in which the solution to a quadratic equation is of value, suppose that ours is a nonprofit institution and that our objective is to earn revenues that are just equal to total cost. Suppose further that the total revenue generated by providing care is given by

$$TR = \$5x$$

where x is the volume of service, as measured in 1,000 units, and \$5 corresponds to the charge per service. Similarly, suppose that the total cost of providing care is given by

$$TC = \$25 + \$5x - \$.25x^2$$

In this case we find that total revenues are equal to total costs when

$$5x = 25 + 5x - .25x^2$$

If we subtract the expression $(25 + 5x - .25x^2)$ from both sides of the equation, we obtain

$$5x - (25 + 5x - .25x^2) = 0$$

$$5x - 25 - 5x + .25x^2 = 0$$

$$.25x^2 - 25 = 0$$

In this case, the value of x that satisfies the equation

$$.25x^2 - 25 = 0 \qquad \textbf{(7.1)}$$

represents the volume of service for which total revenue is equal to total cost. Applying these techniques, we observe that Equation 7.1 might be expressed in the form

$$(.5x - 5)(.5x + 5) = 0$$

Hence, the values of x that equate total revenues and total costs are $x = 10$ and $x = -10$. Given that negative rates of activity are impossible, we find that when 10,000 units of service are provided

$$TR = 5(10)$$
$$= \$50$$

while

$$TC = 25 + 5(10) - .25(10)^2$$
$$= \$50$$

Hence, under the assumption that 10,000 units are provided, neither a net loss nor a net income is realized and the institution is said to break even. As an additional example, suppose that

$$x^2 - 10x + 25 = 0$$

The solution to this expression is obtained by observing that

$$(x - 5)(x - 5) = x^2 - 10x + 25$$

and as a consequence,

$$(x - 5)^2 = 0$$

only when $x = 5$. Hence the value 5 is the root of the equation. Similarly, if

$$x^2 - 10x + 24 = 0$$

we find that

$$(x - 6)(x - 4) = x^2 - 10x + 24$$

and, as a result

$$(x - 6)(x - 4) = 0$$

when $x = 6$ or when $x = 4$. In each of the examples, the process of factoring has led to one or more values of x that solve the quadratic expression.

7.6.4 The Roots of a General Quadratic Equation (R,G,A,F)

Sometimes the factors of a quadratic expression are not obvious. When this situation occurs, we invoke a formula for which we do not give the derivation. In general, if

$$ax^2 + bx + c = 0 \tag{7.2.1}$$

the roots of the equation are given by

$$x = \frac{-b \pm \sqrt{b^2 - 4ac}}{2a} \tag{7.2.2}$$

In the example in which 10,000 units of service represented the break-even volume of service, recall that we required the solution to the quadratic expression

$$.25x^2 + (0)x - 25 = 0$$

Employing Equation 7.2, we find that

$$\begin{aligned} x &= \frac{0 \pm \sqrt{(0)^2 - 4(.25)(-25)}}{2(.25)} \\ &= \frac{0 \pm \sqrt{25}}{.5} \\ &= 0 \pm 5/.5 \\ &= \pm 10 \end{aligned}$$

which, of course, agrees with our earlier results.

As another example, let

$$x^2 + 15x + 54 = 0$$

In this case, the roots of the quadratic expression are given by

$$\begin{aligned} x &= \frac{-15 \pm \sqrt{15^2 - (4 \times 54)}}{2} \\ &= \frac{-15 \pm \sqrt{225 - 216}}{2} \end{aligned}$$

$$= \frac{-15 \pm \sqrt{9}}{2}$$

$$= \frac{-15 \pm 3}{2}$$

$$x = -6 \text{ or } -9$$

This result implies that

$$x^2 + 15x + 54 = (x + 6)(x + 9)$$

Consequently, we find that

$$(x + 6)(x + 9) = 0$$

when x is equal to either -6 or -9.

It also should be noted that many quadratic expressions cannot be solved. As an example, consider

$$3x^2 + 7x + 9 = 0$$

In this case an application of Equation 7.2 yields

$$x = \frac{-7 \pm \sqrt{49 - (12 \times 9)}}{6}$$

$$= \frac{-7 \pm \sqrt{49 - 108}}{6}$$

$$= \frac{-7 \pm \sqrt{-59}}{6}$$

Notice that $\sqrt{-59}$ is an "imaginary" number, and the curve represented by

$$y = 3x^2 + 7x + 9$$

does not cross or touch the x axis.

In the formula

$$x = \frac{-b \pm \sqrt{b^2 - 4ac}}{2a}$$

the expression $b^2 - 4ac$ is called the determinant. When the determinant is

1. greater than zero, there are two real roots;
2. equal to zero, there is only one real root; and
3. less than zero, the roots are imaginary.

Imaginary roots are meaningful but cannot be portrayed in simple graphic form. However, the interpretation of imaginary roots is of limited value to the health administrator and will not be discussed here.

7.7 BREAK-EVEN ANALYSIS: THE GENERAL CASE (O,A,F)

In this section, we extend our discussion of break-even analysis. In general, suppose the revenue function of the institution assumes the form

$$TR = a_1 + b_1 x \tag{7.3}$$

while the total cost function is given by

$$TC = a_2 + b_2 x + b_3 x^2 \tag{7.4}$$

Here, we let

TR correspond to total revenue
TC represent total cost
x represent the volume of service

while the coefficients a_1, a_2, b_1, b_2, and b_3 are known constants. Suppose further that we may superimpose the total revenue curve, as defined by Equation 7.3, on the total cost curve, which was defined by Equation 7.4, as seen in Figure 7-10. When x_1^* or x_2^* units of service are provided, total revenues are equal to total cost and, as a result, neither a net loss nor a net income is realized. However, when the volume of care provided is less than x_1^* or greater than x_2^*, total costs exceed total revenue and a net loss is incurred. When $x_1^* < x < x_2^*$, total revenues exceed total cost and a net income is realized. In the discussion that follows, we use the method developed above to determine the values of x_1^* and x_2^*. We return later to the problem of determining the volume of service that maximizes net income.

Recall that when either x_1^* or x_2^* units of service are provided, total revenues are equal to total costs and, as a result

$$a_1 + b_1 x = a_2 + b_2 x + b_3 x^2$$

Subtracting the expression $(a_2 + b_2x + b_3x^2)$ from both sides of the equation and rearranging slightly we obtain

$$(a_1 - a_2) + (b_1 - b_2)x - b_3x^2 = 0 \quad \textbf{(7.5)}$$

Now, letting $a = -b_3$, $b = (b_1 - b_2)$ and $c = (a_1 - a_2)$ we may rearrange Equation 7.5 in the form of Equation 7.2.1 as

$$ax^2 + bx + c = 0$$

In this case, an application of Equation 7.2.2 yields

$$x = \frac{-(b_1 - b_2) \pm \sqrt{(b_1 - b_2)^2 - 4[(-b_3)(a_1 - a_2)]}}{2(-b_3)}$$

which, of course, results in the value of $x_1{}^*$ and $x_2{}^*$.

7.8 ADDITIONAL EXAMPLES (O,G,F)

As seen previously, the intersection of a linear equation and a quadratic equation may be used to identify the break-even volume of service. To understand better that general approach, suppose we want to find the intersection of

$$y = 18 - .5x$$

and

$$y = x^2 - 4x + 20$$

As before, we find that

$$18 - .5x = x^2 - 4x + 20$$

corresponds to the intersection of the two equations. Subtracting the expression $(x^2 - 4x + 20)$ from both sides of the equation and rearranging the results slightly, we obtain

$$x^2 - (4 - .5)x + (20 - 18) = 0$$

from which we find that

$$x^2 - 3.5x + 2 = 0$$

Figure 7-10 An Example of Break-Even Analysis

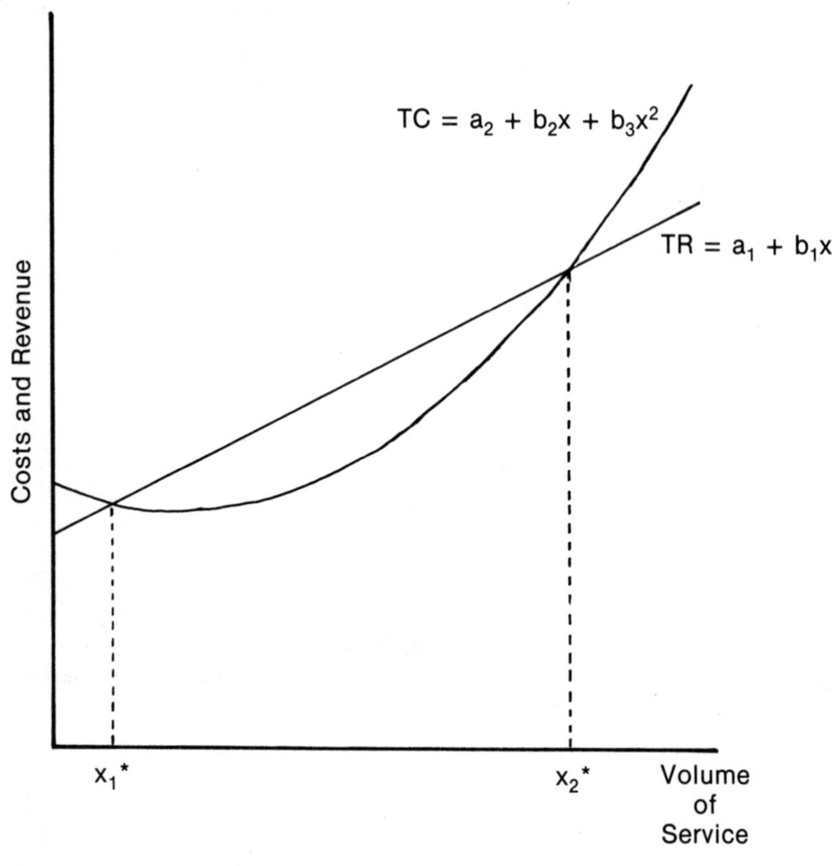

After substituting the coefficients of this equation into

$$x = \frac{-b \pm \sqrt{b^2 - 4ac}}{2a}$$

we obtain

$$x = \frac{3.5 \pm \sqrt{12.25 - 8}}{2}$$

$$= \frac{3.5 \pm \sqrt{4.25}}{2}$$

$$= \frac{3.5 \pm 2.0616}{2}$$

$$\cong .72 \text{ or } 2.78$$

The y values corresponding to these points can be determined by substituting appropriately in the linear equation. Doing so, we obtain the points (.72,17.64) and (2.78,16.61). Substitution in the quadratic verifies the accuracy of the solution. (In this example, the y values have an error of .0016 in each case, because the x values are only accurate to two decimal places.)

As a final example, suppose we require the intersection of

$$y = .02x^2 - .5x + 25 \tag{7.6.1}$$

and

$$y = 5 + 2x \tag{7.6.2}$$

As before, the condition

$$5 + 2x = .02x^2 - .5x + 25$$

represents the intersection of the two curves defined by Equations 7.6.1 and 7.6.2. Employing the approach described earlier, we find that

$$.02x^2 - 2.5x + 20 = 0 \tag{7.6.3}$$

The solution to Equation 7.6.3 is then found by applying Equation 7.2, from which we obtain

$$x = \frac{2.5 \pm \sqrt{6.25 - 1.6}}{.04}$$

$$\cong \frac{2.5 \pm 2.156}{.04}$$

After performing these calculations, we find that the solutions to the quadratic expression are given by $x \cong 8.590$ and $x \cong 116.41$. When these values are substituted into the linear equation, the resulting values of y are 22.18 and 237.82. Verification by substituting $x \cong 8.590$ and 116.41 into the quadratic equation is left as an exercise for the student.

Problems for Solution

1. Find the intersection points and plot the equations of:

a. $y = 3x^2 + 5x + 1$ and $y = 2 + 4x$

b. $y = .5x^2 - 2x + 14$ and $y = 3 + 6x$

c. $y = 2x^2 - x + 10$ and $y = 3 + 2x$

d. $y = x^2 - 3x + 5$ and $y = 2x - 1.25$

Chapter 8

The Mathematics of Finance: Compound Interest and Present Value

Objectives

After completing this chapter you should be able to:

1. Understand and compute simple interest for single and multiple periods;
2. Understand and compute compound interest for multiple periods;
3. Describe and graph the effect of different interest rates for simple and compound interest;
4. Describe and differentiate between compounding and discounting (future and present values);
5. Compute the present value of single and multiple future payments (or receipts);
6. Compute the net present value of a stream of future payments and receipts;
7. Describe and evaluate the effects of different discount rates (5, 10, 15, and 20 percent);
8. Differentiate between current and constant dollars;
9. Evaluate the relative and net effects of inflation and interest (discount) rates;
10. Describe the problems of selecting an appropriate discount rate for analysis of an investment proposal;
11. Describe the problems associated with analysis of intangible or indirect benefits and costs.

Chapter Map

The sections comprising this chapter may be summarized as follows.

Section Number	Required Reading	Optional Reading	Generic Development	Application to Management	Fundamental Principles	Complex Material
	(R)	(O)	(G)	(A)	(F)	(C)
8.1	x		x	x	x	
8.1.1	x		x	x	x	
8.1.2	x		x	x	x	
8.1.3	x		x	x	x	
8.1.4	x		x	x	x	
8.1.5	x			x	x	
8.2	x		x	x	x	
8.3	x		x	x	x	
8.3.1	x			x	x	
8.3.2		x	x	x	x	
8.4	x		x	x	x	
8.4.1	x		x	x	x	
8.4.2	x		x	x	x	
8.5		x		x	x	
8.6		x		x	x	
8.7		x		x	x	

8.1 INTRODUCTION(R,G,A,F)

The administrator of a hospital, clinic, group practice, or any other health care (or nonhealth care) organization must necessarily manage the financial (money) aspects of the enterprise. Borrowing, saving, and investing are important components of money management, and each requires an understanding of the concept of interest. In this chapter, our focus is on the concept and calculation of interest and discounting as well as their relationship to the problem of inflation.

8.1.1 The Concept of Interest(R,G,A,F)

We all are familiar with the fact that if $10 is allowed to remain idle in a demand deposit (i.e., checking) account for one year, usually no interest will be earned during the period. However, if we invest the $10 in a savings account, we expect the bank to pay us *interest* for the privilege of using the money. We also are aware that if we borrow money from the bank, or from someone else, we must repay not only the original amount, which is called the *principal*, but also some interest. Using a system called *simple interest*, the extra amount that must be repaid in the future is given by the product of an *interest rate* (percent per specified period of time), the principal, and the number of periods of time that the principal has been *outstanding*. A much more commonly used system is called *compound interest*. When interest is compounded, the interest payable at the end of a period of time is added to the principal and the sum of these two components earns interest in the next period of time. The process of computing interest on previously earned interest is called *compounding*. Later sections in this chapter will show that compound interest grows much more rapidly than simple interest, and we will examine the effects of different interest rates as well as longer or shorter periods of time (*compounding intervals*).

Different terms are used to identify essentially two points of view that emanate from the process just described. The person or organization who provides the original sum of money is said to be making a loan or an investment (or to be investing) and is called a *lender* or an *investor*. The person who receives and must repay the money is said to be borrowing and is called the *borrower*. It is customary to draw up a written agreement, called a *promissory note*, in which the principal sum, the interest rate, the compounding interval (if compounded), and the period of time or the *due date* are specified.

8.1.2 Future and Present Values(R,G,A,F)

As implied earlier, the investment process may be viewed from essentially two perspectives. The first involves the calculation of the maturity or *future value* of the note, which is simply the sum of money that must be repaid by the borrower. In this case, the concept of future value refers to the principal and any interest that is due and payable on the maturity date. The second perspective involves a concept called *discounting.* When discounting, we begin with the future value and determine the amount, called *present value*, that the lender is willing to provide the borrower today in exchange for the note. When determining the present value of some future amount, the interest rate is called the *discount rate.*

8.1.3 Interest Rates and Rates of Return[R,G,A,F]

Many different interest rates are charged in money markets. Obviously the lender wants as high a rate as possible, while the borrower wants one as low as possible. Observed interest rates are governed by expectations concerning the future and are composed essentially of three components. The first usually is referred to as the basic or risk-free interest rate while the second reflects expectations concerning the economy in general and the value of money (i.e., the inflation rate) in particular. The third component is referred to as a *risk* factor that reflects the likelihood that the borrower will honor all obligations on the due date.

The three components of observed interest rates cannot be measured directly. However, in a country that is politically and economically stable, the risk-free rate usually is thought to be slightly below the rate paid by the central government on its short-term obligations. When central banks such as the Federal Reserve in the United States or the Bank of Canada in Canada set policies concerning the money supply, interest rates frequently respond to changes in the discount rate. Just as commercial banks charge interest on loans to customers, Federal Reserve banks charge interest on loans to commercial banks. In this case, however, the Federal Reserve bank discounts the interest at the time the loan is negotiated rather than collecting the interest when the loan is repaid and, as implied above, the interest rate is called the discount rate. Even though commercial banks do not necessarily borrow from central banks, the discount rate is used as an indicator of the interest rates that are charged on loans to customers. As might be suspected, the interest rates on such loans usually are higher than the current discount rate.

Frequently, borrowed funds are required to implement a project that enables the borrower to produce and sell goods or services whose value is expected to exceed the cost of the investment, including the amount that must be repaid to the lender. The difference between the value of the goods or services associated with the project and the cost of the investment is often specified in decimal fraction or percentage form as a *rate of return.* A priori, the difference is called the *expected* rate of return and after the fact it is called an *actual* rate of return. Traditional economic theory suggests that investors should invest in those projects that earn the highest rates of return in order to maximize net income.

8.1.4 Cash Flows over Time: Net Present Value(R,G,A,F)

Thus far we have not considered a situation in which several payments are made by the lender to the borrower, and vice versa. For example, individuals and economic enterprises deposit and withdraw money from their savings accounts during a given period of time. On the other hand, borrowers may receive a lump sum of money that they repay over many months or several years. An investor may expect to obtain returns over a long period and to incur a series of costs over the same period. More specifically, an *annuity* is a financial arrangement in which one or more payments into a fund are used to purchase a future stream of annual repayments out of the fund. Annuities often are used to provide a retirement income.

As another example, consider an investment project where the initial purchase of equipment or facilities requires a large outlay of funds. In this case, the borrower expects to incur annual operating and maintenance costs, but also to generate revenues from the project over several years. The initial decision to invest in the project hinges on the belief that the eventual return will be greater than the corresponding costs and the return that might be earned by investing the funds in alternate proposals.

Traditionally, the various cash disbursements and cash receipts are expressed in terms of their value at the present time. In turn, the *net present value*, which is simply the difference between the cash receipts and cash disbursements as expressed in present value equivalents, provides the basis for evaluating the economic desirability of the project. It should be noted that, a priori, the cash flows are subject to the uncertainty surrounding the project. Consequently, when evaluating the economic desirability of the project, it is necessary to develop forecasts concerning the cash receipts and cash disbursements expected during the life of the project. The net present value by which the project is evaluated depends not only on the magnitude and timing of the cash flows but also on the discount rate used in translating future cash inflows and outflows into present value equivalents. In general, however, flows in the more distant future exert a progressively lower impact on the present value.

Although the latter point will be discussed later, it deserves brief amplification here. Other things remaining constant, an alteration in the discount rate can exert a dramatic impact on the present value of a given future cash flow. High discount rates imply low present values for cash flows in the distant future while low discount rates treat distant returns more favorably than high rates. Thus, present values depend not only on the size and timing of future cash flows but also on the discount rate imposed on them.

8.1.5 Health Care Organizations and Financial Analysis of Projects(R,A,F)

Historically, hospitals depended on charitable donations for financing—a dependence that led to decisions conforming to the desires of a specific individual or group. In such a situation, a project might be implemented without a formal analysis of its impact on the operational activity of the institution. Recently, however, duplications in service capability and the rising cost of health care have resulted in an increasing need to evaluate proposed projects not only in terms of their impact on operational activity of the institution but also in terms of the system by which health care is delivered to the community. In fact, the need for new equipment or a modification in the physical plant of the institution in most cases must be demonstrated to planning bodies that are external to the hospital.

Such a justification usually requires an evaluation not only of the clinical aspects and potential improvements in patient care associated with the project but also of the economic consequences of the proposal. Management frequently is able to identify the economic consequences of the project by calculating its net present value. For example, when justifying the decision to purchase a new piece of equipment that would permit a health care institution to provide a new service, management must consider both the acquisition and operating costs and the benefits that might be derived. The costs that should be given explicit recognition in this economic analysis are:

1. the initial capital outlay, which includes not only the purchase price but also the costs incurred in preparing the equipment for use; and
2. annual operating costs, which reflect maintenance charges as well as expenditures for labor, heat, power, light, space, and so on.

On the other hand, the potential benefits derivable from the project might be represented by the increase in patient revenues earned by providing the new service.

When estimating relevant costs and revenues for each year of the useful life of the piece of equipment, the resulting data should be presented in tabular form. After expressing costs and revenues in terms of their present value equivalents, we may calculate the net present value of the equipment by

$$NPV = \sum_{t=0}^{n} PVB_t - \sum_{t=0}^{n} PVC_t$$

where

PVB_t = the present value of the benefits in year t; and
PVC_t = the present value of costs in year t.

Here we assume that the equipment is purchased at time $t = 0$ and that the equipment will have a useful life of n years. In this case, assuming that required funds can be obtained and that other factors are favorable for its purchase, the equipment should be acquired if the net present value is greater than zero (i.e., the present value of the benefits exceeds the present value of the costs).

The discount rate used will be at least as great as the cost of financing the purchase. If an outright grant or forgivable loan is used, then the appropriate rate may be debatable, or the agency may impose a specified percentage for evaluation purposes. In a profit-oriented hospital the cost of capital usually is considered to be somewhere between the cost of debt and the desired return on shareholders' equity. By the start of the 1980s decade, interest rates spiralled to the vicinity of 17 to 19 percent, which implies that the cost of capital then was closer to 22 or 24 percent.

8.2 SIMPLE INTEREST[R,G,A,F]

For purposes of illustration, suppose a hospital receives a bequest for $15,000 from a patient who stipulates that interest payments must be used annually for research into the causes of heart disease. Suppose further that the bequest is used to purchase a bond that yields a return of 11 percent per year. On the basis of these assumptions, the purpose of the following discussion is to describe the concept of simple interest and to determine the financial resources generated by the bequest.

Representing the principal by P, the amount of interest by A, the interest rate by r, and the number of years by i, the formula for calculating the interest is given by

$$A_i = P \cdot r \cdot i \tag{8.1}$$

An application of Equation 8.1 yields

$$\begin{aligned} A_1 &= 15{,}000 \times 0.11 \times 1 \\ &= \$1{,}650 \end{aligned}$$

which implies that \$1,650 is generated annually by the research fund. Equation 8.1 also may be employed to determine the amount of interest earned during a period of, say, five years. In this case if we substitute $P = \$15{,}000$; $r = .11$ and $i = 5$ years into Equation 8.1, we obtain

$$A_5 = 15{,}000 \times .11 \times 5$$
$$= \$8{,}250$$

We modify our example slightly and assume that the amount of interest earned by investing the \$15,000 is added to the principal of the bequest rather than expended annually on research. In this case, our objective is to determine the sum of the original principal and the amount of interest earned at the end of a specified period. Using the notation developed earlier, we find that

$$S_i = P + A = P + P \cdot r \cdot i \qquad \textbf{(8.2.1)}$$

where S_i represents the sum of the original principal and the amount of interest earned after i years. Simplifying Equation 8.2.1, we obtain

$$S_i = P(1 + ri) \qquad \textbf{(8.2.2)}$$

Returning to our example, we wish to determine the sum of the principal and the amount of interest earned during a period of five years. After substituting appropriately in Equation 8.2.2, we obtain

$$S_5 = 15{,}000\ [1 + (.11 \times 5)]$$
$$= 15{,}000 \times 1.55$$

or \$23,250. These results are identical to those obtained by summing the principal and the amount of interest earned during the five-year period as calculated by Equation 8.1.

8.3 COMPOUND INTEREST(R,G,A,F)

Suppose that \$78,000 in excess cash is available to the institution for the next three and a half months and we decide to invest this sum in a savings account that earns compound interest at a rate of 1 percent per month. Assume further that we are interested in determining the amount that will be available to satisfy maturing obligations at the end of three months. In this situation, the amount on deposit at the end of the first month is given by the sum of the principal and the interest earned during the month.

Month	Deposit at the Beginning of the Month	×	Interest Rate	=	Interest	+	Deposit at the Beginning of the Month	=	Deposit at the End of the Month
1	\$78,000.00		.01		\$780.00		\$78,000.00		\$78,780.00
2	\$78,780.00		.01		\$787.80		\$78,780.00		\$79,567.80
3	\$79,567.80		.01		\$795.68		\$79,567.80		\$80,363.48

Thus, the interest earned during the first month is given by

$$\$78{,}000 \times .01 = \$780$$

which implies that

$$\$78{,}000 + \$780$$

or \$78,780 represents the amount in the savings account at the *beginning* of the second month. The interest earned during the second month is based not only on the original deposit of \$78,000 but also on the interest earned during the first month. Thus, the interest earned in the second month is given by

$$(\$78{,}000 + \$780) \times .01 = \$787.80$$

The balance of the savings account at the end of the second month is given by the sum

$$\$78{,}000 + \$780 + \$787.80$$

or \$79,567.80. Consequently, the interest earned during the third month is given by

$$(\$78{,}000 + \$780 + \$787.80) \times .01$$

or \$795.68 (which has been rounded to the nearest cent). Therefore, the amount available for satisfying currently maturing obligations at the end of the third month is given by

$$\$78{,}000 + \$780 + \$787.80 + \$795.68$$

or \$80,363.48.

This illustration provides the basis for developing a general formula that permits us to calculate the balance of the account at the end of a specified number of periods. Notice that the balance of our account at the end of the first month may be expressed in the form

$$\begin{aligned} S_1 &= P_0(1 + r) \\ &= 78{,}000(1 + .01) \\ &= \$78{,}780 \end{aligned} \tag{8.3.1}$$

In this case, S_1 represents the balance of the account at the end of the first month while P_0 corresponds to our initial deposit. Similarly, the balance of our account at the end of the second month, which we represent by S_2, might be expressed in the form

$$S_2 = S_1(1 + r) \tag{8.3.2}$$

However, returning to Equation 8.3.1, notice that $S_1 = P_0(1 + r)$ which implies that

$$\begin{aligned} S_2 &= P_0(1 + r)(1 + r) \\ &= P_0(1 + r)^2 \end{aligned} \tag{8.3.3}$$

Returning to our example, we find that an application of Equation 8.3.2 yields

$$\begin{aligned} S_2 &= \$78{,}000(1 + .01)^2 \\ &= \$79{,}567.80 \end{aligned}$$

which, of course, agrees with our earlier results. Similarly, the balance of our account at the end of the third month might be expressed in the form

$$S_3 = S_2(1 + r) \tag{8.3.4}$$

However, referring to Equation 8.3.3, notice that $S_2 = P_0(1 + r)^2$ and, after substituting $P_0(1 + r)^2$ for S_2 in Equation 8.3.4, we obtain

$$\begin{aligned} S_3 &= P_0(1 + r)^2(1 + r) \\ &= P_0(1 + r)^3 \end{aligned} \tag{8.3.5}$$

An application of Equation 8.3.5 to our example yields

$$S_3 = \$80{,}000(1 + .01)^3$$
$$= \$80{,}363.48$$

which, as before, indicates that our account will have a balance of $80,363.48 at the end of the third month.

These results suggest the following theorem. If P dollars are invested in an account that pays a compound rate of interest of r percent per period, the account will contain a total of

$$S_n = P(1 + r)^n \qquad \textbf{(8.4)}$$

dollars at the end of n periods. As an additional example of Equation 8.4, suppose $20,000 are invested for eight years in an account that pays a compound interest rate of 10 percent per year. After substituting $20,000 for P, .10 for r, and 8 for n in Equation 8.4, we find that the account will contain

$$S_8 = \$20{,}000(1 + .10)^8$$
$$= \$42{,}871.78$$

at the end of the eighth year.

8.3.1 The Difference between Simple and Compound Interest(R,A,F)

At this point, it is convenient to examine the difference in operational results obtained when a sum of money is invested in an account that pays simple as contrasted with compound interest. For illustration, let us return to the example in which $78,000 in excess cash is available to our institution. Under the assumption that the sum is invested for three months in an account paying a simple interest rate of 1½ percent per month, we may use Equation 8.2.2 and find that a balance of

$$S_3 = \$78{,}000\,[1 + (.015 \times 3)]$$
$$= \$81{,}510.00$$

will remain in the account after three months. On the other hand, suppose we invested this sum for three months in an account that pays a compound interest rate of 1½ percent per month. In this case, an application of Equation 8.4 reveals that

$$S_3 = \$78{,}000(1 + .015)^3$$
$$= \$81{,}562.91$$

will be in the account at the end of the third month. In this case, an additional \$52.91 (i.e., \$81,562.91 − \$81,510.00) is earned by investing the \$78,000 in the account paying compound interest as opposed to the one that pays simple interest.

In terms of our example, the account that pays compound interest provides our institution with very little extra money for essentially two reasons: (1) the interest rate of 1½ percent is rather low, and (2) only three months are involved. Concerning the latter point, assume that the \$78,000 was invested for 12 months rather than three months. An additional \$1,218.21 is earned by investing the \$78,000 in the account paying compound interest as opposed to the account that pays simple interest. Now, suppose we are able to invest the \$78,000 for 12 months in an account that pays a simple interest rate of 2 percent per month or one paying a compound interest rate of 2 percent per month. In this case, the difference in the interest earned in the two accounts is given by

$$[\$78{,}000(1 + .02)^{12}] - \$78{,}000[1 + (.02 \times 12)]$$

or approximately \$2,202.86.

This illustrates two important aspects that arise from compounded interest. The first is that a longer series of time periods allows the compounding process to accumulate a larger increment of interest relative to the amount obtained by simple interest. The second is that increasing the interest rate causes the compounding effect to be more than proportionally larger.

Using a different example to reinforce these concepts, we assume that the hospital is able to invest \$10,000 at rates of 1 percent, 1¼ percent, or 1½ percent per month for 12, 14, 16, or up to 24 months, both at simple interest and at compound interest. The total accumulated value (principal plus interest) for the three rates at each of the specified intervals, for both simple and compound interest are presented in Table 8-1. This table shows interest has been compounded on a monthly basis. These data also are presented graphically in Figures 8-1 and 8-2. In the latter, we use an expanded scale on the vertical axis to illustrate the difference between the effects of simple and compound interest. The bottom, middle, and top pairs of lines represent 1.00 percent,

Table 8-1 Accumulated Value of $10,000 Invested

Rate and Type of Interest	Accumulated Value (Rounded to Nearest Dollar)						
	Time Periods in Months						
	12	14	16	18	20	22	24
1.00%							
Simple	11,200	11,400	11,600	11,800	12,000	12,200	12,400
Compound	11,268	11,495	11,726	11,961	12,202	12,447	12,697
1.25%							
Simple	11,500	11,750	12,000	12,250	12,500	12,750	13,000
Compound	11,608	11,900	12,199	12,506	12,820	13,143	13,474
1.50%							
Simple	11,800	12,100	12,400	12,700	13,000	13,300	13,600
Compound	11,956	12,318	12,690	13,073	13,469	13,876	14,295

1.25 percent, and 1.50 percent interest rates, respectively. The lower of each pair shows the value accumulated using simple interest.

The two effects show clearly that compound interest generates an increasingly larger amount of interest as time passes, and higher compound interest rates have an effect that is more than proportional to the difference in the interest rates.

8.3.2 Compounding Frequency: Monthly and Annually(O,G,A,F)

Suppose further that our hospital is contemplating a five-year investment in either a note that compounds interest on a monthly basis or in one that compounds interest on an annual basis. To reach a prudent financial decision in such a situation it is necessary to consider the impact of compounding interest at different intervals. To further the illustration, let us examine interest rates of 1, 1¼, and 1½ percent per month, which when translated to annual rates usually are quoted as 12, 15, and 18 percent. These are called the *nominal annual rates* and are obtained by multiplying the monthly rate by 12.

Table 8-2 shows the compound interest factors as calculated on a monthly and annual basis for the first five years. A comparison of the annual and monthly values reveals that the nominally equivalent rates do not yield the same return since interest compounded on a monthly basis gives a larger factor in every case. Thus, the nominal annual rate is not equivalent to the

Figure 8-1 Graphic Representation of Simple and Compound Interest

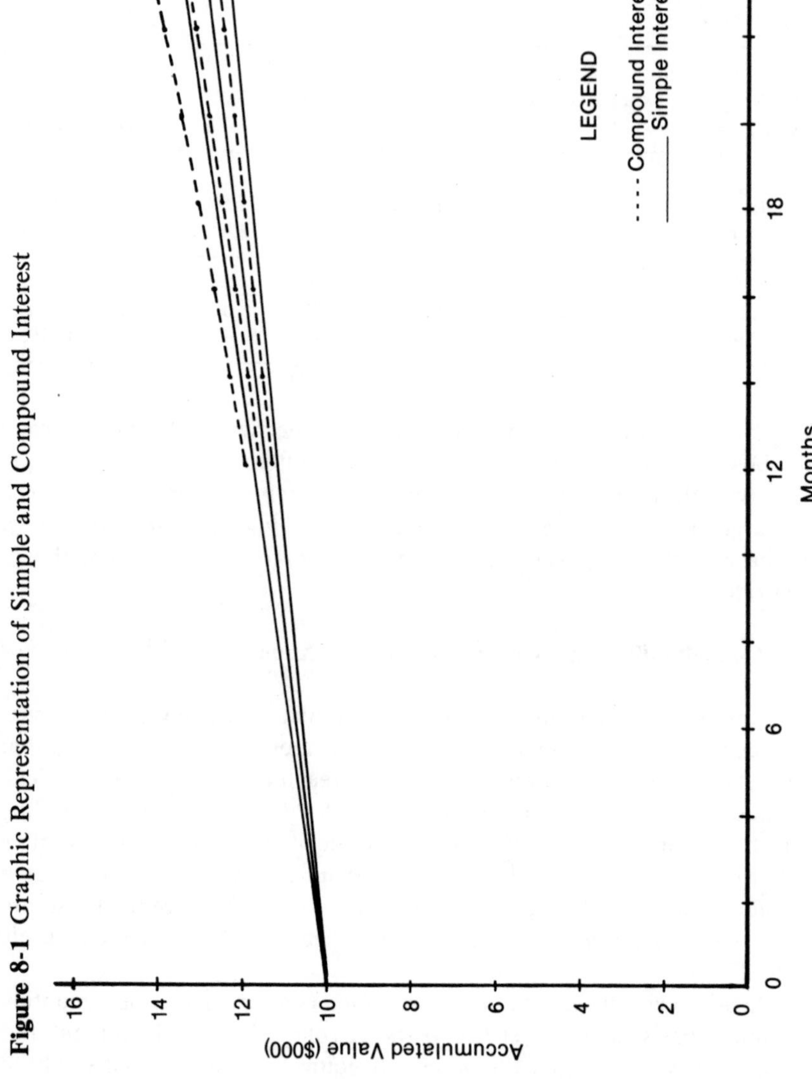

Figure 8-2 Expanded Vertical Scale Shows Differences in Simple and Compound Interest

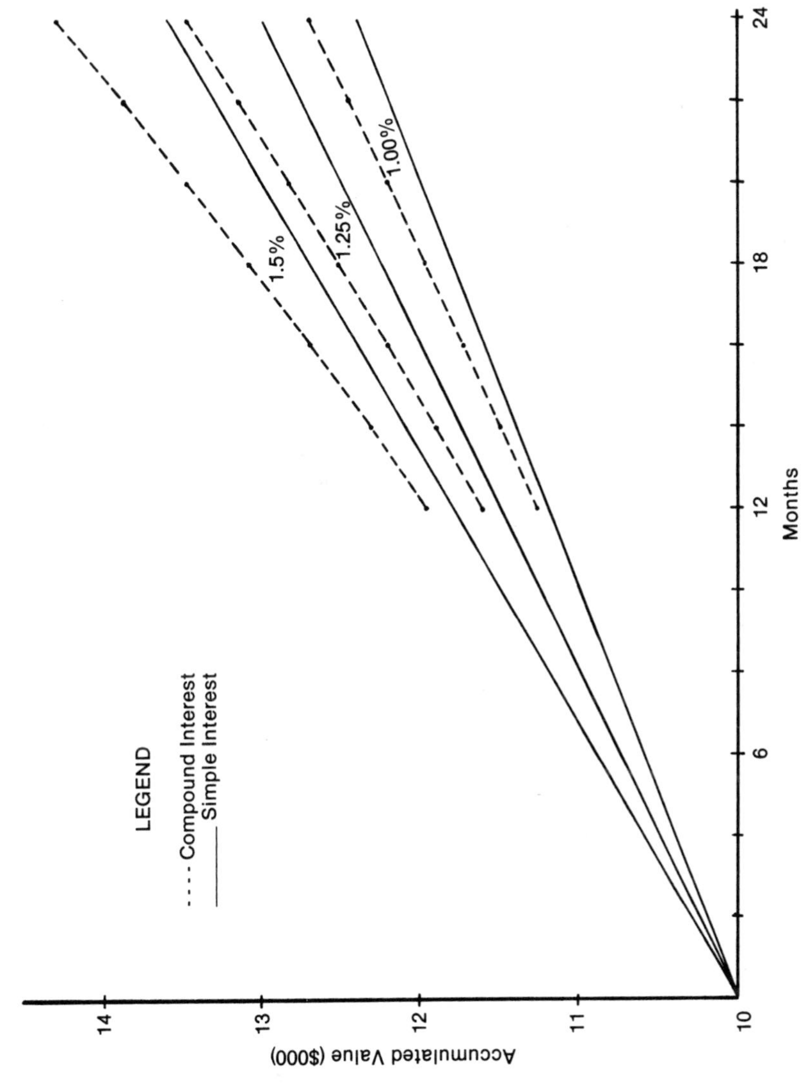

Table 8-2 Annual and Monthly Compound Interest Factors

Compounding Period and Rate	Compound Interest Factor at End of Year				
	1	2	3	4	5
Annual					
12%	1.12000	1.25440	1.40493	1.57352	1.76234
15%	1.15000	1.32250	1.52087	1.74901	2.01136
18%	1.18000	1.39240	1.64303	1.93878	2.28776
Monthly					
1%	1.12683	1.26973	1.43077	1.61223	1.81670
1¼%	1.16075	1.34735	1.56394	1.81535	2.10718
1½%	1.19562	1.42950	1.70914	2.04348	2.44322

monthly rate and the *effective annual rates* for monthly compounding are higher than the nominal rates. For example, a 1 percent monthly rate has an effective rate of 12.683 percent per year, while the monthly rates of 1¼ percent and 1½ percent have effective rates of 16.075 percent and 19.562 percent per year, respectively. Hence, when viewed from the perspective of the investor (or lender), notes or securities for which interest is computed on a monthly basis are preferred to those for which interest is compounded on an annual basis.

8.4 PRESENT VALUE CALCULATIONS(R,G,A,F)

The inverse of the process of compounding interest is that of calculating the present value of, or discounting, a future sum of money. Recall that a sum of money, P, invested at a compound interest rate, r, for i periods of time will have an accumulated value that is given by

$$S_i = P(1 + r)^i \tag{8.5}$$

Finding the present value, PV, of the future sum S_i is accomplished by solving Equation 8.5 for P as follows:

$$PV = P = \frac{S_i}{(1 + r)^i} \tag{8.6}$$

Here we call r the *discount rate.*

Assume that our hospital expects to receive \$15,000 two years hence and that we want to determine the present value of this sum. Employing Equation 8.6 and using a discount rate of 15 percent, the desired present value is given by

$$PV = \frac{15,000}{(1 + .15)^2}$$

$$= \frac{15,000}{1.3225}$$

$$= 15,000 \times .7561437$$

or approximately \$11,342.16.

8.4.1 Present Value Factors(R,G,A,F)

Consider next the present value of \$1 that is received at the end of i years in the future. In this case, we simply substitute \$1 for the term S_i in Equation 8.5 and obtain

$$PV = \frac{1}{(1 + r)^i} \qquad \textbf{(8.7)}$$

As defined by Equation 8.7, the present value of \$1 received at the end of i years is also called a *present value factor.* As seen in Table 8-3, we may employ discount rates of 12, 15, and 18 percent to calculate the present value of \$1 received at the end of year 1, 2, 3, 4, and 5. Notice that we also may use the information presented in the first three rows of Table 8-2 to calculate the present value factors appearing in Table 8-3. For example,

Table 8-3 Present Value Factors

Discount Rate	Present Value of \$1 Received at End of Year				
	1	2	3	4	5
12%	.89286	.79719	.71178	.63552	.56743
15%	.86957	.75614	.65752	.57175	.49718
18%	.84746	.71818	.60863	.51579	.43711

using a discount rate of 12 percent, the present value of $1 received at the end of year 1 is given by the inverse of the compound interest factor 1.12000 (i.e., 1/1.12000 = .89286). The other values in Table 8-3 might be calculated in a similar fashion.

Table 8-3 shows that the present value of $1 received at the end of a given year *decreases* as the discount rate increases. Similarly, for a given discount rate, the present value factors decrease as the number of years in the future grows. Thus, the present value of $1 depends on the discount rate as well as the time horizon.

8.4.2 Present Value of a Series of Future Payments(R,G,A,F)

Suppose we want to evaluate the desirability of acquiring a new piece of equipment that is expected to generate revenue over a series of years. Suppose further that the institution uses a discount rate of 12 percent and that we expect to receive fees during each year of the expected life of the equipment as follows:

Year	Fees
1	$15,000
2	16,000
3	17,000
4	17,000
5	17,000

The present value of any one of these receipts is found easily. For example, the present value of the $16,000 we expect to receive in year 2 is given by

$$PV_2 = \frac{16{,}000}{1.12^2} = \$12{,}755.10$$

while the present value of the $17,000 we expect to receive in year 3 is obtained by

$$PV_3 = \frac{17{,}000}{1.12^3}$$

or approximately $12,000.26. The present value of the series of receipts is the sum of the individual values as follows:

$$PV = \frac{15{,}000}{1.12} + \frac{16{,}000}{1.12^2} + \frac{17{,}000}{1.12^3} + \frac{17{,}000}{1.12^4} + \frac{17{,}000}{1.12^5}$$

$$= 13{,}392.86 + 12{,}755.10 + 12{,}100.26 + 10{,}803.81 + 9{,}646.26$$

$$= \$58{,}698.29$$

As should be expected, the present value of these receipts is substantially less than the undiscounted total of $82,000.

This illustration suggests the following theorem. If we expect to receive $\$R_1, \cdots, \$R_i, \cdots, \$R_n$ at the end of year $1, \cdots, i, \cdots, n$, respectively, the present value of the future stream of receipts is given by

$$PV = \sum_{i=1}^{n} \frac{R_i}{(1 + r)^i} \qquad \textbf{(8.8)}$$

where, as before, r is the discount rate.

8.5 EVALUATION OF AN INVESTMENT PROJECT(O,A,F)

Referring to the example above, assume that the data in Table 8-4 represent the costs and benefits of purchasing the piece of equipment. The table reveals that the initial cost of the new equipment is $30,000 and that its expected salvage value at the end of five years is $10,000. Also notice that our institution expects to incur service costs amounting to $3,000 per year. Finally, notice that in addition to the expected receipts mentioned earlier, operating costs—represented by power, supplies, and wage and salary payments—have been projected and included in the table.

We may now combine these data and evaluate the economic consequences of the proposed acquisition by employing the following criterion:

> The proposal should be accepted if the present value of the receipts is greater than that of the costs; that is, if the net present value (*NPV*) is positive. If the *NPV* is negative, the proposal should be rejected. If the *NPV* is zero, then we would be indifferent to the purchase of the equipment.

Table 8-4 Annual Cash Flows for Testing Equipment

	Cash Flows (Assumed at End of Year)					
Category	0	1	2	3	4	5
Costs						
Purchase	$30,000					
Service	3,000	$ 3,000	$ 3,000	$ 3,000	$ 3,000	$3,000
Salaries		9,200	9,300	9,400	9,400	9,400
Power, Supplies, etc.		900	950	1,000	1,000	1,000
Total Costs	33,000	13,100	13,250	13,400	13,400	13,400
Receipts						
Fees		15,000	16,000	17,000	17,000	17,000
Salvage						10,000
Total Receipts		15,000	16,000	17,000	17,000	27,000

Applying the principles introduced earlier, we find that the present value of the total costs associated with the acquisition is given by

$$\begin{aligned} PV(\text{Costs}) &= \frac{\$33{,}000}{1.12^0} + \frac{\$13{,}100}{1.12^1} + \frac{\$13{,}250}{1.12^2} + \frac{\$13{,}400}{1.12^3} \\ &\quad + \frac{\$13{,}400}{1.12^4} + \frac{\$13{,}400}{1.12^5} \\ &= \$33{,}000 + \$11{,}696.43 + \$10{,}562.82 + \$9{,}537.86 \\ &\quad + \$8{,}515.94 + \$7{,}603.52 \\ &= \$80{,}916.57 \end{aligned}$$

As should be verified, the present value of the anticipated receipts is approximately $64,372.56. Consequently, the net present value of the project is given by

$$\begin{aligned} NPV &= \$64{,}372.56 - \$80{,}916.57 \\ &= -\$16{,}544.01 \end{aligned}$$

Since the net present value is negative, the economic criterion described previously suggests that the proposal should be rejected. If the equipment is purchased for noneconomic reasons, the calculations imply that these factors have been assigned a current value at least equal to $16,544.01. Also notice that if we decide to purchase the equipment, the project must be subsidized from some other source.

That the approach outlined in this section is of importance to the evaluation of investment proposals should be clear. However, other factors must be considered when reaching investment decisions and these are considered next.

8.6 PROBLEMS OF EVALUATING INVESTMENT PROJECTS(O,A,F)

Although the concept of net present value captures many of the economic consequences associated with a given investment proposal, several residual problems remain. The first involves the selection of an appropriate discount rate. Historically, the discount rate that should be used in the evaluation of private, public, and semipublic projects has been the subject of debate. During the mid 1960s, the cost of borrowing ranged from 8 to 14 percent, depending on risk elements. Under such conditions, it was argued that a discount rate between 8 and 25 percent should be used to evaluate proposals that were funded privately. On the other hand, publicly funded projects frequently generate social benefits that are not captured or internalized by the commercial firm. Consequently, it was argued that a low discount rate (approximately 3 percent) should be employed when evaluating a publicly funded project that generates long-term social benefits. Unfortunately, the selection of an appropriate discount rate is still a contentious issue that has not been resolved satisfactorily.

In addition to the problem of selecting an appropriate discount rate, the sharp rise in prices and in interest rates has complicated the process of evaluating investment projects. The major difficulty emanates from the nature of the cash flows associated with most proposals. In this regard, cash flows are either fixed and determined by a contractual agreement or are variable and subject to inflationary pressures.

Consider first the influence of inflation on the interest rate stated on a promissory note. As is well known, an increase in the price level reduces the purchasing power of money. To compensate for the reduced purchasing power of those dollars that are repaid on outstanding loans, interest rates reflect the expectations of the borrower and lender concerning the future rate of inflation. To better understand this observation, consider the following example.

Holding the purchasing power of money constant, we may suppose that the interest rate imposed on a borrower is 5 percent per year. Now, relax the assumption concerning the purchasing power of money and assume that prices are expected to rise at a rate of 9 percent per year. In this situation, we might argue that the interest rate imposed on the borrower is given by the sum of the inflation rate (9 percent) and the "basic" rate of interest (5 percent), or 14 percent per year.

Second, inflationary pressures also complicate the process of evaluating a given investment project. Some of the costs are fixed while others will increase at unknown and possibly different rates during the life of the project. For example, it is possible that the rate of increase in salaries and supply costs will exceed the interest rate charged on an outstanding loan. In such a situation, the fixed loan payments become less expensive relative to the other costs associated with the project. Thus, before calculating the net present value of the project, management should ensure that the effects of inflation are either excluded or included in all of the data incorporated in the analysis. As an alternative, management may elect to apply different discount rates to the different cash flows.

A third problem involves the identification and measurement of cash flows as well as the cash equivalents of other benefits and costs. Direct costs or benefits are those that can be readily identified and measured whereas those that are less easily identified or measured are referred to frequently as the indirect costs and benefits of the project.

Fourth, benefits and costs that cannot be measured are called intangibles. Usually, the net present value of a proposal is based on the direct cost or benefits of the proposal. However, if the intangibles as well as the indirect costs and benefits are ignored completely, the net present value of the proposal may lead to incorrect decisions.

A fifth problem involves the identification of the intangibles and indirect consequences that should be incorporated in the analysis of the proposal. Unfortunately, it is not possible to offer generalizable guidelines that indicate the intangibles or the indirect consequences that should be incorporated in the analysis of a given project. Moreover, once the indirect consequences have been identified, it may not be possible to develop an accurate measure of their importance. In many situations, then, it may only be possible to acknowledge the presence of intangibles and indirect consequences that, in turn, might be evaluated subjectively by management.

In conclusion, it might be argued that the process of evaluation involves the calculation of the net present value of the project as determined by the direct and, to the extent possible, the indirect consequences of the proposal. In part, then, the evaluation of a given project is a scientific undertaking in which known or assumed data are subjected to rational analysis. On the other hand, the treatment of intangibles and the problems posed by inflationary pressures as well as the selection of the discount rate are more of an art than a science. Hence, the mixture of art and science makes the analysis of an investment project an integral part of management.

8.7 CONCLUSIONS[O,A,F]

It should be clear that the financial management of health care institutions requires an understanding of interest and present value. In this chapter, we saw that, with the passage of time, simple and compound interest increase a principal amount, and that compound interest grows more rapidly than simple interest.

The discussion also showed that the calculation of the present value of a future payment is the inverse of the process by which compound interest is calculated. In turn, the economic analysis of an investment proposal consists of identifying and measuring future series of costs and benefits, expressing them as cash flows, and finding the net present value of the project. Although the problems associated with analysis of proposals are difficult to overcome, they reinforce the importance of performing a formal evaluation to ensure that decisions reflect all factors germane to a situation.

Problems for Solution

1. Go to one or more local banks, trust companies, or savings and loans institutions and obtain information about their various savings accounts and savings certificates or notes. How many different names do they use for the accounts or certificates? Do they offer simple or compounding interest? How frequently do they *calculate* the interest? How frequently do they *compound* it? Do they specify nominal or effective annual rates or some other rate? (What do they specify on their loans?) How long a term is required on the deposits or certificates? How many days of notice of withdrawal do they require on the various types of accounts? Which one gives you the best deal? (How did you decide?)
2. Compute simple and compound interest accumulated sums for the following principal amounts and annual interest rates for 2, 4, 6, 8, and 10 years. Plot the accumulated sums on a graph.

   Principal Amount: $2,000, $12,000, $17,500
   Annual Rates: 10%, 15%, 20%

3. Compute the present value of each of the following amounts to be received 2, 4, 6, 8, or 10 years in the future using discount rates as specified.

Amount to be received:	$2,000	$12,000	$17,500
Discount Rates:	11%	16%	21%

4. Find the nominal and effective annual rates of interest for monthly compounded rates of 1.1, 1.2, 1.3, 1.6, and 1.7 percent.
5. A forward-looking hospital is considering investing in the development of a testing facility that will show increasing usefulness over the next 10 years. The estimated costs and benefits are shown in thousands of dollars in Table 8-5. Compute the net present value for discount rates of 2, 6, 12, and 20 percent. What happens to the net present value with increasing discount rates?

Table 8-5 Costs and Benefits of Investing in a Testing Facility

	Year										
	0	1	2	3	4	5	6	7	8	9	10
Construction	1,000	0	0	0	0	10	0	0	0	0	0
Equipment											
purchase	300	50	0	50	0	50	0	0	0	0	0
maintenance	0	16	16	17	17	18	18	18	18	18	18
Building											
maintenance	5	20	20	20	20	25	25	25	30	30	30
heat and power	3	6	6	6	6	7	7	7	7	7	7
Supplies	30	10	11	12	13	15	15	16	16	16	16
Staff Salaries	50	200	200	200	200	250	250	250	250	250	250
Fees for Services	0	378	403	441	500	567	630	693	756	756	756

Part III

Calculus

Chapter 9

Introduction to Calculus

Objectives

After completing this chapter you should be able to:

1. Describe the two branches of calculus;
2. Use the derivative as a measure of the responsiveness of one variable to changes in another;
3. Use the basic rules of differentiation;
4. Use differential calculus in problems of optimization;
5. Use integral calculus to find the area under a curve.

Chapter Map

The sections comprising this chapter may be summarized as follows:

Section Number	Required Reading	Optional Reading	Generic Development	Application to Management	Fundamental Principles	Complex Material
	(R)	(O)	(G)	(A)	(F)	(C)
9.1	x		x		x	
9.2	x		x	x	x	
9.2.1	x		x	x	x	
9.2.2	x		x		x	
9.2.3	x		x	x	x	
9.3	x		x	x	x	
9.3.1	x		x		x	
9.3.2	x		x	x	x	
9.4	x				x	

9.1 AN OVERVIEW(R,G,F)

This part presents the fundamentals of calculus. In general, two distinct but interrelated branches of calculus are identifiable. The first is differential calculus, which permits one to determine the change in the dependent variable that results from a change of one unit in the independent variable. Thus, in order to study rates of change, it is necessary to understand and use the techniques of differential calculus.

The second branch is integral calculus, which has essentially two distinct applications. First, integral calculus permits us to reverse the process of differentiation and, when used in this way, the integral may be viewed as the inverse of the derivative. The second allows us to determine the area under a curve that is defined by the function $f(x)$. This gives the concept of integration central importance when dealing with continuous probability distributions.

The primary purpose of this chapter is to provide an overview of the two branches of calculus. As such, it discusses fundamental concepts and, as a result, ignores much of the mathematical detail that is presented later. Since the development of the basic concepts of integral calculus requires an understanding of the derivative, we begin with a discussion of this branch.

9.2 AN INTRODUCTION TO DIFFERENTIAL CALCULUS(R,G,A,F)

As mentioned, differential calculus may be described as the study of change. An understanding of the changes in one variable in response to changes in another obviously is of vital importance in health administration. For example, management may be interested in the response of costs or revenues to changes in the quantity of care provided by the institution. Management also may be interested in the additional benefits that accrue to the community by increasing the rate of operation in a given area of activity. In each of these examples, we are measuring the change in one variable that results from a change in another. Application of the techniques of differential calculus allows us to measure these changes with precision.

9.2.1 The Derivative(R,G,A,F)

For illustration, consider the problem of analyzing the response of costs to changes in the quantity of care provided by our institution. More specifically, assume that the relation between total costs (y) and the volume of care (x) is given by

$$y = \$50.00 + (\$6.00)x \tag{9.1}$$

where the fixed costs of operation amount to \$50.00 and the variable cost component is represented by the term (\$6.00)x. As the discussion of Section 5.3.1 noted, the slope of the line may be represented in the form

$$\frac{\Delta y}{\Delta x} = \frac{y_2 - y_1}{x_2 - x_1} \tag{9.2}$$

In terms of this example, by increasing the volume of service from, say, 10 units to 11 units, costs increase from \$110 to \$116. Hence, an application of Equation 9.2 yields

$$\frac{\Delta y}{\Delta x} = \frac{116 - 110}{11 - 10}$$
$$= \$6.00$$

which implies that the slope of the line is \$6.00. As a result, each time the service is increased by one unit, total costs increase by \$6.00.

Consider next a slightly more complex situation in which

$$y = \$50 - \$2.00x + .05x^2 \tag{9.3}$$

Figure 9-1 shows that Equation 9.3 is a parabola that reaches a minimum value of \$30.00 when 20 units of service are produced. However, unlike our earlier example, the slope of the curve is not constant. For example, if the provision of service is increased from 25 to 45 units, total costs increase from \$31.25 to \$61.25. In this case, an application of Equation 9.2 yields

$$\frac{\Delta y}{\Delta x} = \frac{\$61.25 - \$31.25}{45 - 25}$$
$$= \$30/20$$
$$= \$1.50$$

On the other hand, if the volume of service is increased from 25 units to 35 units, total costs increase from \$31.25 to \$41.25 and an application of Equation 9.2 yields

$$\frac{\Delta y}{\Delta x} = \frac{\$41.25 - \$31.25}{35 - 25}$$
$$= \$10/10$$
$$= \$1.00$$

Figure 9-1 The Total Cost Function

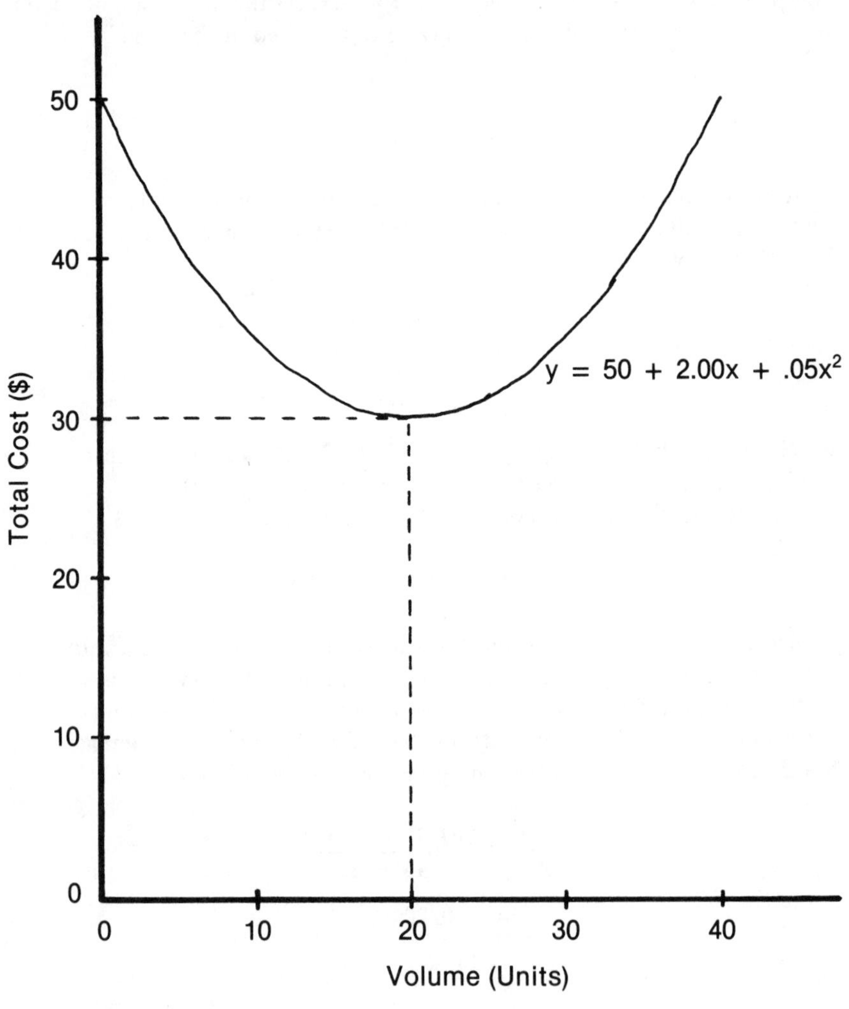

Finally, if the service is increased from 25 units to 26 units, costs increase from \$31.25 to \$31.80, which implies that

$$\frac{\Delta y}{\Delta x} = \frac{\$31.80 - \$31.25}{26 - 25}$$

$$= \$.55$$

These calculations illustrate several important points. First, when Equation 9.2 is used to calculate the slope, which is simply a numerical measure of the steepness of the line, the resulting value depends on the magnitude of Δx. In terms of our example, notice that

(1) $\Delta y/\Delta x = \$1.50$ when $\Delta x = 20$ units;
(2) $\Delta y/\Delta x = \$1.00$ when $\Delta x = 10$ units; and
(3) $\Delta y/\Delta x = \$.55$ when $\Delta x = 1$ unit.

In this case, we are faced with the possibility that, by employing an even smaller change in x, we might obtain different results.

The techniques of differential calculus are designed to deal with this possibility. Conceptually, we proceed by calculating the ratio $\Delta y/\Delta x$ for smaller and smaller values of Δx. Referring to the following calculations, we see that, as Δx becomes smaller and smaller, the value of the ratio $\Delta y/\Delta x$ approaches \$.50.

x	Δx	Δy	$\Delta y/\Delta x$
25	1.00	\$.55	\$.55
25	.75	\$.40	\$.54
25	.50	\$.26	\$.53
25	.25	\$.13	\$.51

Employing standard mathematical terminology, we say that as Δx approaches zero, the ratio $\Delta y/\Delta x$ approaches the limiting value of \$.50. (However, we cannot make $\Delta x = 0$ in Equation 9-2 because the division would not be possible.) The limiting value normally is represented by the symbol dy/dx, which is the *first derivative* of the variable y with respect to the variable x. In fact, the limiting value that is approached by an infinite sequence of numbers, when coupled with the definition of the derivative, is the foundation of differential calculus.

9.2.2 Basic Rules of Differentiation(R,G,F)

The primary purpose of this section is to investigate the use of the definitions that have been presented. For illustration, suppose that

$$y = 50{,}000 \quad \textbf{(9.4.1)}$$

$$y = 7x \quad \textbf{(9.4.2)}$$

$$y = 4x^2 \tag{9.4.3}$$

represent three rather implausible relationships between cost and the volume of service. In Equation 9.4.1, we assume that costs are unaffected by volume, Equation 9.4.2 assumes that costs are proportional to the volume of service, and Equation 9.4.3 represents a second-degree relation between cost and volume.

These three equations provide the basis for introducing four rules of differentiation that are of fundamental importance to many managerial problems.

Rule 1: The derivative of any variable with respect to a *constant* is zero.

The first rule is intuitively plausible when viewed from the perspective of Equation 9.4.1, where it is assumed that all costs are fixed and invariant with respect to output. In this case, we represent Rule 1 by observing that if

$$y = k$$

where k is any constant, $dy/dk = 0$.

Rule 2: In a first degree relationship of the form $y = bx$, where b is any number, the derivative of y with respect to x is b.

Equation 9.4.2 demonstrates that the cost of providing x units of service is $7x$ and that the cost of producing $x + 1$ units of service is $7(x + 1) = 7x + 7$. Hence, for each unit increase in the provision of service, costs increase by \$7.00.

Rule 3: If $y = ax^n$, where a and n are any constants, the derivative of y with respect to x is given by $n(a)x^{n-1}$.

Although the validity of this rule is demonstrated later, we may refer to Equation 9.4.3 and apply Rule 3 to obtain

$$\begin{aligned} dy/dx &= 2(4)x^{2-1} \\ &= 8x \end{aligned}$$

Notice that dy/dx is given by the product of the coefficients n and a and the variable x which is raised to the power $n - 1$. Such an expression allows us to find the slope (i.e., the instantaneous rate of change in y relative to x) of the curve at the point $x = c$. Notice that the first derivative of y with respect to x is a function of x. If we let $x = 4$, we find that

$$\begin{aligned} dy/dx &= 8(4) \\ &= 32 \text{ units} \end{aligned}$$

which represents the rate of change in y with respect to x at the point $x = 4$.

> **Rule 4:** The derivative of a sum of several terms is the sum of the derivatives of the terms.

For example, suppose that we combine Equations 9.4.1, 9.4.2, and 9.4.3 to form

$$y = 50{,}000 + 7x + 4x^2$$

Referring to Rule 1, recall that the derivative of a constant (50,000) is zero while Rule 2 suggests that the derivative of $7x$ is 7. Also notice that we may employ Rule 3 to obtain the derivative of $4x^2$, which is $8x$. Hence, applying Rule 4, we find that

$$\begin{aligned} dy/dx &= 0 + 7 + 8x \\ &= 7 + 8x \end{aligned}$$

Let us return to our example in Equation 9.3 (and Figure 9-1) in which we assumed that the cost-volume relationship might be represented by

$$y = \$50 - \$2.00x + \$.05x^2$$

In this case, we apply the rules presented in this section to obtain

$$dy/dx = 0 - \$2.00 + \$.10x$$

Also recall that we calculated the value of the ratio $\Delta y/\Delta x$ for the point $(25 + \Delta x)$ as the value of Δx became smaller and smaller. In the limit of this process, then, we find that

$$dy/dx = 0 - \$2.00 + \$.10(25)$$
$$= \$.50$$

which agrees with our earlier results.

9.2.3 The Derivative and Problems of Optimization[R,G,A,F]

At this point it is useful to translate the derivative into graphic terms. Consider Figure 9-2, where the curve defined by

$$y = \$50 - 2.00x + .05x^2$$

has been reproduced. The slope of the curve changes as the volume of service is increased. As a consequence, the value of the ratio $\Delta y/\Delta x$ is ambiguous which, in turn, requires the use of differential calculus. Earlier, we calculated values of $\Delta y/\Delta x$ for the general point $(25 + \Delta x)$. Notice that as Δx approaches zero, the general point $(25 + \Delta x)$ approaches the point 25. Ultimately, however, continued reductions in the value Δx will result in a state in which Δy and Δx are both zero.

Unfortunately, the slope of the curve at the point C cannot be interpreted in the same way as the slope of a finite interval since there is neither a Δy nor a Δx. Rather, as will be seen later, the derivative of the curve at C is given by the slope of the tangent line *TT* that has been constructed at that point.

We use this result to describe the general problem of optimization that involves the identification of the value of x that corresponds to the minimum or maximum point on the curve. Figure 9-2 demonstrates that when more than 20 units of service are provided, total costs rise, since the slope of the curve is positive to the right of the point B. Similarly, when fewer than 20 units of service are provided, total costs are higher. (The slope of the curve is negative to the left of the point B.) The minimum point on the curve (i.e., the point B) occurs when the slope of the curve is zero.

In such a situation, we identify the value of x that corresponds to the minimum point on the curve by

1. finding the first derivative,
2. forcing the first derivative to equal zero,
3. solving for x, and
4. ensuring that the value of the first derivative changes from negative to positive in the region surrounding the minimum point.

Following these procedures, we find that, as before,

$$dy/dx = -2.00 + .10x$$

Forcing this result to equal zero, we obtain

$$-2.00 + .10x = 0$$

and solving for x yields

$$.10x = 2.00$$
$$x = 20 \text{ units}$$

Finally, observe that 19 < 20 units and

$$dy/dx = -2.00 + .10(19)$$
$$= -.10$$

which is less than zero. Similarly, notice that 21 > 20 units and that

$$dy/dx = -2.00 + .10(21)$$
$$= \$.10$$

which is positive. On the basis of these calculations we conclude that the costs of care are minimized when 20 units of service are provided.

The process of identifying the value of x that corresponds to the maximum point on a curve is similar. For example, Figure 9-3 shows that x^* corresponds to the maximum value of y. Also notice that the slope of the curve is

1. positive to the left of point B;
2. zero at point B; and
3. negative to the right of point B.

As in the discussion above, we identify the value of x that corresponds to the maximum value of y by

1. finding the first derivative dy/dx;
2. forcing dy/dx to equal zero;
3. solving the resulting equation for x; and
4. ensuring that the slope of the curve changes from positive to negative in the region surrounding the maximum point.

Figure 9-2 Cost of Service

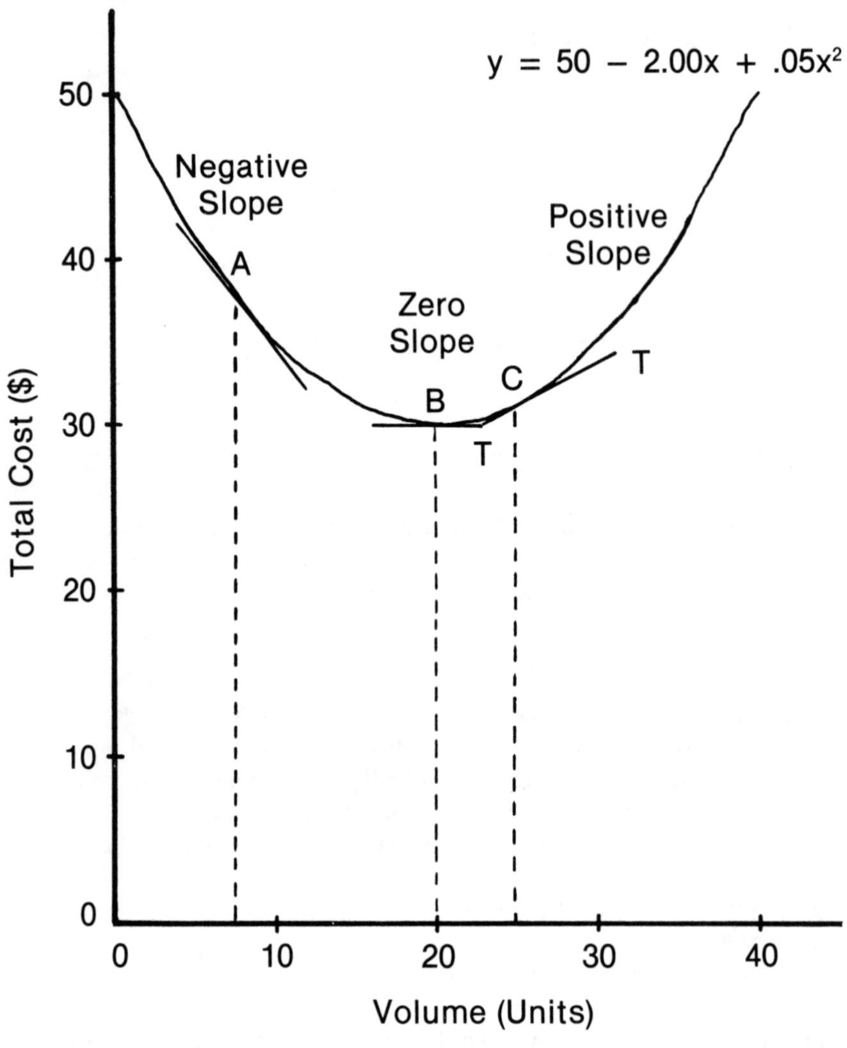

Figure 9-3 Example of a Maximum Point

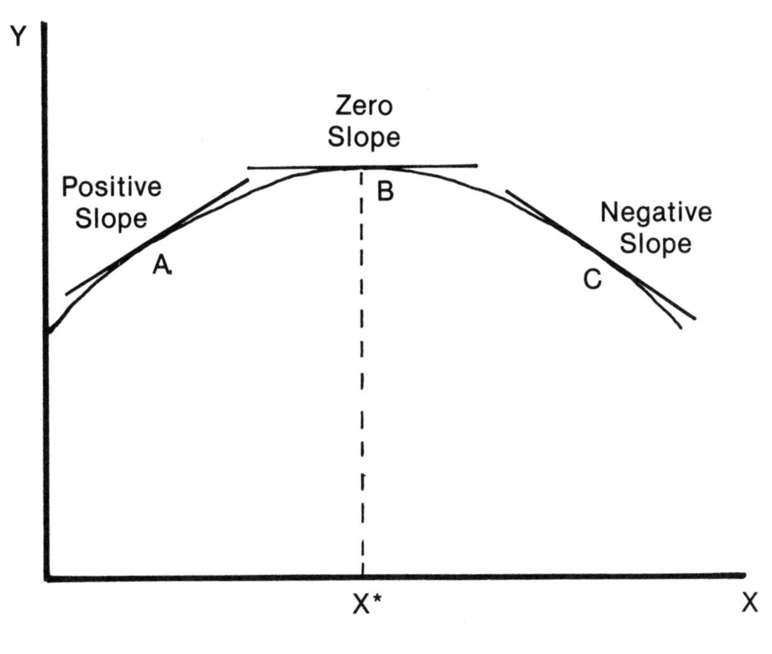

As will be seen later, the process of identifying the minimum or maximum point on a curve is greatly simplified when the second derivative is employed.

9.3 INTRODUCTION TO INTEGRAL CALCULUS(R,G,A,F)

As mentioned earlier, integral calculus has two distinct uses. The first is to reverse the process of differentiation and the second involves the problem of determining the area under a curve. In this section we consider the role played by integral calculus in reversing the process of differentiation first and then consider the use of the integral in finding the area under a curve.

9.3.1 The Inverse of Differentiation(R,G,F)

To illustrate the use of integration in reversing the process of differentiation, suppose that $3x$ represents the first derivative of an unknown equation that we require. The rules developed in the previous section must have been employed in obtaining the first derivative. Hence, if we reverse these rules and apply a similar process, we may obtain the desired equation. In this case, we let

$$f(x) = 3x$$

We require a function $F(x)$ that has as its first derivative the function $f(x)$.

Referring to Rule 3 presented earlier, recall that, if $y = ax^n$,

$$dy/dx = n(a)x^{n-1}$$

To reverse the process of differentiation, we simply increase the power to which x is raised by one and then divide the result by the new power. In general, if $f(x) = ax^x$ for $n \neq -1$, we find that

$$F(x) = \frac{a}{n+1}x^{n+1} + C \qquad \textbf{(9.5)}$$

where C represents a *constant of integration* and may assume any real value. The presence of the constant of integration comes from an application of the first rule of differentiation, which indicated that the derivative of a constant is zero. Hence, any constant value that appeared in the original equation was eliminated when the process of differentiation was performed. To compensate for this fact, we attach the constant of integration to the term $[a/(n + 1)]x^{n+1}$ when finding the function $F(x)$. Also, since C may assume any real value, the function $F(x)$ represents an infinite number of equations.

Returning to the example in which $f(x) = 3x$, we find that an application of Equation 9.5 yields

$$\begin{aligned} F(x) &= \frac{3}{2}x^2 + C \\ &= 1.5x^2 + C \end{aligned}$$

Hence, the expression $1.5x^2 + C$ represents a family of curves, each of which has the slope $3x$.

At this point, it is convenient to introduce the *indefinite integral.* The indefinite integral of the function $f(x)$ is given by

$$\int f(x)dx = F(x) + C \qquad \textbf{(9.6)}$$

where C is the constant of integration. Since C can take on *any* value the integral is called *indefinite.*

9.3.2 The Area under a Curve(R,G,A,F)

The purpose of this section is to illustrate the use of the *definite integral* in finding the area under a curve. As will be demonstrated, we may define the definite integral of $f(x)$ for $x = a$ and $x = b$ by

$$\int_a^b f(x)dx = F(x)\Big|_a^b = F(b) - F(a) \tag{9.7}$$

where $F(b)$ refers to the value of $F(x)$ that is obtained when we substitute the value b for x in $F(x)$. Similarly, when we substitute the value a for x in $F(x)$ we obtain $F(a)$. In turn, the difference between $F(b)$ and $F(a)$ corresponds to the area under the curve defined by $f(x)$ between the points $x = a$ and $x = b$ for $b > a$.

To illustrate, suppose a health care institution is considering a project that has a life expectancy of ten years and results in *annual* savings that may be approximated by

$$y = 50.00 + 4.00x$$

where x represents the number of years of operation. Assuming that the project requires an initial capital outlay of \$1,000, our objective is to determine whether the total savings resulting from the project are greater than, less than, or equal to the initial investment. Presumably, if the total savings are at least equal to or greater than the initial outlay, management should implement the project. On the other hand, if the initial capital outlay is greater than total savings, the proposed project probably should be rejected.

The objective in Figure 9-4 is to find the area under the line defined by $y = 50.00 + 4.00x$ for $x = 0$ and $x = 10$ years. Employing Equation 9.7, we find that the desired area is given by

$$\begin{aligned}\int_0^{10} (50 + 4x)dx &= 50x + 2x^2 \Big|_0^{10} \\ &= [50(10) + 2(10)^2] - [50(0) + 2(0)^2] \\ &= \$700\end{aligned}$$

Figure 9-4 The Use of the Definite Integral To Find the Area under the Line $y = 50.00 + 4.00x$

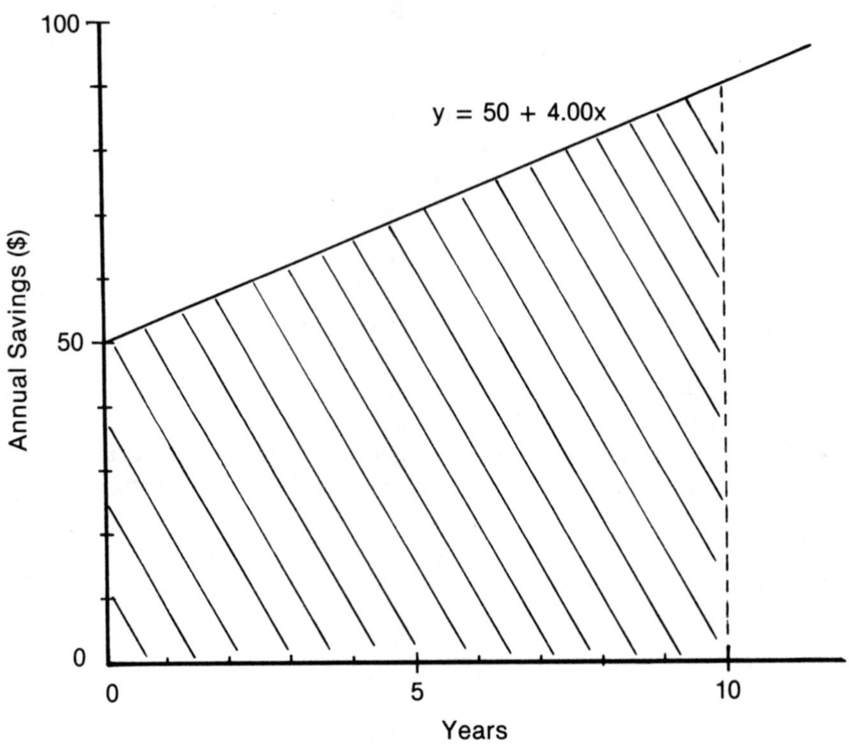

Given that the initial investment of $1,000 is greater than the total savings generated by the project (i.e., $700), these calculations suggest that the proposed project should be rejected. Notice that the expressions $[5(10) + 2(10)^2]$ and $[50(0) + 2(0)^2]$ correspond to the terms $F(a)$ and $F(b)$, respectively.

This approach also may be used to determine the savings that are realized between, say, the fifth and tenth years of operation. In this case, we expect to realize savings amounting to

$$\int_5^{10} (50 + 4x)dx = 50x + 2x^2 \Big|_5^{10}$$

$$= [50(10) + 2(10)^2] - [50(5) + 2(5)^2]$$

$$= \$400$$

during this period. Notice that the symbol $|_5^{10}$ simply specifies the limits of integration and indicates that the integral of $f(x)$ between the points $x = 5$ and $x = 10$ is given by $F(10) - F(5)$.

9.4 CONCLUSIONS[(R,F)]

In general the purpose of this chapter has been to present the fundamental concepts of the two branches of calculus and to indicate the potential usefulness of differential and integral calculus in the management of the health care facility. In the next chapter we consider differential calculus in greater detail. In Chapter 11 we examine the use of the partial derivative in determining changes in the dependent variable that result from altering the value of one of several independent variables while holding the others constant. Finally, our examination of calculus concludes in Chapter 12 with a more detailed discussion of the antiderivative, the indefinite integral, and the definite integral.

Chapter 10

The Derivative of a Function

Objectives

After completing this chapter, you should be able to:

1. Distinguish between average and marginal rates of change;
2. Define and understand the derivative of a function;
3. Find the derivative of:
   a. a constant
   b. x^n
   c. a constant and a function
   d. a sum or a difference;
4. Find second and higher order derivatives;
5. Use the first and second derivative to identify maximum and minimum points on a curve.

Chapter Map

The sections comprising this chapter may be summarized as follows:

Section Number	Required Reading	Optional Reading	Generic Development	Application to Management	Fundamental Principles	Complex Material
	(R)	(O)	(G)	(A)	(F)	(C)
10.1	x		x	x	x	
10.2		x	x	x		x
10.2.1		x	x	x		x
10.2.2		x	x			x
10.2.3		x	x	x		x
10.3	x	x	x	x		x
10.3.1		x	x			x
10.3.2	x		x	x		x
10.4		x	x			x
10.4.1		x	x			x
10.4.2		x	x			x
10.4.3		x	x			x
10.4.4		x	x			x
10.5	x			x	x	
10.5.1	x		x		x	
10.6		x	x	x		x
10.7		x		x		x
10.8	x			x		x
Appendix		x	x			x

10.1 INTRODUCTION[R, G, A, F]

To illustrate the usefulness of differential calculus, let us focus on the problem of containing the cost of health care and improving the operational performance of the institution. As will be recalled, total cost may be defined as the sum of total fixed and total variable costs. Also, when used in this context, the term *fixed* refers to costs that are invariant with respect to changes in the volume of care provided. On the other hand, the term *variable* refers to cost components that increase (decrease) as the volume of care provided increases (decreases). More specifically, many writers argue that variable costs vary in direct proportion to changes in the volume of care provided. However, for the purposes of discussion here we shall assume that the rate of change in total variable costs may be less than, equal to, and greater than the rate of change in the volume of service provided.

These observations suggest that:

1. if the rate of change in the volume of care exceeds the rate of change in total variable cost, average variable costs decline as the volume of service provided expands;
2. if the rate of change in the volume of care is equal to the rate of change in total variable cost, average variable costs remain constant as the quantity of service is increased or decreased; and
3. if the rate of increase in total variable cost exceeds the rate of increase in the volume of service, average variable costs will increase as the amount of care provided is increased.

On the other hand, since the fixed cost component is constant for all rates of activity, average fixed cost must decline continuously as the volume of care provided is expanded.

Since the cost per unit of service is influenced by the behavior of average fixed costs and average variable costs, it seems reasonable to argue that the average total cost curve assumes the configuration presented in Figure 10-1. Here it will be observed that as the volume of care provided is initially increased, the average total cost curve, (ATC), falls, reaches a minimum at V^*, and rises continuously for those volumes of service that exceed V^*. Notice that the unit costs are *minimized* when the institution provides the volume of care specified by V^*. Thus, identifying the volume of service that minimizes the unit costs of the care provided is useful in achieving the institution's goal of containing the costs of health services.

In this regard, differential calculus is particularly useful in determining the points at which a specific function assumes a maximum or minimum value. Hence, the derivative should be viewed as a managerial tool that facili-

Figure 10-1 Effects on Service of Fixed and Variable Costs

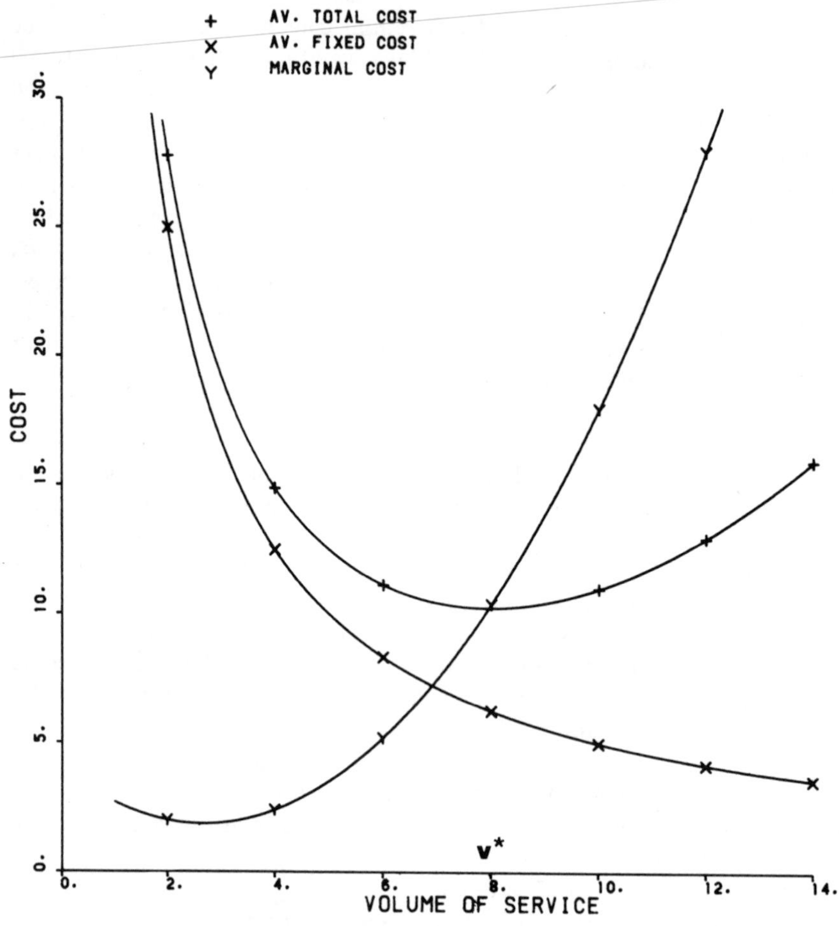

tates the development of plans designed to minimize or maximize the value assumed by a given variable or phenomenon.

In addition to the concept of optimization (i.e., minimizing or maximizing the value of a function), the derivative enables us to examine the response of one variable to changes in another. For instance, consider the marginal cost curve (*MC*) in Figure 10-1. Such a curve measures the change in total cost that results from altering the volume of care by one unit.

To illustrate the usefulness of these data, suppose management is considering the adoption of alternative A that would permit the substitution of outpatient care that would have otherwise been provided to 380 inpatients under the status quo. We also assume that the substitution of outpatient care for inpatient care does not result in a reduction in the quality or effectiveness of the services provided. Suppose further that the following data are available to management. These data reveal that the reduction in the flow of inpatients results in cost savings that amount to $95,000. On the other hand, the treatment of an additional 380 patients on an ambulatory basis results in a $20,000 increase in outpatient costs. These data suggest that, by adopting alternative A, the institution's management is able to provide required services at an acceptable level of quality while realizing a net saving of $75,000 (i.e., $95,000 − $20,000). In this case, the derivative is a managerial tool that may be used to obtain data on which these and similar decisions might be reached.

	Patient Flow		Costs	
	Inpatient Flow (# of Patients)	Outpatient Flow (# of Patients)	Inpatient Costs	Outpatient Costs
Alternative A	17,000	14,380	$3,100,000	$120,000
Status Quo	17,380	14,000	$2,195,000	$100,000
Increase (Decrease)	(380)	380	($95,000)	$20,000

The potential importance of changes in one variable (costs) in response to changes in another (flow of patients) was exemplified by the foregoing illustration. On the other hand, calculus was described earlier as the study of change. For example, recall that the expression

$$y = f(x) \tag{10.1}$$

means that y is a function of x. Our earlier discussion indicated that, if y is a single valued function of x, the value of y is determined by the values assumed by x. In this chapter we extend our understanding of functional relationships and examine methods of comparing the *changes* in y resulting from changes in x as well as determining the rate of change in y when x increases or decreases.

In our discussion of linear functions (Chapter 5) we limited our analysis to equations of the form

$$y = a + bx \tag{10.2}$$

where a is the intercept of the equation (i.e., the value assumed by y when $x = 0$) and b is the slope (i.e., the change in y resulting from a unit change in the variable x). In this chapter we examine differential calculus as a tool that allows us to extend our analysis so as to accommodate *nonlinear*, or *curvilinear*, functions.

To understand the role of the derivative in optimization, it first is necessary to discuss the process of differentiation. As a result, we turn initially to a consideration of the derivative as a mathematical tool. The purpose of this discussion is to define and measure the rate of change in one variable that results from a change in the value assumed by one or more other factors. On the basis of this, we may then consider a diagrammatic interpretation of the derivative, which in turn provides the foundation for applying the derivative in problems requiring optimization. Finally, we consider additional applications of these techniques in health care administration and research.

10.2 THE DERIVATIVE OF A FUNCTION(O,G,A,C)

In this section, we introduce the derivative by considering first the average rate of change in a function and then the derivative of linear and curvilinear functions.

10.2.1 The Average Rate of Change(O,G,A,C)

Consider the general function given by $y = f(x)$. We may define the average rate of change (ARC) by

$$ARC = \frac{\Delta y}{\Delta x} \tag{10.3}$$

where Δy is the change in y that results from a given change in x (i.e., Δx). With respect to Δx, suppose that x is increased from x_1 to x_2. In this case, we have

$$\Delta x = x_2 - x_1 \tag{10.4.1}$$

Recalling that y_i is the value assumed by y when $x = x_i$, we find that $y_2 = f(x_2)$ while $y_1 = f(x_1)$. As a consequence, the change in y is given by

$$y_2 - y_1 = f(x_2) - f(x_1) \tag{10.4.2}$$

Substituting $f(x_2) - f(x_1)$ for Δy and $x_2 - x_1$ for Δx in Equation 10.3, we find that the average rate of change becomes

$$ARC = \frac{\Delta y}{\Delta x} = \frac{f(x_2) - f(x_1)}{x_2 - x_1} \tag{10.4.3}$$

To illustrate the use of Equation 10.4.3, in finding the average rate of change, assume that the total cost of providing ambulatory care in our institution is related to the volume of outpatient visits by

$$y = 3 + 5x$$

where y represents total cost and x corresponds to the volume of service. On the basis of our earlier discussion, we see that $y = 3 + 5x$ is the equation of a straight line with an intercept of 3 and a slope of 5. In terms of our example, it also will be recognized that the y intercept corresponds to the fixed costs of providing outpatient care while the term $5x$ represents the variable cost component of total costs.

Assume that x increases from one unit of care to, say, five units of care. Employing Equation 10.4.3, we find that the average rate of change in total cost is given by

$$\frac{\Delta y}{\Delta x} = \frac{[3 + 5(5)] - [3 + 5(1)]}{5 - 1} = \frac{28 - 8}{4} = \$5 \text{ unit}$$

Alternatively, suppose the volume of care increases from one unit to, say, ten units. In this case, the average rate of change in the total cost of outpatient care is found to be

$$\frac{\Delta y}{\Delta x} = \frac{[3 + 5(10)] - [3 + 5(1)]}{10 - 1} = \frac{53 - 8}{9} = \$5 \text{ unit}$$

This illustrates that the slope (i.e., the average rate of change in y) of a linear equation is constant.

We now want to transform Equation 10.4.3 into a more useful form. Recall that we defined Δx by

$$\Delta x = x_2 - x_1$$

If we solve for x_2, we find that

$$x_2 = x_1 + \Delta x \tag{10.5.1}$$

and substituting $x_1 + \Delta x$ for x_2 in Equation 10.4.2 yields

$$\Delta y = f(x_1 + \Delta x) - f(x_1) \tag{10.5.2}$$

If we then divide both sides of Equation 10.5.2 by Δx, we obtain

$$\frac{\Delta y}{\Delta x} = \frac{f(x_1 + \Delta x) - f(x_1)}{\Delta x} \tag{10.6}$$

which yields results that are identical to those obtained by Equation 10.4.3.

Example—Substitution of Outpatient for Inpatient Care—Quadratic Equation

To illustrate the use of Equation 10.6, consider a slightly more complex example and assume that the relation between total inpatient costs and the number of days of care provided in our institution is given by

$$y = 400 + 3x + x^2$$

where, as before, y corresponds to total cost while x represents the volume of care. In this case, however, the fixed costs of providing inpatient care are \$400 while the variable cost component is given by the expression $3x + x^2$.

Assume that management is contemplating the maintenance of the status quo or the introduction of an alternative mode of treatment, which we represent by the letter A. Under the status quo, suppose that management expects to provide 400 outpatients visits and 1,200 days of inpatient care. On the other hand, the implementation of proposal A is expected to permit the substitution of outpatient for inpatient care in the treatment of 50 individuals, without a reduction in the effectiveness or the quality of care provided. Suppose further that

1. each of the 50 patients will require five days of care if proposal A is not implemented (i.e., maintenance of the status quo); and

2. if proposal A is implemented, each of the 50 patients will require three outpatient visits.

In this case, we find that the introduction of proposal A reduces the inpatient load of our institution by 250 days (50 patients × 5 days per patient) and increases the outpatient load by 150 visits (50 patients × 3 visits per patient). Our objective then, is to evaluate the relative costliness of implementing proposal A as contrasted with maintaining the status quo.

At this point, we may use Equation 10.6 to calculate the average rate of change expected in the costs of inpatient and outpatient care. In this example, we assume that the status quo will result in the provision of 1,200 days of care ($x = 1200$) and that the introduction of proposal A will reduce the inpatient load by 250 days ($\Delta x = -250$). The corresponding average rate of change in $y = 400 + 3x + x^2$, is given by

$$\frac{f(x + \Delta x) - f(x)}{\Delta x} = \frac{[400 + 3(1200 - 250) + (1200 - 250)^2] - [400 + 3(1200) + (1200)^2]}{-250} = \$2{,}153$$

By contrast, proposal A would increase the outpatient load from 400 visits ($x = 400$) to 550 visits ($\Delta x = 150$). Hence, using Equation 10.6, we find that the average rate of change in $y = 3 + 5x$ is given by

$$\frac{f(x + \Delta x) - f(x)}{\Delta x} = \frac{[3 + 5(400 + 150)] - [3 + 5(400)]}{150} = \$5$$

These findings suggest that, in the range $950 \leq x \leq 1200$, reducing the inpatient load of our institution by one day of care results, on the average, in a cost savings of \$2,153. Since proposal A is expected to reduce the inpatient load by 250 days of care, the total cost savings associated with altering the volume of inpatient care is given by 250 × \$2,153 or \$538,250. On the other hand, the calculations also suggested that, on the average, each additional outpatient visit increases costs by \$5. Since proposal A would increase the outpatient load by 150 visits, the resulting increase in cost is given by 150 × \$5 or \$750. The use of the average rate of change thus allows us to assert that the introduction of proposal A results in a net savings of \$537,500 (\$538,250 − \$750). Other things remaining constant, our calculations suggest that proposal A is preferred to the status quo.

The Changing Rate of Change

The use of Equation 10.6 was of particular importance in this example because the average rate of change in $y = 400 + 3x + x^2$ does not remain constant. To illustrate, let us limit our analysis to the finite interval defined by $1 \leq x \leq 5$. Under this assumption, we may portray our cost function as seen in Figure 10-2. To calculate the average rates of change, we require the following data:

Point in Figure 10-2	Volume of Care	Total Cost
A	1	\$404
B	2	\$410
C	3	\$418
D	4	\$428
E	5	\$440

Notice that the points A, B, C, D, and E in Figure 10-2 correspond to the coordinates (1, 404), (2, 410), (3, 418), (4, 428), and (5, 440), respectively. Also notice that points A and E have been connected by a straight line that is called a secant. The slope of secant AE is given by

$$\frac{f(x + \Delta x) - f(x)}{\Delta x} = \frac{\$440 - \$404}{5 - 1} = \$9$$

Similarly, if we mentally construct a straight line through the points B and E, we find that the slope of the resulting secant is given by

$$\frac{f(x + \Delta x) - f(x)}{\Delta x} = \frac{\$440 - \$410}{5 - 2} = \$10$$

As a final example, the slope of the secant constructed through the points C and E is given by

$$\frac{f(x + \Delta x) - f(x)}{\Delta x} = \frac{\$440 - \$418}{5 - 3} = \$11$$

These findings suggest that the average rate of change in $y = 400 + 3x + x^2$ and the net savings realized by substituting outpatient for inpatient care depend on the range of values assumed by the variable x.

Let us consider a slightly different situation and calculate the average rate of change in $y = 400 + 3x + x^2$ when x is increased from $x = 2$ to $x + \Delta x$

Figure 10-2 Inpatient Costs versus Days of Care

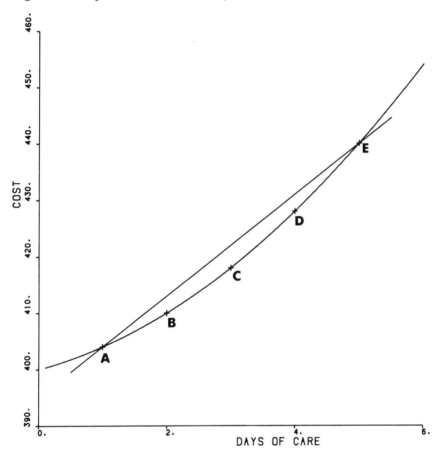

and Δx becomes smaller and smaller. Table 10-1 shows that the average rates of change, as defined by Equation 10.6, are obtained by dividing the values appearing in column (4) by the corresponding value appearing in column (2). Also note that as Δx becomes smaller and approaches zero, the average rate of change approaches \$7.00.

Given that division by zero is not defined, we are not able to use the ratio $\Delta y/\Delta x$ to determine the average rate of change when $\Delta x = 0$. Before introducing the derivative, then, it is necessary to consider the limit of a function.

10.2.2 The Limit of a Function[O,G,C]

In this section, we want to find the value (if one exists) that the function $f(x)$ approaches as the value of x comes closer and closer to a given number b. In this regard, the equation

$$\lim_{x \to b} f(x) = L \tag{10.7}$$

indicates that the value of the function $f(x)$ approaches the limit L as x approaches the value given by b. It is important to note that x never reaches b, but, rather, Equation 10.7 indicates that the value of $f(x)$ approaches the limit L as x approaches the value b.

By way of illustration, suppose we are given the function

$$y = f(x) = 3 + 2x$$

and we wish to find the limit of $f(x)$ as x approaches 4. In this case, we find that the values of y that correspond to $x = 1, 2, 3, 3.5, 3.9, 3.98$, and 4.01 are as follows:

Value of x	Value of y
1	5
2	7
3	9
3.5	10
3.9	10.8
3.98	10.96
4.01	11.02

These calculations reveal that, as x approaches 4, the value of y seems to approach 11.0. As a result, we may write

$$\lim_{x \to 4} (3 + 2x) = L = 11.0$$

since

$$f(4) = 3 + 2(4) = 11 = \lim_{x \to 4} 3 + 2(x)$$

Table 10-1 The Average Rate of Change in $f(x) = 400 + 3x + x^2$ When x Is Increased from $x = 2$ to $x = x + \Delta x$

x (1)	Δx (2)	$x + \Delta x$ (3)	Δy (4)	Average Rate of Change (5)
2	3	5	30	10
2	2	4	18	9
2	1	3	8	8
2	.1	2.1	.71	7.1
2	.01	2.01	.0701	7.01
2	.001	2.001	.007001	7.001

Thus, the limit of $y = 3 + 2x$ is 11.0 when x approaches 4.0. This is an example in which the limit may be determined by substitution in the expression.

Consider next a slightly more complex example. Suppose we are given

$$y = f(x) = \frac{1}{3 - x} \tag{10.8}$$

for which we wish to find the limit as x approaches 3. Replicating our earlier work we find that the values assumed by y as x approaches 3 are as follows:

Value of x	Value of y	Value of x	Value of y
1	.5	3.05	−20.0
2	1.0	3.1	−10.0
2.5	2.0	3.2	−5.0
2.9	10.0	4.0	−1.0
2.95	20.0	5.0	−.5

In this case the function $f(x)$ does not approach a single value as x approaches 3 and

$$\lim_{x \to 3} \frac{1}{3 - x}$$

does not exist. However, as x approaches any value other than 3, the limit of $f(x) = 1/(3 - x)$ is defined. For example, as can be verified, the limit of $1/(3 - x)$ as x approaches 2 is 1.

10.2.3 The Derivative(O,G,A,C)

At this point, we may use the material developed in previous sections to define the derivative. Let us return to the cost function given by $y = 400 + 3x + x^2$ that, as will be recalled, pertained to the provision of inpatient care. As shown in Figure 10-3, we let the points A, B, and C correspond to the coordinates (1, 404), (2, 410), and (3, 418), respectively. We consider first a line constructed tangent to the point A—i.e., the coordinate (1, 404). As before, we find that the slope of secant AC is given by

$$\frac{f(x + \Delta x) - f(x)}{\Delta x} = \frac{\$418 - \$404}{3 - 1} = \$7$$

while the slope of secant AB is given by

$$\frac{f(x + \Delta x) - f(x)}{\Delta x} = \frac{\$410 - \$404}{2 - 1} = \$6$$

If we now select values of x that are closer and closer to one, we obtain these results.

Value of $x + \Delta x$	Value of y	Resulting Point	Slope of Line Through Resulting Point and (1, 404)
1.1	404.51	(1.1, 404.51)	5.1
1.01	404.0501	(1.01, 404.0501)	5.01
1.001	404.005001	(1.001, 404.005001)	5.001

These calculations suggest that the slopes of the secant lines approach a limit of \$5. To verify that the limiting value of these slopes is \$5, we calculate the slope of the line constructed through the point (1, 404) and the general point $(1 + \Delta x, f(1 + \Delta x)$ where $1 + \Delta x$ represents the value of x and

$f(1 + \Delta x)$ represents the corresponding value of y. Using the results of previous sections, we find that

$$\frac{\Delta y}{\Delta x} = \frac{f(1 + \Delta x) - f(1)}{1 + \Delta x - 1} = \frac{f(1 + \Delta x) - f(1)}{\Delta x}$$

Focusing on the numerator $f(1 + \Delta x) - f(1)$, recall that $f(x) = 400 + 3x + x^2$. As a result, we find that

$$\begin{aligned} f(1 + \Delta x) &= 400 + 3(1 + \Delta x) + (1 + \Delta x)^2 \\ &= 400 + 3 + 3\Delta x + 1 + 2\Delta x + \Delta x^2 \\ &= 404 + 3\Delta x + 2\Delta x + \Delta x^2 \\ &= 404 + 5\Delta x + \Delta x^2 \\ &= 404 + \Delta x(5 + \Delta x) \end{aligned}$$

Consequently, after substituting appropriately, we find that

$$\begin{aligned} \frac{\Delta y}{\Delta x} = \frac{f(1 + \Delta x) - f(1)}{\Delta x} &= \frac{[404 + \Delta x(5 + \Delta x)] - [400 + 3(1) + (1)^2]}{\Delta x} \\ &= \frac{404 - 404 + \Delta x(5 + \Delta x)}{\Delta x} \\ &= \frac{\Delta x(5 + \Delta x)}{\Delta x} \\ &= 5 + \Delta x \end{aligned}$$

As Δx approaches zero, the general point $(1 + \Delta x)$, $f(1 + \Delta x)$ approaches the point (1, 404) and the secant approaches the line that is tangent to point A. Thus, if we let Δx approach zero, we find that

$$\lim_{\Delta x \to 0} \frac{f(1 + \Delta x) - f(1)}{\Delta x} = \lim_{\Delta x \to 0}(5 + \Delta x) = \$5$$

These findings indicate that, as Δx approaches zero, the slope of the line that is tangent to point A has a slope of \$5. The tangent line derived above is shown graphically in Figure 10-4.

The limit used to find the slope of the line constructed tangent to the curve defined by $y = 400 + 3x + x^2$ is of particular importance and is useful in

Figure 10-3 Use of the Derivative in Cost/Care Analysis

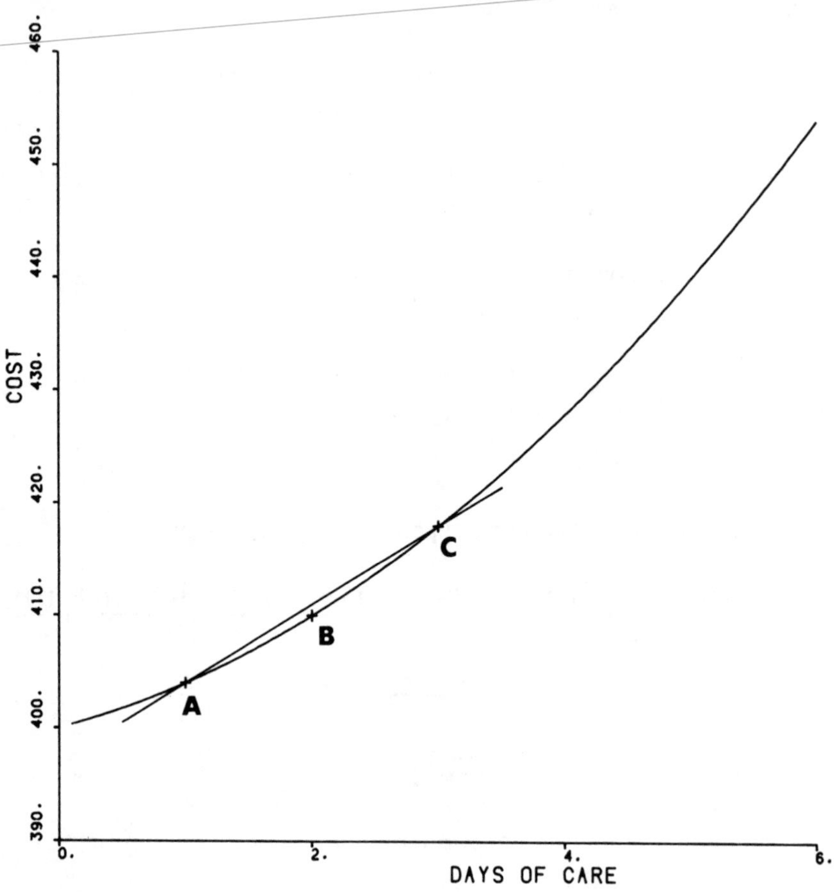

many practical applications. As a result, it is given a special name. We refer to such a limit as the derivative of $f(x)$. The derivative of $f(x)$ is defined as

$$dy/dx = \lim_{\Delta x \to 0} \frac{f(x + \Delta x) - f(x)}{\Delta x} \quad \textbf{(10.9)}$$

assuming that the limit exists. On the other hand, if the limit does not exist, the derivative does not exist. In addition to dy/dx, the derivative may also be represented by y'; $f'(x)$, $df(x)/dx$, $D_x y$, and $D_x f(x)$.

Figure 10-4 Tangent Line Solution in Cost/Care Determination

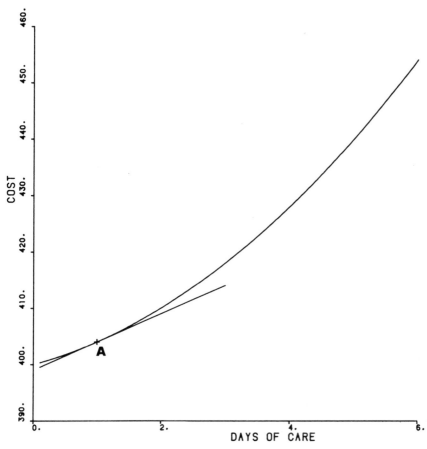

Example—Substitution of Outpatient for Inpatient Care—Revisited

Returning to our example, let us use Equation 10.9 to find the derivative of the cost function defined by $y = 400 + 3x + x^2$ when $x = 1$. In this case we find that

$$\frac{f(x + \Delta x) - f(x)}{\Delta x} = \frac{400 + 3(x + \Delta x) + (x + \Delta x)^2 - (400 + 3x + x^2)}{\Delta x}$$

$$= \frac{400 + 3x + 3\Delta x + x^2 + 2x\Delta x + \Delta x^2 - 400 - 3x - x^2}{\Delta x}$$

$$= \frac{3\Delta x + 2x\Delta x + \Delta x^2}{\Delta x}$$

$$= \frac{\Delta x(3 + 2x + \Delta x)}{\Delta x}$$

$$= 3 + 2x + \Delta x$$

Therefore

$$\lim_{\Delta x \to 0} \frac{f(x + \Delta x) - f(x)}{\Delta x} = 3 + 2x$$

On the basis of these findings we find that $dy/dx = 3 + 2x$ and the slope of the line constructed tangent to the point (1, 404) is simply $3 + 2(1)$ or \$5. This result agrees with our earlier findings in which we used the point $(1 + \Delta x)$, $f(1 + \Delta x)$ to find the slope of the line constructed tangent to the point A.

10.3 OBSERVATIONS CONCERNING THE DERIVATIVE(O, R, G, A, C)

At this point, several observations concerning the derivative are appropriate. The first set of observations concerns the technical conditions that must be satisfied before the derivative exists, while the second involves the interpretation of the derivative.

10.3.1 The Existence of the Derivative(O, G, C)

As defined previously, the derivative applies only to single valued functions of a continuous variable. The function $y = f(x)$ is said to be continuous at the point b if:

(1) $\lim_{x \to b} f(x)$ exists;

(2) $\lim_{x \to b} f(x) = f(b)$ and

(3) $f(b)$ exists

On the other hand, if these three conditions are not satisfied, the function is said to be discontinuous and the point $x = b$ is referred to as a point of discontinuity. For example, if $f(x) = 1/(7 - x)$, the function is discontinuous at the point $x = 7$ since neither $\lim f(x)$ nor $f(7)$ exist.

10.3.2 Interpreting the Derivative[R, G, A, C]

It should be clear now that the sign of the derivative indicates the direction of relationship between the dependent variable y and the independent variable x. Thus, if dy/dx is positive, the dependent variable y is positively related to the independent variable x. Conversely, if dy/dx is negative, the variable y is negatively related to x.

As seen earlier, the derivative may also be viewed as the slope of a line constructed tangent to a point on the curve defined by $f(x)$. For example, suppose that we measure potential benefits derived from a given program on the vertical axis and various levels of expenditure on the horizontal axis of Figure 10-5. The slope of the line that is drawn tangent to point A is greater than the

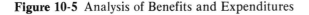

Figure 10-5 Analysis of Benefits and Expenditures

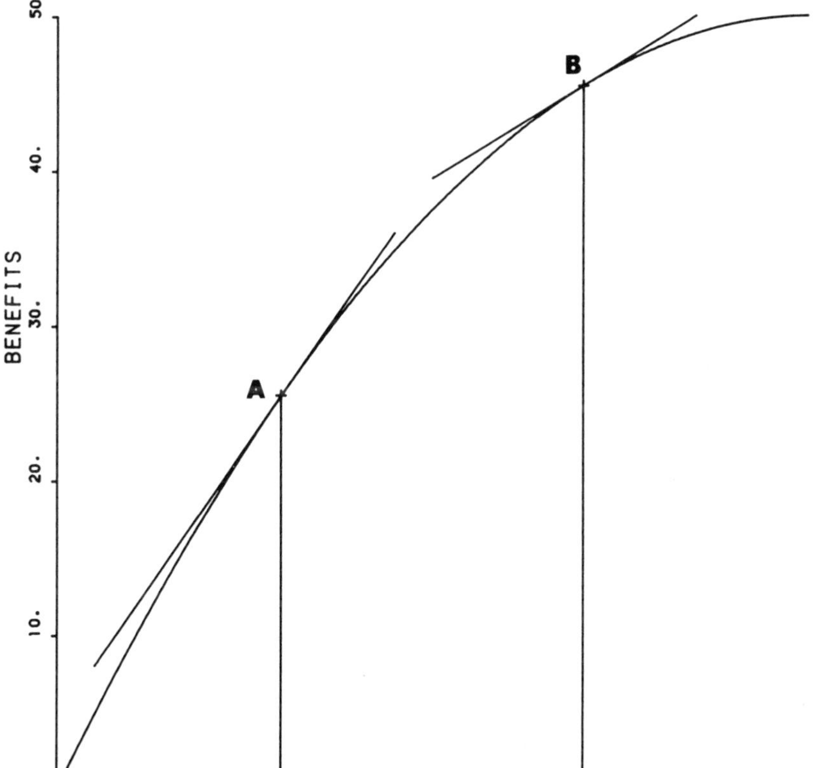

slope of the line constructed tangent to the point B. In this situation, we say that the marginal rate of change in benefits at point A is greater than the marginal rate of change in benefits at point B. Thus, as the rate of activity in the program is expanded from E_1 to E_2, the additional benefit obtained from the last dollar of expenditure declines.

10.4 DIFFERENTIATION FORMULAS(O,G,C)

The primary purpose of this section is to review and verify the rules of differentiation presented initially in Chapter 9. As mentioned previously,

(1) $f(x) = a$, where a is any constant;

(2) $f(x) = x^n$;

(3) $f(x) = a \cdot g(x)$; and

(4) $f(x) = g(x) + h(x)$;

represent the functional relationships most frequently encountered in health care management. As a consequence, we shall limit our discussion in this section to the rules of differentiation that pertain to these functional forms. On the other hand, the reader is directed to the appendix at the end of this chapter for rules pertaining to the differentiation of the functions:

(1) $f(x) = g(x) \cdot h(x)$

(2) $f(x) = g(x)/h(x)$

and more complicated functions, using the *chain rule*.

10.4.1 The Derivative of a Constant(O,G,C)

As seen earlier, we might express a linear equation by

$$y = a + bx \tag{10.10}$$

where a is the y intercept (i.e., the value of y when $x = 0$) and b is the slope of the line. Such a line might assume the form presented in Figure 10-6. If the slope of the linear function given by Equation 10.10 is zero, the expression

$$y = a$$

Figure 10-6 The Linear Equation and the Constant

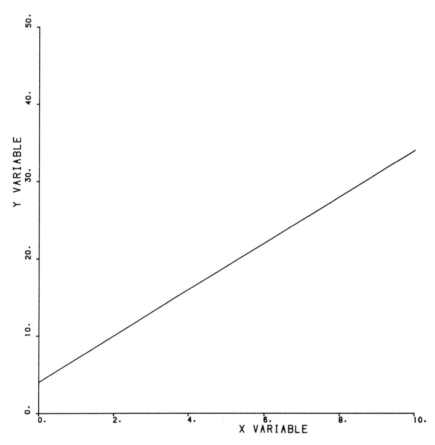

assumes the form shown in Figure 10-7. Here, it will be observed that $y = a$ is a straight line that is horizontal (parallel to the x axis) and there is no variation or change in the variable y over the interval. As a result, the derivative of $f(x) = a$, where a is a constant, is zero. Employing functional notation, we find that:

> if $f(x) = a$, where a is a constant, then the derivative $f'(x) = 0$.

This rule implies that, since the function is not dependent on x, the derivative is equal to zero.

10.4.2 The Derivative of x^n (O,G,C)

As seen earlier the functional relation

$$y = f(x) = x^n \tag{10.11}$$

implies that the value of y is obtained by raising x to the power n, where n is assumed to be a rational number. The general form for the derivative of $f(x) = x^n$ is given by

$$f'(x) = nx^{n-1} \tag{10.12}$$

where n is any rational number. In the following discussion, we demonstrate the application of the rule to

$$f(x) = x^3$$

$$f(x) = \frac{1}{x} = x^{-1}$$

and

$$f(x) = \frac{1}{x^2} = x^{-2}$$

and illustrate that the results are identical to those obtained when the general procedure $\text{Lim}_{\Delta x \to 0} [f(x + \Delta x) - f(x)]/\Delta x$ is employed.

Consider first the function $f(x) = x^3$. Applying the rule $f'(x) = nx^{n-1}$, we find that

$$f'(x) = 3x^2$$

We also may use Equation 10.9, which defined the derivative to verify these results. In this case, we find that

$$\begin{aligned} \text{Lim}_{\Delta x \to 0} \frac{f(x + \Delta x) - f(x)}{\Delta x} &= \text{Lim}_{\Delta x \to 0} \frac{(x + \Delta x)^3 - x^3}{\Delta x} \\ &= \text{Lim}_{\Delta x \to 0} \frac{(x^2 + 2x\Delta x + \Delta x^2)(x + \Delta x) - x^3}{\Delta x} \end{aligned}$$

$$= \lim_{\Delta x \to 0} \frac{x^3 + 2x^2 \Delta x + x(\Delta x)^2 + x^2(\Delta x) + 2x(\Delta x)^2 + (\Delta x)^3 - x^3}{\Delta x}$$

$$= \lim_{\Delta x \to 0} \frac{x^3 + 3x^2 \Delta x + 3x(\Delta x)^2 + (\Delta x)^3 - x^3}{\Delta x}$$

$$= \lim_{\Delta x \to 0} (3x^2 + 3x\Delta x + (\Delta x)^2)$$

$$= 3x^2$$

Figure 10-7 Absence of Change Over an Interval

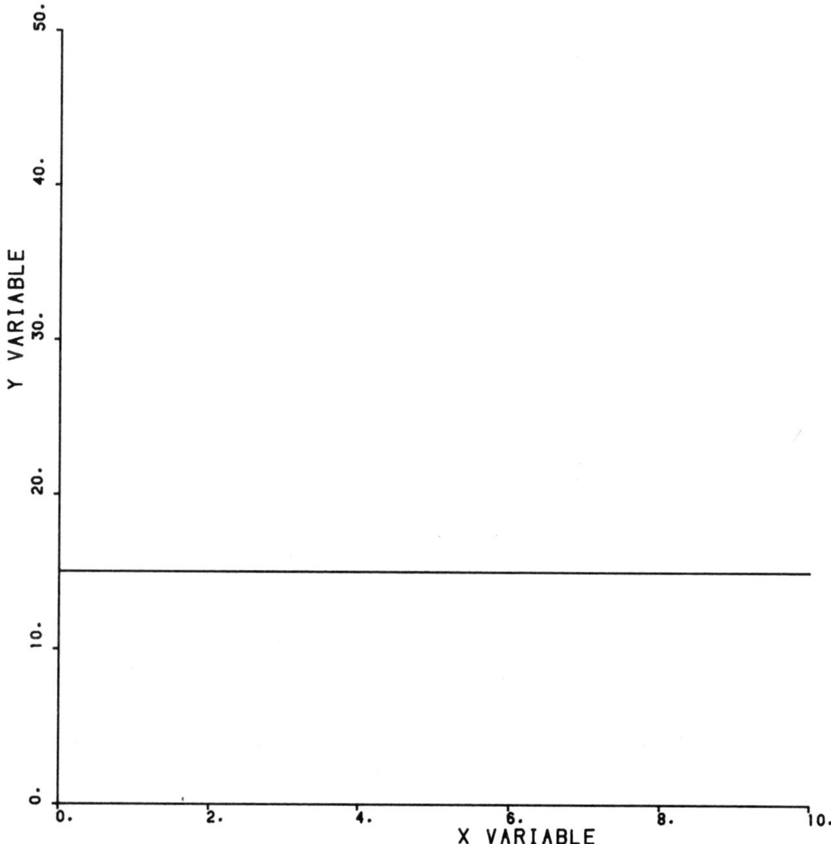

Consequently, the rule $f'(x) = nx^{n-1}$ yields results that are identical to those obtained by the general procedure $\text{Lim}_{\Delta x \to 0} [f(x + \Delta x) - f(x)]/\Delta x$.

Consider next the function $f(x) = 1/x = x^{-1}$. In this case, an application of the rule $f'(x) = nx^{n-1}$ yields

$$f'(x) = -x^{-2} = -\frac{1}{x^2}$$

Applying the definitional form of the derivative, we find that

$$\begin{aligned}
\lim_{\Delta x \to 0} \frac{f(x + \Delta x) - f(x)}{\Delta x} &= \lim_{\Delta x \to 0} \frac{\dfrac{1}{x + \Delta x} - \dfrac{1}{x}}{\Delta x} \\
&= \lim_{\Delta x \to 0} \frac{1}{\Delta x}\left(\frac{1}{x + \Delta x} - \frac{1}{x}\right) \\
&= \lim_{\Delta x \to 0} \frac{1}{\Delta x}\left(\frac{x - (x + \Delta x)}{x(x + \Delta x)}\right) \\
&= \lim_{\Delta x \to 0} \frac{1}{\Delta x}\left(\frac{-\Delta x}{x(x + \Delta x)}\right) \\
&= \lim_{\Delta x \to 0} - \frac{1}{x^2 + x\Delta x} \\
&= -\frac{1}{x^2}
\end{aligned}$$

As before, the general procedure and the rule $f'(x) = nx^{n-1}$ yield identical results.

Finally consider the derivative of $f(x) = 1/x^2 = x^{-2}$. Applying the rule $f'(x) = nx^{n-1}$, we find that

$$f'(x) = -2x^{-3} = -2/x^3$$

Similarly, when the general procedure represented by Equation 10.9 is employed, we obtain

$$\begin{aligned}
\lim_{\Delta x \to 0} \frac{f(x + \Delta x) - f(x)}{\Delta x} &= \lim_{\Delta x \to 0} \frac{\dfrac{1}{(x + \Delta x)^2} - \dfrac{1}{x^2}}{\Delta x} \\
&= \lim_{\Delta x \to 0} \frac{1}{\Delta x}\left(\frac{1}{x^2 + 2x\Delta x + (\Delta x)^2} - \frac{1}{x^2}\right)
\end{aligned}$$

$$= \lim_{\Delta x \to 0} \frac{1}{\Delta x} \left(\frac{x^2 - [x^2 + 2x\Delta x + (\Delta x)^2]}{x^2 [x^2 + 2x\Delta x + (\Delta x)^2]} \right)$$

$$= \lim_{\Delta x \to 0} \frac{1}{\Delta x} \left(- \frac{2x\Delta x + (\Delta x)^2}{x^4 + 2x^3\Delta x + x^2(\Delta x)^2} \right)$$

$$= \lim_{\Delta x \to 0} - \frac{2x + \Delta x}{x^4 + 2x^3\Delta x + x^2(\Delta x)^2}$$

$$= - \frac{2x}{x^4} = - \frac{2}{x^3}$$

which, of course agrees with our earlier results.

10.4.3 The Derivative of the Product of a Constant and a Function[(O, G, C)]

The discussion in the previous section may be extended to accommodate expressions that assume the general form

$$f(x) = a \cdot g(x)$$

where a is a constant and $g(x)$ is a function of x. As might be suspected, if the values of a function, say $g(x)$, are multiplied by the constant a, the derivative of the function is also multiplied by the constant.

For example, consider the equation

$$y = f(x) = 10x^2$$

for which the first derivative is found by

$$\lim_{\Delta x \to 0} \frac{f(x + \Delta x) - f(x)}{\Delta x} = \lim_{\Delta x \to 0} \frac{10(x + \Delta x)^2 - 10x^2}{\Delta x}$$

$$= \lim_{\Delta x \to 0} \frac{10x^2 + 20x\Delta x + 10(\Delta x)^2 - 10x^2}{\Delta x}$$

$$= \lim_{\Delta x \to 0} \frac{\Delta x(20x + 10\Delta x)}{\Delta x}$$

$$= 20x$$

Employing functional notation, this result suggests that

if $f(x) = a \cdot g(x)$, where a is a constant, $f'(x) = a \cdot g'(x)$.

Applying this rule, we find that if

$$f(x) = 5x^2; \qquad f'(x) = 10x$$

$$f(x) = 7x^3; \qquad f'(x) = 21x^2$$

$$f(x) = 3\left(\frac{1}{x^2}\right); \qquad f'(x) = -6x^{-3} = -\frac{6}{x^3}$$

As can be verified, the marginal rates of change in each of these functions at $x = 2$ are 20, 84, and $-3/4$ respectively.

10.4.4 The Derivative of a Sum or a Difference(O,G,C)

As seen earlier, two or more functions may be added. The derivative of the sum (or difference) of two functions is simply the sum (or difference) of the separate derivatives. Thus,

if $f(x) = h(x) + g(x)$, then $f'(x)$ is given by $h'(x) + g'(x)$.

Similarly,

if $f(x) = h(x) - g(x)$, then $f'(x)$ is given by $h'(x) - g'(x)$.

By way of illustration, consider the function

$$f(x) = g(x) + h(x)$$

where $g(x) = 5x^2$ and $h(x) = 4x^3$. Observe that

$$g'(x) = 10x$$

while

$$h'(x) = 12x^2$$

Thus,

$$f'(x) = 10x + 12x^2$$

In this case, the marginal rate of change in the function at $x = 3$ is given by

$$10(3) + 12(3)^2 = 138$$

Alternatively, if

$$f(x) = g(x) - h(x)$$

where $g(x) = 7x^2$ and $h(x) = x^3$, we find that

$$f'(x) = 14x - 3x^2$$

Hence the rate of change in $f(x)$ at $x = 2$ is given by

$$14(2) - 3(2)^2 = 16$$

In the next section we return to the use of these rules in identifying the maximum or minimum point on the curve defined by the function $f(x)$.

10.5 THE PROBLEM OF OPTIMIZATION REVISITED(R, A, F)

The purpose of this section is twofold. The first objective is to review the use of differential calculus when addressing the general problem of optimization (i.e., the determination of the value of the independent variable that results in either a maximum or minimum value of the dependent variable). The second major objective is to express the criterion by which we identify a maximum or minimum point on a curve in terms that are slightly more analytical or formal than those employed in Chapter 9.

As seen earlier, the derivative of the single valued function $f(x)$ may be viewed as the rate of change of the function at the point $x = b$. Also recall that the derivative of a function may be found by determining the slope of a straight line constructed tangent to the point $x = b$ on the curve defined by $y = f(x)$.

Employing the notation developed earlier, we find the value of the derivative of the function at the point $x = b$ is given by $f'(b)$. Thus, if $f'(b)$ is positive, it follows that the rate of change of $f(x)$ at $x = b$ is positive. Conversely, if $f'(b)$ is negative, the rate of change of $f(x)$ at $x = b$ is negative.

Consequently, when $f'(b) > 0$, we find that the function $f(x)$ increases as x increases and that the curve defined by $y = f(x)$ rises from left to right at the point $x = b$. Conversely, when $f'(b) < 0$, we find that $f(x)$ decreases as x increases and that the curve defined by $y = f(x)$ falls from left to right at the point $x = b$. Thus, the numerical value of $f'(b)$ measures how rapidly the function $f(x)$ rises or falls at the point $x = b$.

In this section we focus on the special situation in which $f'(b) = 0$. When $f'(b) = 0$, the tangent constructed to the curve at point b is parallel to the x axis, which implies that $f(x)$ is neither rising nor falling. When $f'(b)$ is zero, the point b is referred to as a stationary value of the function $f(x)$, and our main concern in this section involves the so-called stationary values of the function.

Example—Maximum Productivity

Let us consider the relationship between the productivity of labor and the level of employment. In this case, we might rely on economic theory that suggests that, if other factors are held constant, initial increases in the employment of labor expands the amount of output or service per worker. After a point, however, additional increases in the amount of labor results in a reduction in productivity.

We assume that, ceteris paribus, the function

$$y = f(x) = 8x - 2x^2 \tag{10.13}$$

represents the relationship between productivity and the amount of labor, x, employed in a given department of our health care facility. In this example, we measure x in terms of full-time equivalents that allow us to examine the labor input in decimal form. We also assume that management wishes to identify the level of employment that maximizes the productivity of the workers in the unit. In this case, the first derivative of $y = f(x) = 8x - 2x^2$ is given by

$$f'(x) = 8 - 4x = -4(x - 2)$$

which implies that

(1) $f'(x) > 0$ when fewer than two workers are employed in the department,

(2) $f'(x) = 0$ when two workers are employed in the department, and
(3) $f'(x) < 0$ when more than two workers are employed in the department.

To verify these conclusions, the value assumed by $f'(x)$ when one, two, and three workers are employed in the department is given by

$$f'(1) = -4(1 - 2) = 4$$

$$f'(2) = -4(2 - 2) = 0$$

$$f'(3) = -4(3 - 2) = -4$$

respectively. As a result, the function increases for the values $0 \leq x \leq 2$, becomes stationary at $x = 2$, and decreases when more than two full-time equivalents are employed. Figure 10-8 demonstrates that the function $y = f(x) = 8x - 2x^2$ is a parabola that reaches a maximum value of 8 when $x = 2$. In terms of our example, a maximum of eight units per worker is obtained when two full-time equivalents are employed in the department.

Example—Minimum Cost

Consider next a slightly different situation in which we are interested in minimizing the unit cost of providing a given service. As before, we might rely on the economic theory that suggests that, as the volume of service is initially increased, unit costs decline. However, after a point, increases in the provision of the service result in rising unit costs.

We let y represent the unit costs of the service and x represent the volume of service as expressed in terms of 1,000 units. Assume that historical data provide the basis for the estimation of

$$y = f(x) = x^2 - 4x + 8 \quad \textbf{(10.14)}$$

As can be verified, the derivative of the unit cost function is given by

$$f'(x) = 2x - 4 = 2(x - 2)$$

Recalling that the volume of service is measured in terms of 1,000 units, we find that

(1) $f'(x) < 0$ when the volume of care provided is less than 2,000 units,
(2) $f'(x) = 0$ when the volume of care provided is equal to 2,000 units, and
(3) $f'(x) > 0$ when more than 2,000 units of care are provided.

Figure 10-8 Relationship Between Productivity and Employment

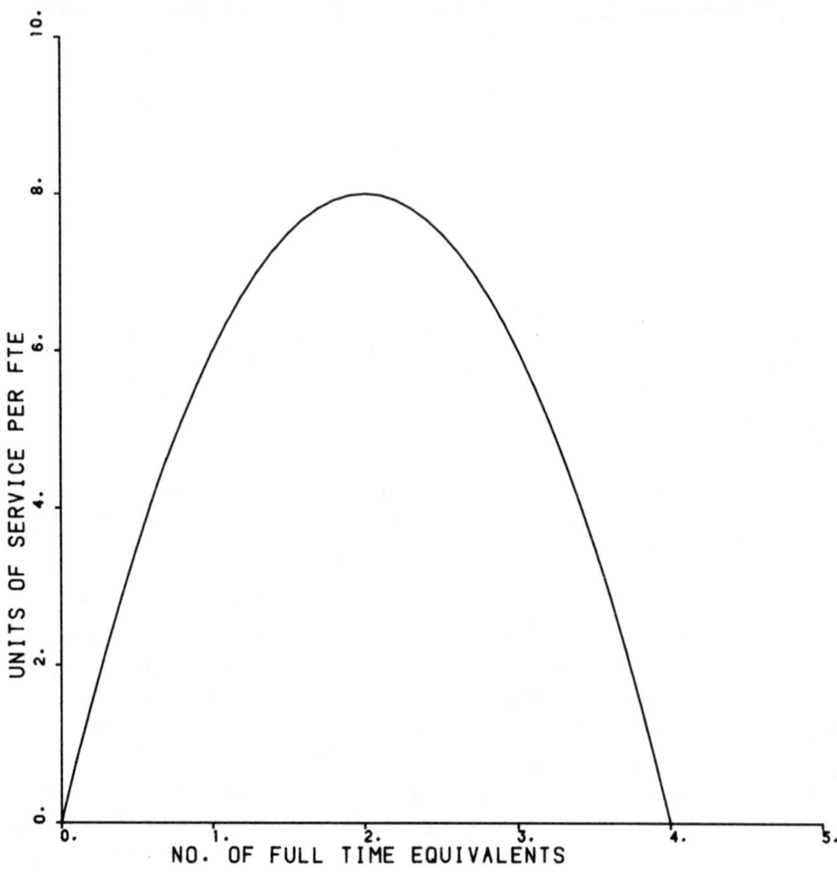

The function $y = x^2 - 4x + 8$ is portrayed graphically in Figure 10-9, which reveals that the minimum value of $y = 4$ occurs when $x = 2$. Hence, when 2,000 units of service are provided, average total costs are minimized and equal to \$4 per service.

Example—Maximizing Net Benefits

Consider next a slightly more complex example. Suppose that the benefits accruing to the community amount to \$125 on each occasion that a given procedure is performed. Consequently, letting x represent the number of ser-

Figure 10-9 Average Total Costs in Relation to Service Volume

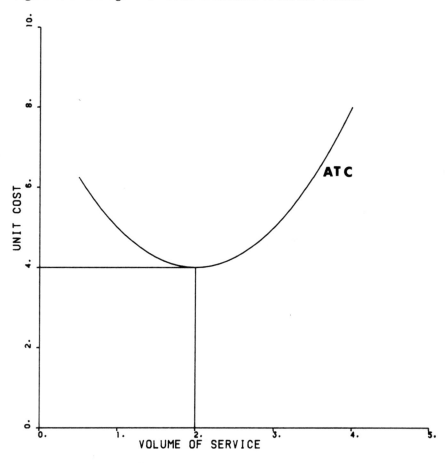

vices that are provided by our health care institution, the total benefit (TB) accruing to the community is given by

$$TB = 125x$$

On the other hand, suppose that the total cost of providing the procedure may be approximated by

$$TC = 30 + 5x + 6x^2$$

In this example, the basic objective of management is to maximize the net benefits (y) accruing to the community that is given by

$$y = TB - TC$$

In terms of our example, we find that

$$\begin{aligned} y &= 125x - (30 + 5x + 6x^2) \\ &= 120x - 6x^2 - 30 \end{aligned}$$

for which we require the value of x (the volume of service) that yields a maximum value of y (the net benefits accruing to the community).

As before, our first task is to find the first derivative of y with respect to x, which is given by

$$dy/dx = 120 - 12x$$

We then force dy/dx to equal zero and obtain

$$120 - 12x = 0$$

which, of course, defines a stationary value of the function $y = f(x)$. Solving for x, we find that

$$\begin{aligned} 12x &= 120 \\ x &= 10 \end{aligned}$$

units of service maximize the net benefits accruing to the community.

To verify that ten units of service maximize net benefits, we simply substitute $x = 9$, 10, and 11 into $f'(x)$ and obtain

$$\begin{aligned} f'(9) &= +\$12 \\ f'(10) &= 0 \\ f'(11) &= -\$12 \end{aligned}$$

These findings may be interpreted as follows. When the value of $f'(b)$ is positive, providing an additional unit of service adds more to total benefits than to total costs and, as a result, the expansion of the amount of care provided increases the net benefits accruing to the community. On the other hand, when $f'(b) < 0$, an additional unit of service adds more to total costs than to total benefits, so that increasing the volume of service reduces the net benefits for the community. Thus, to maximize the difference between total benefits and total costs, health care management should provide the procedure up to the point at which the marginal benefits of $125 are just equal to the additional costs of providing the last unit of service.

These observations may be verified as follows. Notice that the value of the benefits generated by providing ten units of the service is given by 10 units × $125 per unit or $1,250. By way of contrast, the total cost of providing this volume of care is $680 [$30 + 5(10) + 6(10)^2$]. As a result, the value of the net benefits generated by providing ten units of the service is $570 ($1,250 − $680). As should be verified, the provision of more or less than ten units of the service results in a lower level of net benefits. For example, the value of the net benefits generated by providing either 9 or 11 units of the service is $564.

Example—Optimum Staffing of Credit Collection Department

As another example, consider the operation of the business office of a large hospital that engages in credit and collection activities. We may argue that the hospital's credit and collection efforts represent a costly undertaking and, to minimize costs, should be eliminated. On the other hand, by increasing its credit and collection efforts, management may improve the cash position of the hospital by increasing the realization of outstanding accounts receivable.

Let us assume that management's objective is to maximize the cash inflows relative to the costs of operating the credit and collection department of the hospital. Suppose further that

$$TC = 30 + 4x + x^2$$

represents the total cost function of the credit and collection department. In this example, we let x represent the efforts of this department as measured by the number of manhours devoted to this activity per day. In addition, we assume that the cash inflows realized by payments on outstanding receivables also are a function of the effort devoted to credit and collection activities. Specifically, assume that

$$CI = 24.2x - .01x^2$$

represents the functional relation between cash inflows, CI, and the number of manhours employed in this unit per day. In this case, the objective of management is to select the value of x (i.e., the number of manhours devoted to credit and collection efforts) that maximizes

$$\begin{aligned} y &= CI - TC \\ &= 24.2x - .01x^2 - (30 + 4x + x^2) \\ &= 20.2x - 1.01x^2 - 30 \end{aligned}$$

As before, we find the first derivative of y with respect to x, which in this example is given by

$$dy/dx = 20.2 - 2.02x$$

Forcing this result to equal zero, we obtain

$$20.2 - 2.02x = 0$$

which corresponds to a stationary value of the function $f(x)$. Solving for x, we find that the use of

$$\begin{aligned} 2.02x &= 20.2 \\ x &= 10 \end{aligned}$$

manhours in this area per day maximizes the difference between CI and TC.

Similar to our earlier discussion, we may verify that the use of ten manhours per day in credit and collection activities maximizes the difference between CI and TC when we inspect the signs of the first derivative when $x = 9$, 10, and 11 manhours. In this case, we find that

$$\begin{aligned} f'(9) &= +\$2.02 \\ f'(10) &= \$0.00 \\ f'(11) &= -\$2.02 \end{aligned}$$

which implies that $f(10)$ is the maximum value of y. Notice that when ten manhours per day are employed in credit and collection, the corresponding cash inflows amount to

$$\begin{aligned} CI &= 24.2(10) - .01(10)^2 \\ &= \$241 \end{aligned}$$

per day. On the other hand, the costs of ten manhours are

$$TC = 30 + 4(10) + (10)^2$$
$$= \$170$$

per day. Hence, the difference between the daily cash inflows and the daily costs is \$70 (\$241.00 − \$170.00). On the other hand, if either 9 or 11 manhours per day are used, it can be verified that

$$CI - TC = \$69.99$$

which is less than the net returns associated with devoting ten staff hours per day to credit and collection.

10.5.1 Stationary Values(R, G, F)

These examples illustrate the meaning of minimum and maximum values. More formally, we say that the function $f(x)$ has a maximum (minimum) value at a point at which the value of $f(x)$ is greater (less) than all other values in the vicinity of the point. Minimum and maximum points on a curve defined by $f(x)$ are called *extreme values* of the function. At such points the function also is *stationary*, which means it is neither increasing nor decreasing.

Although stationary values include cases other than maximum and minimum values, we limit this discussion to the so-called extreme values of the function. To present this discussion in somewhat more analytic terms, we consider first the function portrayed diagrammatically in Figure 10-10. This figure shows that: (1) the slope of the tangent constructed at the point $x = a$ is positive, which implies that $f'(a) > 0$; (2) the tangent constructed at the point $x = b$ is parallel to the x axis, which implies that $f'(b) = 0$; and (3) the slope of the tangent constructed at the point $x = c$ is negative, implying that $f'(c) < 0$. Hence, when the slope of the curve is positive immediately to the left of point b and negative immediately to the right of point b, we find that the stationary value given by $f'(b) = 0$ defines a maximum value of the function. Stated differently, when the derivative of the function changes from positive to zero to negative, and $f'(b) = 0$, the point is a maximum value of the function $f(x)$.

Consider next a different function $f(x)$, as presented in Figure 10-11. In this case, the slope of the tangent constructed at point a is negative while the slope of the tangent at point c is positive. As before, the tangent constructed at the point b is parallel to the x axis, which implies that $f'(b) = 0$. Thus, when the derivative changes from negative to zero to positive and $f'(b) = 0$, it follows that $f(b)$ is a minimum value of the function $f(x)$.

Figure 10-10 Analysis of the Extreme Values of a Function

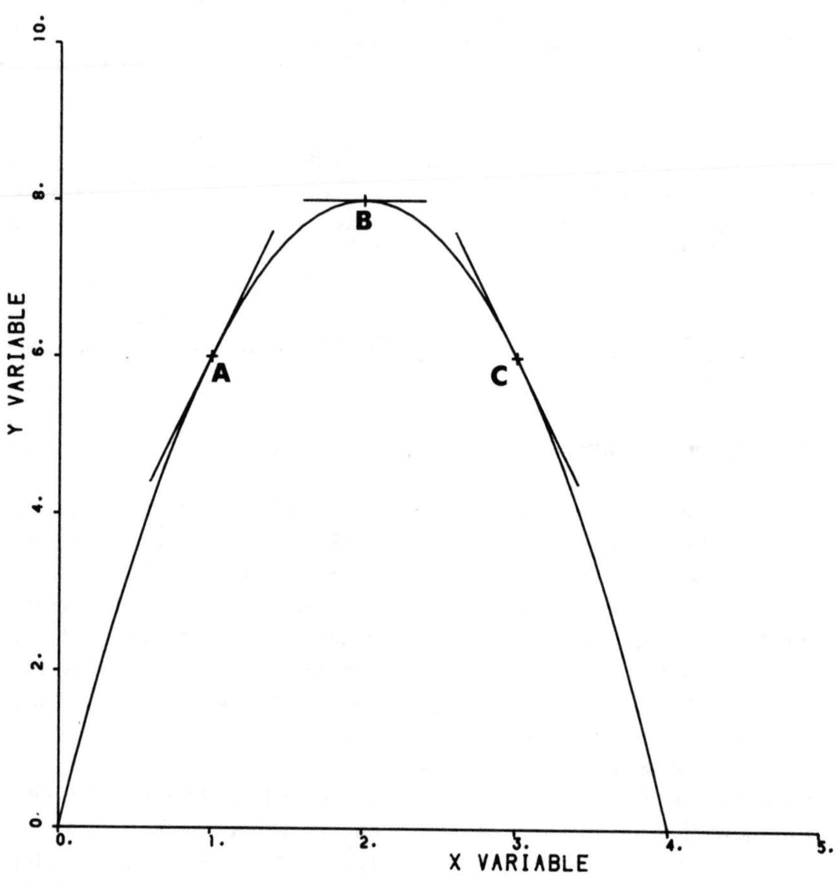

The results may be summarized as follows:

(1) Extreme values of a function are stationary values and occur when $f'(x) = 0$;
(2) $f(b)$ is a maximum value of $f(x)$ if $f'(b)$ is zero and $f'(x)$ changes in sign from *positive* to *negative* as x passes through the point $x = b$;
(3) $f(b)$ is a minimum value of $f(x)$ if $f'(b)$ is zero and $f'(x)$ changes in sign from *negative* to *positive* as x increases through the point $x = b$.

Figure 10-11 Tangents and Minimum Value

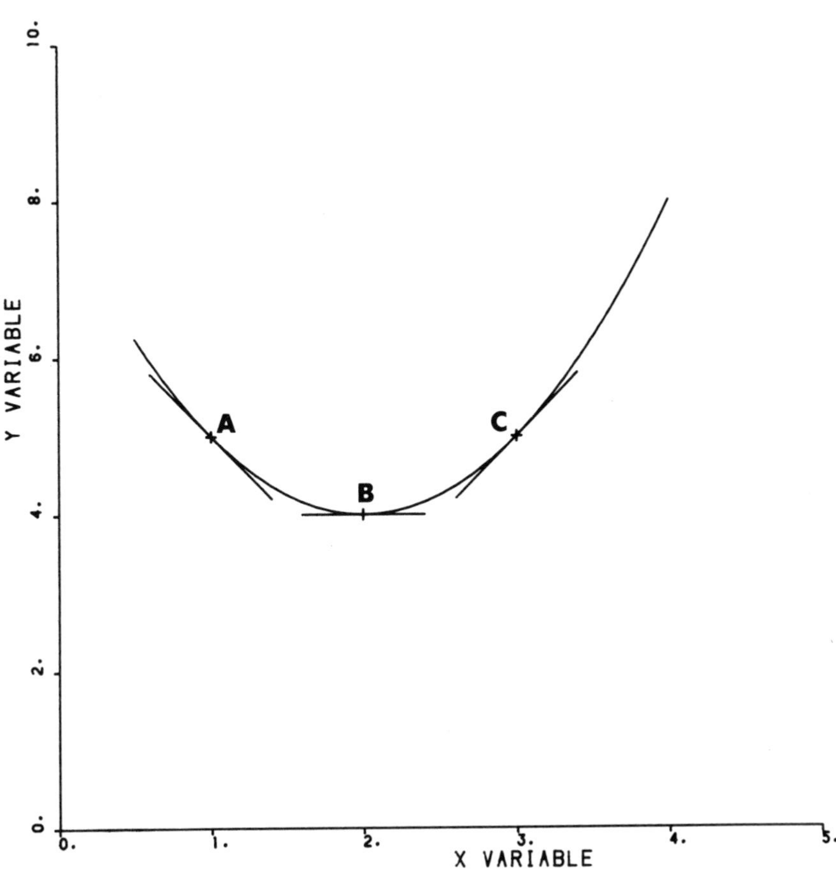

Observe that, as before, these criteria apply to functions and derivatives that are continuous.

10.6 SECOND AND HIGHER ORDER DERIVATIVES(O, A, G, C)

There are many situations in which it is of interest to determine the rate of change in the slope of a curve defined by a given function. For example, in managing the health care facility, it is desirable to provide service at minimum cost. In this situation, the second derivative may be employed to identify the minimum or maximum point on a given curve.

The methods of obtaining second and higher order derivatives require nothing that is new. Once we have found the derivative of the function by applying the rules, the second derivative, $f''(x)$, is obtained by an application of these rules to $f'(x)$ where the first derivative is a function of x. The third, fourth, and higher order derivatives are obtained in succession in the same manner.

For example, consider the function

$$f(x) = 3x^4 + 6x^3 - 2x^2 - 4x$$

In this case we find that the first derivative of this function is given by

$$f'(x) = 12x^3 + 18x^2 - 4x - 4$$

Recognizing that $f'(x)$ is itself a function of x, we find the second derivative by applying the rules cited in Section 10.4.4. Thus, we find that the second derivative is given by

$$f''(x) = 36x^2 + 36x - 4$$

Recognizing that the second derivative is also a function of x, we find that the third derivative is given by

$$f'''(x) = 72x + 36$$

This illustration serves to indicate the general process of finding second and higher order derivatives, the use of which is discussed in the next section.

10.7 THE GENERAL PROBLEM OF OPTIMIZATION AND THE SECOND DERIVATIVE(O, A, C)

The minimum and maximum values of a function also may be identified by the use of the second derivative. Given that the second derivative of the function $f(x)$ is the derivative of the first derivative, $f''(x)$ measures the rate of increase or decrease in $f'(x)$. Stated in another way, the second derivative measures the rate of change in the slope of the tangent to the curve $y = f(x)$ as x passes through the point of interest.

The sign of $f''(x)$ is of considerable importance to the problem of identifying minimum and maximum values. If $f''(b)$ is positive, then $f(x)$ is changing at an *increasing* rate and the slope of the tangent is increasing as x passes through the point $x = b$. In this case, the slope of the tangent rotates in counterclockwise direction, which suggests that the curve is concave when viewed from above at the point $x = b$. Conversely, if $f''(b) < 0, f(x)$ changes at a decreasing rate as x increases through the point $x = b$. Hence, the slope of the tangent turns in clockwise direction that implies that the curve is concave when viewed from above at the point $x = b$. (Note that this discussion of concavity and convexity assumes that *above* is *outside*. Concavity means bending inward, while convexity means bending outward.) These results are portrayed in Figure 10-12, where it will be observed that our interpretation of $f''(x)$ does not depend on either the value or the sign of $f'(b)$. It also will be noted that the numerical value of $f''(b)$ indicates how rapidly the value of $f(x)$ is accelerating as well as the extent of the curvature of $y = f(x)$ at the point $x = b$.

Figure 10-12 Uses of the Second Derivative

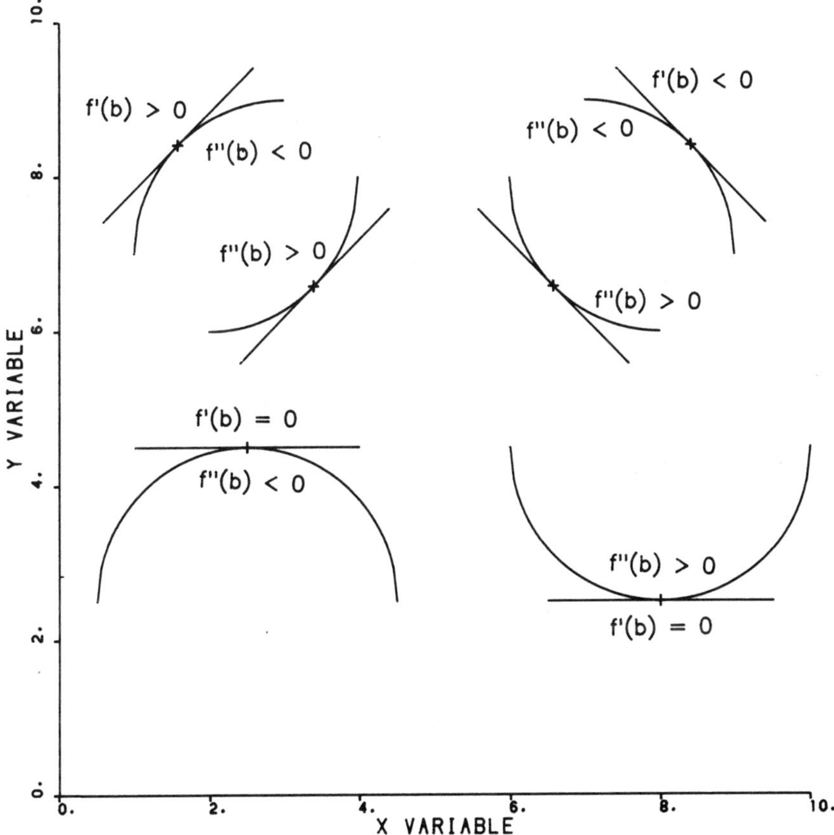

These observations provide an alternate set of criteria for identifying minimum and maximum values of the function $f(x)$. Assuming the function $f(x)$ is continuous and possesses continuous first and second derivatives, the discussion suggests the following criteria:

(1) If $f'(b)$ is zero and $f''(b)$ is negative, $f(b)$ is a maximum value of $f(x)$;
(2) If $f'(b)$ is zero and $f''(b)$ is positive, $f(b)$ is a minimum value of $f(x)$.

With respect to maximum values, we saw earlier that the sign of the slope changes from positive to zero to negative. As a consequence, the rate of change in the slope at the maximum point should be negative. Conversely, when identifying the minimum point, we found that the sign of the slope changes from negative to zero to positive. Hence, the rate of change of the slope should be positive.

To illustrate these criteria let us return to the example in which we wanted to find the level of employment that maximized the amount of service provided per worker. Recall that

$$y = 8x - 2x^2$$

represented the relationship between productivity and the number of full-time equivalents employed in the department. Also recall that $dy/dx = 8 - 4x$, which implied that $8 - 4x$ was equal to zero only when $x = 2$. As a result, the stationary value of the function is defined by $f'(2)$. Also notice that

$$f''(x) = -4$$

and since $f''(x) < 0$, we find that $f(2)$ is a maximum value of the function

$$y = f(x) = 8x - 2x^2$$

Consider next the example in which our objective was to identify the volume of service that minimized unit costs. In this case, we assumed that

$$y = x^2 - 4x + 8$$

represented the relationship between unit costs and the volume of service. Recall that

$$f'(x) = 2x - 4$$

which implied that $f'(x) = 0$ when $2x - 4 = 0$. Thus, the stationary value of the function occurs when $x = 2$. Also note that

$$f''(x) = 2$$

Since $f'(2) = 0$ and $f''(x) > 0$, we conclude that $f(2)$ is a minimum value of the function

$$y = f(x) = x^2 - 4x + 8$$

At this point in the analysis the reader should employ the second derivative to verify the other results obtained in Section 10.5.

Local and Global Maximums and Minimums

The discussion thus far has assumed that a function has either one maximum or one minimum. However, there may be none, one, two, several or an infinite number of minimum or maximum points.

A straight line with zero slope has an infinite number of maximum (also minimum) points. A straight line with nonzero slope has no finite maximum or minimum point.

A quadratic equation (i.e. one that has x^2 as the highest power) has a single maximum *or* a single minimum point. This is called a *global* maximum (or minimum) because no other point on the curve is higher (or lower).

A cubic equation (i.e. one which has x^3 as the highest power) may have either zero, one or two stationary points at which the first derivative, a quadratic, is zero. (Recall from section 7.6 that a quadratic may have imaginary roots, or one or two real roots. The determinant $b^2 - 4ac$ indicates which situation prevails.) Figure 10-13 shows examples of the three possibilities. Part (a) of the figure shows a curve with no stationary point. Part (b) shows one with a single stationary point, but it is neither a maximum nor a minimum. Part (c) shows a curve with two stationary points. One of them is a *local maximum* and the other is a *local minimum.* The reason that the stationary points are *local* is that the outer arms of the curve extend to positive and negative infinity.

Figure 10-13 Example of zero, one, and two stationary points for cubic equations

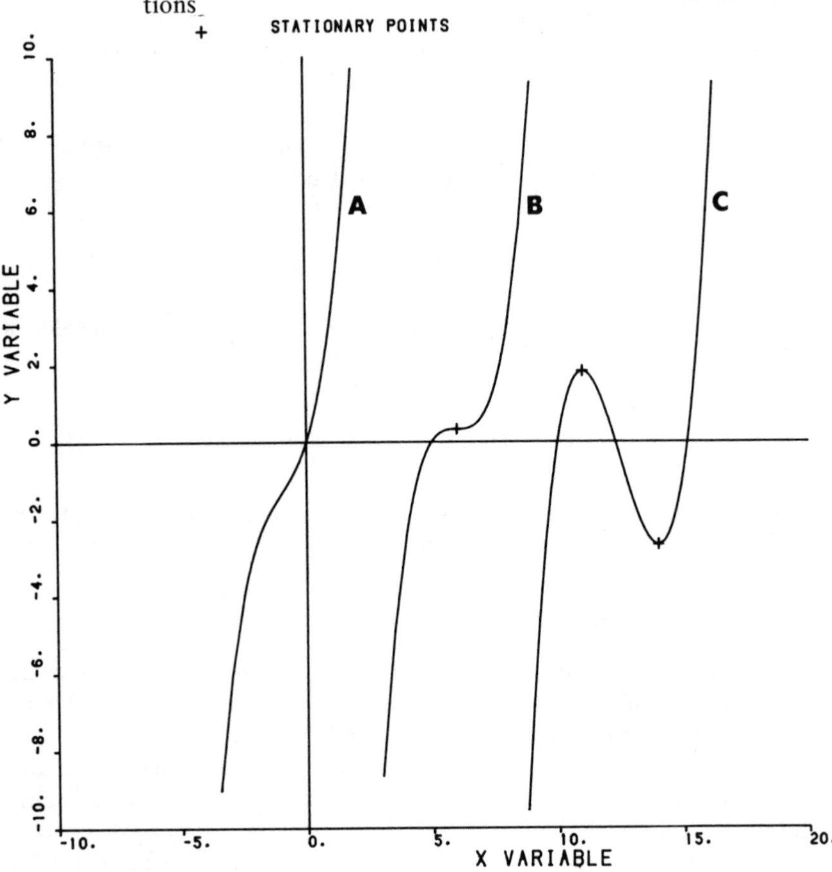

As an example of a cubic equation with a local minimum and a local maximum, consider the function

$$y = f(x) = 2x^3 - 6x^2 + 5 \qquad \textbf{(10.15.1)}$$

In this case, we find that

$$f'(x) = 6x^2 - 12x = 6x(x - 2) \qquad \textbf{(10.15.2)}$$

Thus, we find that $f'(x) = 0$ when $x = 0$ and $x = 2$, which of course defines the stationary values of the function. As seen earlier, the identification of minimum and maximum values requires an evaluation of the second derivative that in this example is given by

$$f''(x) = 12x - 12 \qquad \textbf{(10.15.3)}$$

Substituting $x = 0$ and $x = 2$, which correspond to the condition $f'(b) = 0$, we find that

$$f''(0) = -12 < 0$$

and

$$f''(2) = 12 > 0$$

Thus, $x = 0$ corresponds to a maximum value since $f'(0) = 0$ and $f''(0) < 0$ while $x = 2$ corresponds to a minimum value since $f'(2) = 0$ and $f''(2) > 0$. The values of y that correspond to $x = 0$ and $x = 2$ are given by substituting appropriately into Equation 10.15.1. After substituting and solving for y, we find that the minimum and maximum values assumed by the function $f(x)$ are -3 and 5, respectively.

10.8 APPLICATIONS TO HEALTH CARE MANAGEMENT(R, A, C)

Changes in one variable that result from changes in another variable are of considerable importance to health care management. Examples of situations in which the application of differential calculus and the identification of extreme values are of value include the problems of determining: (1) the rate of activity that corresponds to minimum average cost, (2) the incremental changes in costs that result when activity is increased by one unit, (3) the change in the use of care that results when the characteristics of the population at risk change, and (4) the scale of operations that results in minimum unit costs.

For example, suppose that the number of days of care provided by an institution of fixed size during a given period are used as a measure of hospital activity. Suppose also that the relation between the average total costs (*ATC*) of the institution and the rate of activity, as measured by number of days of care (x) provided per period, may be represented by

$$f(x) = ATC = 80{,}200 - 800x + 2x^2$$

As can be verified, $f'(x)$ is given by $4x - 800$, and setting this result to zero yields

$$4(x - 200) = 0$$

Observe that $f'(x)$ equals zero when $x = 200$ while

$$f'(x) < 0 \text{ when } x < 200$$
$$f'(x) > 0 \text{ when } x > 200$$

Thus, these calculations suggest that unit costs are minimized when 200 days of care are provided per period and that the corresponding cost per day is $200.

Consider a slightly different example in which the institution is divided into four departments, B_1, B_2, B_3, and B_4. Assume that these departments produce services s_1, s_2, s_3, and s_4, respectively, and that each department operates independently of the others. If the functional relation between the average total costs (ATC_i for $i = 1, 2, 3$, and 4) and the volume of services are given by

$$ATC_1 = f(s_1)$$
$$ATC_2 = f(s_2)$$
$$ATC_3 = f(s_3)$$
$$ATC_4 = f(s_4)$$

the conditions

$$f'(s_1{}^*) = 0; \qquad f''(s_1{}^*) > 0$$
$$f'(s_2{}^*) = 0; \qquad f''(s_2{}^*) > 0$$
$$f'(s_3{}^*) = 0; \qquad f''(s_3{}^*) > 0$$
$$f'(s_4{}^*) = 0; \qquad f''(s_4{}^*) > 0$$

indicate that $s_1{}^*$, $s_2{}^*$, $s_3{}^*$, and $s_4{}^*$ correspond to the rates of activity that minimize the unit costs of operating departments B_1, B_2, B_3, and B_4, respectively. These results are of rather obvious value when formulating policies designed to reduce the costs of care or the rate at which such costs have risen.

Finally, suppose that management wants to maximize the net income generated in a given area of activity. Suppose further that

$$TR = 202x$$

represents the total revenue earned by providing x units of service. Similarly, suppose that the total cost of providing care may be approximated by

$$TC = 500 + 2x + .1x^2$$

Hence, the objective of management is to maximize

$$\begin{aligned} y &= TR - TC \\ &= 202x - (500 + 2x + .1x^2) \\ &= 200x - .1x^2 - 500 \end{aligned}$$

In this case, we proceed as before and find that the first derivative is given by

$$\frac{dy}{dx} = 200 - .2x$$

Forcing this result to equal zero, we obtain

$$200 - .2x = 0$$

which, of course, represents a stationary value of the net income or profit function. We may now solve this equation for x and obtain

$$\begin{aligned} .2x &= 200 \\ x &= 1{,}000 \end{aligned}$$

That $x = 1{,}000$ corresponds to a maximum value of y may be verified by considering the signs of $f'(999)$, $f'(1{,}000)$, and $f'(1{,}001)$. In this example, we find that

$$\begin{aligned} f'(999) &= +.2 \\ f'(1{,}000) &= 0 \\ f'(1{,}001) &= -.2 \end{aligned}$$

which verifies that $x = 1{,}000$ is a maximum value of the function $f(x)$. Thus, when 1,000 units of service are provided, a net income of

$$\begin{aligned} y &= 202(1{,}000) - [5 - 0 + 2(1{,}000) + .1(1{,}000)^2] \\ &= \$99{,}500 \end{aligned}$$

is realized. In this example, then, the use of differential calculus permits the identification of the volume of service that corresponds to the maximum net income that might be earned by management.

Appendix 10A

Rules of Differentiation and the Chain Rule[O,G,C]

In this appendix, we consider the rules of differentiation that pertain to products and quotients as well as the chain rule that allows us to find the derivative of more complicated functions.

THE DERIVATIVE OF A PRODUCT

Occasionally two functions may be multiplied such that

$$f(x) = g(x) \cdot h(x)$$

In this case, the product rule states that,

> if $f(x) = g(x) \cdot h(x)$, the derivative $f'(x)$ is equal to the product of the first function and the derivative of the second *plus* the product of the second function and the derivative of the first. As a result, we find that
>
> $$f[g(x) \cdot h(x)] = g(x) \cdot h'(x) + g'(x) \cdot h(x) \qquad \textbf{(10.16)}$$

To illustrate the use of Equation 10.16, suppose that $f(x)$ is given by

$$f(x) = (3x^2 + 3)(4x^2 + 7)$$

In this example, it will be observed that $g(x) = 3x^2 + 3$ and $g'(x)$ is found to be $6x$. Similarly, $h(x) = 4x^2 + 7$ and $h'(x) = 8x$. Thus, we find that

$$\begin{aligned} f'(x) &= g(x) \cdot h'(x) + g'(x) \cdot h(x) \\ &= (3x^2 + 3)(8x) + (6x)(4x^2 + 7) \\ &= 24x^3 + 24x + 24x^3 + 42x \\ &= 48x^3 + 66x \end{aligned}$$

The reader should verify by expanding $f(x)$ to $12x^4 + 33x^2 + 21$ and differentiating.

THE DERIVATIVE OF A QUOTIENT

Consider next a function that assumes the form

$$f(x) = \frac{g(x)}{h(x)}$$

In this case, the derivative of $f(x)$ is equal to

> the product of the denominator and the derivative of the numerator *minus* the product of the derivative of the denominator and the numerator, all divided by the square of the denominator.

Employing functional notation and letting $f(x) = g(x)/h(x)$, we find that

$$f'(x) = \frac{h(x) \cdot g'(x) - g(x) \cdot h'(x)}{[h(x)]^2} \qquad \textbf{(10.17)}$$

To illustrate the use of Equation 10.17, let

$$f(x) = \frac{g(x)}{h(x)} = \frac{x^2 - 3}{x^3}$$

In this case, we find that $g'(x) = 2x$ and $h'(x) = 3x^2$. Thus, $f'(x)$ is given by

$$\frac{h(x) \cdot g'(x) - g(x) \cdot h'(x)}{[h(x)]} = \frac{x^3(2x) - (x^2 - 3)(3x^2)}{(x^3)^2}$$

$$= \frac{2x^4 - (3x^4 - 9x^2)}{x^6}$$

$$= \frac{9x^2 - x^4}{x^6}$$

In this example, it will be observed that when $x = 0$, the derivative is not defined.

THE CHAIN RULE

The *chain rule* is a special method of differentiation that allows us to find the derivative of complicated functions with relative ease. For example, suppose we were given the function

$$f(x) = (6 + x^2)^3$$

for which we wish to find $f'(x)$. We begin by letting $u = 6 + x^2$ such that

$$y = u^3$$

Applying the rule developed in Section 10.4.2, we find that

$$dy/du = 3u^2$$

where the notation dy/du is read as the derivative of y with respect to u. The next step is to return to $u = 6 + x^2$ and differentiate u with respect to x, which is found to be

$$\frac{du}{dx} = 2x$$

The chain rule states that

$$\frac{dy}{dx} = \frac{dy}{du}\left(\frac{du}{dx}\right) \tag{10.18}$$

which in our example is given by

$$3u^2(2x)$$

Substituting $6 + x^2$ for u we find that

$$\frac{dy}{dx} = 3(6 + x^2)^2 2x$$

which reduces to

$$\begin{aligned} 6x(6 + x^2)^2 &= 6x(36 + 12x^2 + x^4) \\ &= 216x + 72x^3 + 6x^5 \end{aligned}$$

Consider a slightly more complicated application of Equation 10.18. Suppose we are given the function

$$f(x) = 3(7 + 2x^2)^4 - 5(12x + 4x^2)^{10}$$

and we wish to find $f'(x)$. Using Equation 10.18, we find that

$$\begin{aligned} f'(x) &= (4)(3)(7 + 2x^2)^3(4x) - 10(5)(12x + 4x^2)^9(12 + 8x) \\ &= 12(4x)(7 + 2x^2)^3 - 50(12 + 8x)(12x + 4x^2)^9 \\ &= 48x(7 + 2x^2)^3 - (600 + 400x)(12x + 4x^2)^9 \end{aligned}$$

Problems for Solution

1. Suppose that the sizes of several hospitals are measured by the active bed complement and that

$$f(x) = 90{,}200 - 60x + .1x^2$$

represents the relation between scale of operations and annual average cost/bed. What scale of activity minimizes average cost? What is the minimum average cost?

2. Suppose that the relation between average productivity (output per unit of service) and the number of factor inputs employed by an institution is given by

$$f(x) = 10 + 12x - .3x^2$$

   a. Find the value of x that gives the maximum $f(x)$.
   b. What is the value of $f''(x)$ corresponding to $f'(x) = 0$?
   c. Find the maximum amount of service per input.

3. Suppose that the relation between total costs and the rate of operation as measured by the number of visits is given by

   a. $f(x) = 6.5x - 4x^2 + \dfrac{1}{2}x^3$

   b. $f(x) = \dfrac{1}{10}x^2 - 5x + 200$

   Find $f'(x)$, and the rate of operation for $f'(x) = 0$, in each case. What interpretation is given to these results?

4. Suppose that departments D_1, D_2, and D_3 provide services S_1, S_2, and S_3 and that the relation between unit cost and the rate of activity is given by

   a. $ATC_1 = f(x_1) = .12x^2 - 48x + 5000$

   b. $ATC_2 = f(x_2) = .02x^2 - 16x + 3400$

   c. $ATC_3 = f(x_3) = .03x^2 - 12x + 1500$

What are the rates of activity that minimize the unit cost in each of these departments? What is the minimum average total cost of operating each unit?

5. Find the extreme values for

   a. $f(x) = x^3 - 3x + 10$

   b. $f(x) = 3x^4 - 10x^3 + 6x^2 + 30$

   c. $f(x) = 6x^3 + 3x^2 - 10$

   Illustrate graphically.

6. Economic theory suggests that the use of care, x, is inversely related with the price, p, of care. Show that

$$(x + b)(p + c) = a$$

   is downward sloping and concave from above where a, b and c are constant values.

7. Show that the total cost functions

$$TC = (ax + b)^{1/2} + c$$

$$TC = ax\frac{x + b}{x + c} + a \qquad (b > c)$$

   yield average and marginal cost curves that fall continuously as the rate of activity increases.

8. Suppose that an institution has a total cost function $TC = f(x)$ and the fee charged per unit of service is fixed at p. Show that the rate of output resulting in a maximum net revenue corresponds to the condition $f'(x) = p$, provided that the total costs are covered.

9. Referring to Problem 8, let $f(x) = ax^2 + bx + c$; show that the marginal cost function is linear. Also show that p must exceed $b + 2\sqrt{ac}$ if total costs are to be covered but p need only exceed b if variable costs are to be covered.

Chapter 11

Partial Derivatives

Objectives

After completing this chapter you should be able to:

1. Define a partial derivative;
2. Find second and higher order partial derivatives;
3. Interpret partial derivatives;
4. Use first and second order partial derivatives to find relative maxima and minima.

Chapter Map

The sections comprising this chapter may be summarized as follows:

Section Number	Required Reading	Optional Reading	Generic Development	Application to Management	Fundamental Principles	Complex Material
	(R)	(O)	(G)	(A)	(F)	(C)
11.1	x		x	x		x
11.2	x		x	x	x	
11.3	x		x			x
11.4	x		x			x
11.5	x		x			x
11.6		x	x	x		x
11.7		x	x	x		x
11.8	x			x		x

11.1 INTRODUCTION(R,G,A,C)

In the previous chapter we limited our analysis to a function of the general form $y = f(x)$. We now extend this analysis to accommodate problems in which y is a function of more than one independent variable. In such a situation, we might be interested in an equation of the form

$$y = f(x_1, x_2, \cdots, x_i, \cdots, x_n)$$

More specifically, our objective is to define and measure the change in y that is derived from changes in x_i, holding the variables $x_j\,(j \neq i)$ constant.

Many phenomena of interest to the health care administrator are influenced by more than one variable. For example, consider the problem of planning and controlling the operational activity of the health care facility. In general, it might be argued that the facility exists for the purpose of providing services required by the population at risk. In turn, the "demand" for service is a dynamic phenomenon that is influenced by factors such as the incidence of disease and injury, the severity of medical conditions treated in the institution, the economic and demographic characteristics of the population at risk, and characteristics of the physician population. Changes in any one of these factors, holding other variables constant, can influence the demand for care.

On the other hand, the amount of service provided depends on the use of real resources. For example, the quantity of care provided is related to the amount of labor, capital equipment, and consumable supplies employed in a process that results in patient services. Management might be interested in the increase in service that results from employing an additional unit of labor, while holding the amount of capital equipment and consumable supplies constant.

These examples suggest that in discharging the planning and control function, it frequently is necessary for health care administrators to understand the relation between the phenomenon of interest and one of the "independent" variables while holding other factors constant. It is in this regard that the partial derivative represents a managerial tool that may be used productively by the health administrator.

11.2 PARTIAL DERIVATIVES OF A FUNCTION OF TWO VARIABLES(R,G,A,F)

Assume that the amount of service provided is a function of the amount of labor, capital, and consumable supplies employed by the health care facility. Further, if during a reasonably short period of time the stock of capital equipment is held constant, we might argue that the amount of service provided is a function of the quantity of labor and consumable supplies used by the institution.

Letting y represent the amount of service, x_1 the amount of labor, and x_2 the amount of consumable supplies, we assume that

$$y = f(x_1, x_2) \tag{11.1}$$

In this institution, we may hold variable x_2 constant and measure the changes in y that are derived from changes in x_1. As a consequence, we may define

1. y as a function of x_1, holding x_2 constant; or
2. y as a function of x_2, holding x_1 constant.

The derivative of each of these functions may be determined using techniques similar to those described earlier. However, the resulting derivatives of the function $y = f(x_1, x_2)$ are referred to as *partial* derivatives. Here, the term partial refers to the fact that the resulting derivative applies only when special assumptions concerning the variation in the variables x_1 and x_2 are satisfied. More specifically, one partial derivative is applicable when only x_1 is allowed to vary while x_2 is held constant. Another partial derivative applies when only x_2 is allowed to vary and x_1 is fixed.

Assume that we want to find the change in the amount of service that results from increasing the use of labor, x_1, by one unit while holding the use of consumable supplies constant and equal to a given number of units. Assuming that $y = f(x_1, x_2)$ is single valued, the partial derivative of y with respect to x_1 at the point defined by $(x_1 = a, x_2 = b)$ is given by the limiting value of the ratio

$$\frac{f(x_1 + \Delta x_1, x_2) - f(x_1, x_2)}{\Delta x_1}$$

In this case, $f(x_1 + \Delta x_1, x_2) - f(x_1, x_2)$ represents the change in the amount of service (i.e., Δy) obtained by changing the use of labor (Δx_1) while holding factor input x_2 constant and equal to b units. Letting Δx_1 approach zero, we may express the partial derivative of y with respect to x_1 by employing the notation $\partial y/\partial x_1$, $\partial f(x_1, x_2)/\partial x_1$, or $f_{x1}'(x_1, x_2)$. We use the symbol "∂" (pronounced "die") to indicate that we are finding a partial derivative.

Similarly, the notation $\partial y/\partial x_2$ indicates that we are finding the partial derivative of y with respect to x_2. In terms of our example, the partial derivative $\partial y/\partial x_2$ at the point $(x_1 = a, x_2 = b)$ indicates the change in the amount of service resulting from a change in the use of consumable supplies while holding the employment of labor constant and equal to a units.

The partial derivative of y with respect to x_1 at the point $(x_1 = a, x_2 = b)$ is given by

$$\partial y/\partial x_1 = \lim_{\Delta x_1 \to 0} \frac{f(x_1 + \Delta x_1, x_2) - f(x_1, x_2)}{\Delta x_1} \qquad \textbf{(11.2)}$$

while the partial derivative of y with respect to x_2 at the point $(x_1 = a, x_2 = b)$ is given by

$$\partial y/\partial x_2 = \lim_{\Delta x_2 \to 0} \frac{f(x_1, x_2 + \Delta x_2) - f(x_1, x_2)}{\Delta x_2} \qquad \textbf{(11.3)}$$

The similarity between these ratios and those used previously should be readily apparent. In fact, nothing new is involved in the definitions advanced by Equations 11.2 and 11.3. As a consequence, partial derivatives are found by using the techniques described in the previous chapter.

By way of illustration, suppose that the relationship between the amount of service and the factor inputs x_1, x_2 is given by the simple expression,

$$y = f(x_1, x_2) = 4x_1 + 7x_2 \qquad \textbf{(11.4.1)}$$

Given that y is linear with respect to x_1 and x_2, $f(x)$ defines a flat plane or surface. In this case, we find that

$$\begin{aligned} y_1 &= 4(10) + 7(20) \\ &= 180 \end{aligned}$$

units of service are obtainable when 10 units of x_1 and 20 units of x_2 are employed by management. Now suppose that the use of resource x_1 is increased by one unit while the use of resource x_2 is held constant and equal to 20 units. In such a situation,

$$\begin{aligned} y_2 &= 4(10 + 1) + 7(20) \\ &= 184 \end{aligned}$$

which implies that, by increasing the use of resource x_1 by one unit and holding the use of resource x_2 constant and equal to 20 units, the quantity of patient care has increased by four units (i.e., 184 − 180 units of service).

In general, let us consider the partial derivative $\partial y/\partial x_1$ at the point for which $x_1 = a$ units of labor and $x_2 = b$ units of consumable supplies. In this case, we know that x_2 (the quantity of consumable supplies) is held constant and equal to b units, which implies that the term $7x_2$ is constant and equal to $7(b)$. As a result, we may let c represent the constant value $7(b)$. In such a situation, we may rewrite Equation 11.4.1 in the form

$$y = 4x_1 + c$$

Now, if we want to find the derivative of y with respect to x_1, we employ the rules developed in the previous chapter and obtain

$$dy/dx_1 = 4$$

since the derivative of a constant is zero. When finding the partial derivative, we usually eliminate the explanatory steps and simply write

$$\partial y/\partial x_1 = 4$$

These findings suggest that an increase of four units of service results when labor is increased by one unit while holding factor input x_2 constant and equal to b units. These results also agree with our earlier findings.

Now consider the partial derivative $\partial y/\partial x_2$ at the point defined by a units of labor and b units of consumable supplies. When finding the partial derivative $\partial y/\partial x_2$ at the point $(x_1 = a, x_2 = b)$, the term $4x_1$ must be constant and equal to $4(a)$. As a result, the partial derivative $\partial y/\partial x_2$ is given by

$$\partial y/\partial x_2 = 7$$

which implies that the volume of service is increased by seven units when the amount of consumable supplies is increased by one unit while the quantity of x_1 is held constant and equal to a units.

Consider a slightly more complex relation between amount of service y and resources x_1, x_2. In this situation we might assume that

$$y = x_1^2 + 3x_1x_2 + 2x_2 \qquad \textbf{(11.4.2)}$$

where the term $3x_1x_2$ contains both x_1 and x_2. Suppose further that we want to find the increase in the volume of service resulting from an increase in the use of labor at the point $(x_1 = a, x_2 = b)$. In this case, we might express Equation 11.4.2 in the form

$$y = x_1^2 + 3(b)x_1 + 2(b)$$

where we substitute the constant value b for x_2. Employing the techniques developed earlier, we find that

$$dy/dx_1 = 2x_1 + 3(b)$$

As before, when dealing with partial derivatives, we may eliminate the explanatory comments and simply write

$$\partial y/\partial x_1 = 2x_1 + 3x_2$$

where x_2 is held constant and equal to b units. As should be verified, the partial derivative $\partial y/\partial x_2$ is given by

$$\partial y/\partial x_2 = 3x_1 + 2$$

As a general rule, if

$$y = \alpha x_1 + \beta x_1 x_2 + \omega x_2$$

we find that

$$\partial y/\partial x_1 = \alpha + \beta x_2$$

while

$$\partial y/\partial x_2 = \beta x_1 + \omega$$

where α, β, and ω are known and fixed parameters (constants).

Now, consider a slightly more complex situation in which

$$y = Ax_1{}^{\alpha} x_2{}^{\beta}$$

In this case, it will be observed that the partial derivative of y with respect to x_1 is given by

$$\partial y/\partial x_1 = \alpha A x_1{}^{\alpha-1} x_2{}^{\beta}$$

which implies that

$$\partial y/\partial x_1 = \alpha y/x_1$$

Similarly, the partial derivative of y with respect to x_2 is given by

$$\partial y/\partial x_2 = \beta A x_1{}^{\alpha} x_2{}^{\beta-1}$$

from which we find that

$$\partial y/\partial x_2 = \beta y/x_2$$

In this formulation, the parameters α and β are elasticity coefficients of output with respect to x_1 and x_2, respectively. As such, the coefficients α and β indicate the percentage change in the amount of service resulting from a percentage change in the use of the inputs x_1 and x_2, respectively.

As an example, suppose that

$$\begin{aligned} x_1 &= 200 \text{ units} \\ x_2 &= 400 \text{ units} \\ A &= 10 \\ \alpha &= .45 \\ \beta &= .55 \end{aligned}$$

In this case, we find that

$$\begin{aligned} y &= 10(200)^{.45}(400)^{.55} \\ &= 2928.17 \end{aligned}$$

units of service are obtained from 200 and 400 units of labor and consumable supplies, respectively. Further we find that

$$\begin{aligned} \frac{\partial y}{\partial x_1} &= .45(2928.17/200) \\ &\cong 6.59 \text{ units} \end{aligned}$$

while

$$\begin{aligned} \frac{\partial y}{\partial x_1} &= .55(2928.17/200) \\ &\cong 4.03 \text{ units} \end{aligned}$$

These results should be verified using $\partial y/\partial x_1 = \alpha A x^{\alpha-1} x_1{}^{\beta}$ and $\partial y/\partial x_2 = A x_1{}^{\alpha} x_2{}^{\beta-1}$.

11.3 A GRAPHIC PRESENTATION OF THE PARTIAL DERIVATIVE(R,G,C)

The meaning of the partial derivative may be illustrated by using Figure 11-1. Letting $x_1 = a$ and $x_2 = b$, we observe that Q corresponds to a point on the surface defined by $y = f(x_1, x_2)$. As seen in this figure, two sections have been constructed through the point Q. One of these sections is constructed so that it is perpendicular to the x_1 axis and shows how y changes in response to variation in x_2, holding x_1 constant and equal to the value a. The other section is constructed perpendicular to the x_2 axis and illustrates the variation in y as x_1 changes, holding x_2 constant and equal to the value b. Also note that line x_1* is tangent to the point Q, which appears on the section that is constructed perpendicular to the x_2 axis. As a consequence, the

Figure 11-1 Example of the Partial Derivative

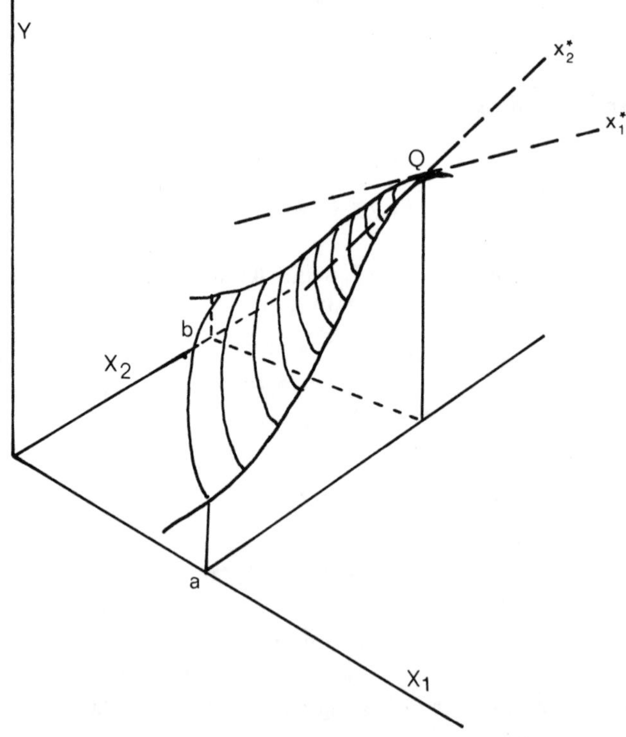

slope of line $x_1{}^*$ is represented by the partial derivative $\partial y/\partial x_1$, which is a numerical measure of the relation between y and x_1 while holding the variable x_2 constant and equal to the value b. Similarly, line $x_2{}^*$ has been constructed so that it is tangent to the other section at the point Q. Hence, the slope of line $x_2{}^*$ is represented by $\partial y/\partial x_2$, which is a numerical measure of the relation between y and x_2 while holding the variable x_1 constant and equal to the value a.

It must be emphasized that the value of $\partial y/\partial x_1$ depends not only on the value originally assumed by x_1 in the limiting process but also on the value of x_2 that is held constant. Similar remarks are appropriate when considering $\partial y/\partial x_2$. Thus, the values of $\partial y/\partial x_1$ and $\partial y/\partial x_2$ are functions of both x_1 and x_2. This may be seen by moving the point Q in any direction and observing the changes in the corresponding slopes of the tangent lines constructed at the new point on the surface.

11.4 SECOND AND HIGHER ORDER PARTIAL DERIVATIVES(O,G,C)

We may define and use a second partial derivative in much the same way that we did for a second derivative. As seen earlier, the two partial derivatives of $y = f(x_1, x_2)$ are functions of x_1 and x_2. Thus, in order to obtain the second partial derivatives, we must derive two partial derivatives from *each* of the partial derivatives of y. In this case, we shall use $\partial^2 y/\partial x_1{}^2$, $\partial^2 y/\partial x_2 \partial x_1$, $\partial^2 y/\partial x_2{}^2$, and $\partial^2 y/\partial x_1 \partial x_2$ to represent the second partial derivative. Employing this notation we may write

$$\frac{\partial^2 y}{\partial x_1{}^2} = \frac{\partial}{\partial x_1}\left(\frac{\partial y}{\partial x_1}\right); \qquad \frac{\partial^2 y}{\partial x_2 \partial x_1} = \frac{\partial}{\partial x_2}\left(\frac{\partial y}{\partial x_1}\right);$$

$$\frac{\partial^2 y}{\partial x_2{}^2} = \frac{\partial}{\partial x_2}\left(\frac{\partial y}{\partial x_2}\right) \quad \text{and} \quad \frac{\partial^2 y}{\partial x_1 \partial x_2} = \frac{\partial}{\partial x_1}\left(\frac{\partial y}{\partial x_2}\right)$$

This notation indicates the order in which partial derivatives are found. For example, $\partial^2 y/\partial x_2 \partial x_1$ is the second partial derivative obtained by first finding the partial derivative of y with respect to x_1 and then with respect to x_2. Similarly, $\partial^2 y/\partial x_1 \partial x_2$ is the second partial derivative obtained by first finding the partial derivative of y with respect to x_2 and then with respect to x_1. The other second-order partial derivatives are found in a similar fashion.

As an example of the method by which second-order partial derivatives are found, consider the function

$$y = 3x_1^2 + x_1^3x_2^2 - 3x_2x_1 + 4x_2^2$$

Applying the rules developed earlier, we find that the partial derivative of y with respect to x_1 is given by

$$\begin{aligned}\frac{\partial y}{\partial x_1} &= \frac{\partial}{\partial x_1}(3x_1^2 + x_1^3x_2^2 - 3x_2x_1 + 4x_2^2)\\ &= 6x_1 + 3x_1^2x_2^2 - 3x_2\end{aligned}$$

while the partial derivative of y with respect to x_2 is given by

$$\begin{aligned}\frac{\partial y}{\partial x_2} &= \frac{\partial}{\partial x_2}(3x_1^2 + x_1^3x_2^2 - 3x_2x_1 + 4x_2^2)\\ &= 2x_1^3x_2 - 3x_1 + 8x_2\end{aligned}$$

Consider next the four second-order partial derivatives. As seen, the second-order partial derivative $\partial^2y/\partial x_1^2$ is given by $\partial/\partial x_1(\partial y/\partial x_1)$ where in terms of our example, $\partial y/\partial x_1 = 6x_1 + 3x_1^2x_2^2 - 3x_2$. Thus, we find that

$$\begin{aligned}\frac{\partial}{\partial x_1}\left(\frac{\partial y}{\partial x_1}\right) &= \frac{\partial}{\partial x_1}(6x_1 + 3x_1^2x_2^2 - 3x_2)\\ &= 6 + 6x_1x_2^2\end{aligned}$$

Similarly, we noted that the second-order partial derivative $\partial^2y/\partial x_2\partial x_1$ is given by $\partial/x_2(\partial y/\partial x_1)$. In terms of our example, we find that

$$\begin{aligned}\frac{\partial^2 y}{\partial x_2\partial x_1} &= \frac{\partial}{\partial x_2}(6x_1 + 3x_1^2x_2^2 - 3x_2)\\ &= 6x_1^2x_2 - 3\end{aligned}$$

Consider next the second-order partial derivative $\partial^2y/\partial x_1\partial x_2$ that is given by $\partial/\partial x_1(\partial y/\partial x_2)$. In terms of our example, we find that $\partial y/\partial x_2$ is $2x_1^3x_2 - 3x_1 + 8x_2$ and, as a result

$$\frac{\partial}{\partial x_1}\left(\frac{\partial y}{\partial x_2}\right) = \frac{\partial}{\partial x_1}(2x_1^3x_2 - 3x_1 + 8x_2)$$

$$= 6x_1^2x_2 - 3$$

Finally, consider $\partial^2y/\partial x_2^2$ that is given by $\partial/\partial x_2(\partial y/\partial x_2)$. In this case we find that

$$\frac{\partial}{\partial x_2}\left(\frac{\partial y}{\partial x_2}\right) = \frac{\partial}{\partial x_2}(2x_1^3x_2 - 3x_1 + 8x_2)$$

$$= 2x_1^3 + 8$$

Referring to our numeric illustration, observe that $\partial^2y/\partial x_1\partial x_2$ is equal to $\partial^2y/\partial x_2\partial x_1$. This result is obtained whenever $\partial y/\partial x_1$ and $\partial y/\partial x_2$ both exist and $\partial y/\partial x_1\partial x_2$ is continuous. Under these conditions, then, the second-order partial derivatives we usually require are $\partial^2y/\partial x_1^2$, $\partial^2y/\partial x_2^2$, and $\partial^2y/\partial x_1\partial x_2$.

11.5 INTERPRETING PARTIAL DERIVATIVES(R,G,C)

As mentioned earlier, $\partial y/\partial x_1$ measures the change in $f(x_1, x_2)$ relative to the point $(x_1 = a, x_2 = b)$ as x_1 varies and x_2 is held constant. Similarly, $\partial y/\partial x_2$ measures the change in $f(x_1, x_2)$ relative to the point $(x_1 = a, x_2 = b)$ as x_2 varies and x_1 is held constant. Thus, these partial derivatives measure the slope of the surface defined by $f(x_1, x_2)$ in two perpendicular directions. The meaning of the signs attached to $\partial y/\partial x_1$ and $\partial y/\partial x_2$ are interpreted as follows:

When $\partial y/\partial x_1$ is greater than zero at the point (a, b), the function $y = f(x_1, x_2)$ increases as x_1 increases from the value a while x_2 remains constant and equal to the value b. Conversely, if $\partial y/\partial x_1$ is less than zero at the point (a, b), the function $y = f(x_1, x_2)$ decreases as x_1 is increased relative to the value a and x_2 is held constant.

The sign of $\partial y/\partial x_2$ is interpreted in a similar way. For example, if $\partial y/\partial x_2$ is greater than zero, the function $y = f(x_1, x_2)$ increases as x_2 is increased and x_1 is held constant. On the other hand, when $\partial y/\partial x_2$ is less than zero, the function $f(x_1, x_2)$ decreases as x_2 is increased relative to b and x_1 is held constant and equal to a.

Consider next the interpretation of the second-order partial derivatives $\partial^2y/\partial x_1^2$ and $\partial^2y/\partial x_2^2$. Similar to our earlier discussion of $f''(x)$, the second-order partial derivative $\partial^2y/\partial x_1^2$ measures the rate of change in $\partial y/\partial x_1$ from the point (a, b) while holding x_2 constant and equal to b. Thus, in terms of Figure 11-1, $\partial^2y/\partial x_2$ measures the rate of change in the slope of the surface defined by $y = f(x_1, x_2)$ from the point Q. As a result, if $\partial^2y/\partial x_1^2$ is positive at the point (a, b), the function $f(x_1, x_2)$ changes at an increasing rate as x_1

moves from the value a while x_2 is held constant and equal to b. Conversely, if $\partial^2y/\partial x_1{}^2$ is negative, the function $f(x_1, x_2)$ changes at a decreasing rate as x_1 moves away from a while x_2 remains constant and equal to b. The interpretation of $\partial^2y/\partial x_2{}^2 \gtrless 0$ also is similar to the meaning attached to the sign of $\partial^2y/\partial x_1{}^2$.

The interpretation of the second-order partial derivatives $\partial^2y/\partial x_1\partial x_2$ and $\partial^2y/\partial x_2\partial x_1$ is slightly more complex. For the purposes of this text, it is necessary to recognize only that $\partial^2y/\partial x_1\partial x_2$ measures the rate of change in $\partial y/\partial x_2$ as x_1 increases, holding x_2 constant, as well as the rate of change of $\partial y/\partial x_1$ as x_2 increases, holding x_1 constant. Referring to point Q in Figure 11-1, $\partial^2y/\partial x_1\partial x_2$ measures changes in the slope of the line $x_2{}^*$ in response to increases in the value of x_2, holding x_1 constant, as well as changes in the slope of the line $x_1{}^*$ in response to increases in x_1, holding x_2 constant. If $\partial^2y/\partial x_1\partial x_2$ is positive at the point Q, we find that:

1. the slope of the line $x_2{}^*$ increases as x_1 is increased, holding x_2 constant, and
2. the slope of line $x_1{}^*$ increases as x_2 is increased, holding x_1 constant.

Once it is recognized that $\partial^2y/\partial x_1\partial x_2$ is a "cross" second-order partial derivative, this interpretation becomes more intuitively plausible.

11.6 MINIMUM AND MAXIMUM POINTS: THE PROBLEM OF OPTIMIZATION(O,G,A,C)

On the basis of the previous discussion, we may now examine the use of first and second partial derivatives to determine the minimum and maximum points on the surface defined by $y = f(x_1, x_2)$. In this case, we define a local maximum or minimum in a manner similar to the way these terms were used earlier. A point for which $x_1 = a$ and $x_2 = b$ is called a critical point for the function $f(x_1, x_2)$, if

$$\partial y/\partial x_1 = 0 \quad \textit{and} \quad \partial y/\partial x_2 = 0$$

However, it is possible for the point ($x_1 = a$, $x_2 = b$) to be neither a local maximum nor a local minimum even though $\partial y/\partial x_1 = 0$ and $\partial y/\partial x_2 = 0$.

To determine the presence of a relative minimum or maximum, we define the number K by

$$K = \frac{\partial(\partial y/\partial x_1}{\partial x_1} \times \frac{\partial(\partial y/\partial x_2)}{\partial x_2} - \left[\frac{\partial(\partial y/\partial x_1)}{\partial x_2}\right]^2 \qquad \textbf{(11.5)}$$

where the values of the partial derivatives are determined at the point ($x_1 = a$, $x_2 = b$). In the following discussion we shall assume that: (1) $y = f(x_1, x_2)$ is continuous, (2) the second-order partial derivatives exist, and (3) a critical point occurs at the point ($x_1 = a$, $x_2 = b$). If these assumptions are satisfied, the point ($x_1 = a$, $x_2 = b$) is

(1) a relative minimum when $[\partial(\partial y/\partial x_1)]/\partial x_1 > 0$ and $K > 0$;
(2) a relative maximum if $[\partial(\partial y/\partial x_1)]/\partial x_1 < 0$ and $K > 0$.

On the other hand, when K is less than zero, $f(x_1 = a, x_2 = b)$ is neither a maximum nor minimum value of y. Rather, no information concerning a relative maximum or minimum is given when the value of K is less than zero.

To illustrate this technique, assume that a department in our health care institution uses labor of types x_1 and x_2 in providing medical service. Ceteris paribus, assume that the cost (in hundreds of dollars) of employing labor of types x_1 and x_2 is given by

$$y = f(x_1, x_2) = 1.5x_1^2 + x_2^2 - 4x_1 - 2x_2 - 2x_1x_2 + 93 \qquad \textbf{(11.6.1)}$$

Also assume that our objective is to determine the values of x_1 and x_2, as expressed in 100 staff hours, which minimize the labor cost of the department.

When identifying a relative minimum or maximum, we first require that $\partial y/\partial x_1$ and $\partial y/\partial x_2 = 0$ at the point ($x_1 = a$, $x_2 = b$). In this example, we find that

$$\partial y/\partial x_1 = 3x_1 - 4 - 2x_2$$

while

$$\partial y/\partial x_2 = 2x_2 - 2 - 2x_1$$

Rearranging slightly and setting $\partial y/\partial x_1$ and $\partial y/\partial x_2$ equal to zero, we obtain

$$\frac{\partial y}{\partial x_1} = 3x_1 - 2x_2 - 4 = 0 \qquad \textbf{(11.6.2)}$$

$$\frac{\partial y}{\partial x_2} = -2x_1 + 2x_2 - 2 = 0 \qquad \textbf{(11.6.3)}$$

Now we must solve for x_1 and x_2.

If we add Equations 11.6.2 and 11.6.3 we obtain

$$x_1 - 6 = 0$$

Solving for x_1 we find that

$$x_1 = 6$$

Substituting this result into Equation 11.6.2, we obtain

$$\begin{aligned} 3(6) - 2x_2 - 4 &= 0 \\ -2x_2 &= -14 \end{aligned}$$

which implies that $x_2 = 7$. As a result, the only critical point of the function is given by ($x_1 = 6$, $x_2 = 7$).

The next step in the process is to find the second-order partial derivatives required to calculate K. In this example, we find that

$$\frac{\partial\left(\frac{\partial y}{\partial x_1}\right)}{\partial x_1} = \frac{\partial(3x_1 - 2x_2 - 4)}{\partial x_1} = 3$$

while

$$\frac{\partial\left(\frac{\partial y}{\partial x_2}\right)}{\partial x_2} = \frac{\partial(-2x_1 + 2x_2 - 2)}{\partial x_2} = 2$$

Finally, the calculation of K requires the cross partial derivative that is given by

$$\frac{\partial(\partial y/\partial x_1)}{\partial x_2} = \frac{\partial(3x_1 - 2x_2 - 4)}{\partial x_2} = -2$$

As a result, we find that the value of K for the point ($x_1 = 6$, $x_2 = 7$) is given by

$$\begin{aligned} K &= 3 \times 2 - (2)^2 \\ &= 2 \end{aligned}$$

Since K and $[\partial(\partial y/\partial x_1)]/\partial x_1$ are both greater than zero, the critical point defined by $(x_1 = 6, x_2 = 7)$ is a relative minimum. Hence, the use of 600 manhours of type x_1 labor and 700 manhours of type x_2 labor minimizes the labor costs of the department.

11.7 PARTIAL DERIVATIVES INVOLVING MORE THAN TWO VARIABLES(O,G,A,C)

We now extend our discussion of partial derivatives to accommodate functions involving more than two independent variables. For example, if

$$y = f(x_1, x_2, x_3)$$

is a single valued function of three variables (i.e., x_1, x_2, and x_3), we may form an equation of one variable while holding the two other variables constant. Thus, we may express y as a function of:

1. x_1, holding x_2 and x_3 constant,
2. x_2, holding x_1 and x_3 constant, and
3. x_3, holding x_1 and x_2 constant.

As a consequence, we may define three first-order partial derivatives $\partial y/\partial x_1$, $\partial y/\partial x_2$, and $\partial y/\partial x_3$ at any point defined by $(x_1 = a, x_2 = b, x_3 = c)$. Similar to our earlier work, the partial derivatives $\partial y/\partial x_1$, $\partial y/\partial x_2$, and $\partial y/\partial x_3$ are defined by the limiting value of

$$\lim_{\Delta x_1 \to 0} \frac{f(x_1 + \Delta x_1, x_2, x_3) - f(x_1, x_2, x_3)}{\Delta x_1}$$

$$\lim_{\Delta x_2 \to 0} \frac{f(x_1, x_2 + \Delta x_2, x_3) - f(x_1, x_2, x_3)}{\Delta x_2}$$

and

$$\lim_{\Delta x_3 \to 0} \frac{f(x_1, x_2, x_3 + \Delta x_3) - f(x_1, x_2, x_3)}{\Delta x_3}$$

respectively.

Following the procedure outlined in the previous sections, we may then form three second-order partial derivatives from each of the first-order partial derivatives, which results in a total of nine second-order partial derivatives. However, assuming that the continuity conditions mentioned earlier are satisfied, we find that

$$\frac{\partial^2 y}{\partial x_1 \partial x_2} = \frac{\partial^2 y}{\partial x_2 \partial x_1}; \qquad \frac{\partial^2 y}{\partial x_1 \partial x_3} = \frac{\partial^2 y}{\partial x_3 \partial x_1}$$

and

$$\frac{\partial^2 y}{\partial x_2 \partial x_3} = \frac{\partial^2 y}{\partial x_3 \partial x_2}$$

As a result we need only determine $\partial^2 y/\partial x_1{}^2$, $\partial^2 y/\partial x_2{}^2$, $\partial^2 y/\partial x_3{}^2$, $\partial^2 y/\partial x_1 \partial x_2$, $\partial^2 y/\partial x_1 \partial x_3$, and $\partial^2 y/\partial x_2 \partial x_3$.

We now extend this discussion to accommodate the general function

$$y = f(x_1, \cdots, x_n)$$

for which there are n first-order partial derivatives of the form

$$\frac{\partial y}{\partial x_1} \cdots \frac{\partial y}{\partial x_n}$$

Here, $\partial y/\partial x_1$ indicates the change in y that is derived by a unit change in the variable x_1 while $x_2, \cdots, x_n$ remain constant.

On the other hand, when we consider ordinary continuous functions, the second-order partial derivatives that are not redundant may be represented by

$$\frac{\partial^2 y}{\partial x_1{}^2}, \frac{\partial^2 y}{\partial x_2{}^2}, \cdots, \frac{\partial^2 y}{\partial x_n{}^2}$$

and by

$$\frac{\partial^2 y}{\partial x_1 \partial x_2}, \frac{\partial^2 y}{\partial x_1 \partial x_3}, \cdots, \frac{\partial^2 y}{\partial x_1 \partial x_n}, \frac{\partial^2 y}{\partial x_2 \partial x_3}, \cdots, \frac{\partial^2 y}{\partial x_2 \partial x_n}, \cdots$$

The interpretation of these first- and second-order partial derivatives is as follows. The first-order partial derivative measures the rate of change in the function as one of the variables changes and the other $n - 1$ variables remain constant. Similarly, $\partial^2 y/\partial x^2$, $\cdots$, $\partial^2 y/\partial x_n{}^2$ measure the rates of change in $\partial y/\partial x_1, \cdots, \partial y/x_n$ as one of the variables increases from the given value and the others remain constant.

11.8 APPLICATIONS TO HEALTH CARE MANAGEMENT(R,A,C)

The use and provision of health services in any setting is influenced by a complex set of interrelated factors. For example, the use of health services is influenced by such factors as the incidence of disease or injury in the population at risk, the complexity and severity of presenting diagnoses, the sociodemographic characteristics of the population, and economic issues. When predicting the use of care, the administrator must realize that a change in any one of these factors may be reflected in the utilization of care by the population at risk.

Similarly, the provision of service may be viewed as a process in which factor inputs such as labor, supplies, and capital equipment are combined in order to offer care. The relationship between the provision of service and the use of factor inputs may be represented in the form

$$y = f(K, L, CS)$$

where

y represents output or the quantity of service provided,
K represents capital equipment,
L represents labor,
CS represents consumable supplies.

From a theoretic perspective, the production function may be defined as a technological statement of the maximum amount of output attainable when exactly L^* units of labor, K^* units of capital equipment, and CS^* units of consumable supplies are employed in the process of providing care. When expressed in specific form, an analysis of the production function allows an examination of the change in output that results from a change in the use of one of the factor inputs.

In both of these situations, we are interested in examining the change in the dependent variable in response to a change in one of the independent variables while holding other factors constant. As such, the technique of partial differentiation is an appropriate method for examining these changes.

Returning to the first example introduced, suppose that the use of care by primary diagnosis is related to the age, income, and education of the patient.

Let U represent the use of care, as measured by number of visits per year; A represent age, as measured in years; I represent income, as measured in dollars per year; and E represent education, as measured in years. Suppose further that

$$U = 100 + .2A^2I + .001I^2E + .0002EA$$

represents the functional relation between use and the set of independent variables. As can be verified

$$\frac{\partial U}{\partial I} = .2A^2 + .002IE$$

$$\frac{\partial U}{\partial E} = .001I^2 + .0002A$$

$$\frac{\partial U}{\partial A} = .4AI + .0002E$$

indicate the rate of change in U associated with a change in I, E, and A respectively while holding the other factors constant. Clearly these findings are of considerable value to solving the problem of estimating the use of service by a population that is expected to exhibit changes in its age, income, and educational distribution.

Concerning the second example, management may be interested in examining the rate of chage in the quantity of service provided that results from changes in the level of resources employed by the facility. For example, suppose that the relation between the quantity of services provided, Y, and the amount of labor, consumable supplies, and capital is given by

$$Y = AL^{\alpha}K^{\beta}CS^{\gamma}$$

In this case, the coefficients α, β, and γ correspond to elasticities of output with respect to labor, capital, and consumable supplies, respectively. The change in output resulting from changing the quantity of labor by one unit is given by

$$\frac{\partial Y}{\partial L} = \alpha AL^{\alpha-1}K^{\beta}CS^{\gamma}$$

$$= \alpha\frac{Y}{L}$$

while the change in output resulting from a change in the use of capital equipment is given by

$$\frac{\partial Y}{\partial K} = \beta A L^{\alpha} K^{\beta - 1} CS^{\gamma}$$

$$= \beta \frac{Y}{K}$$

Similarly, the change in output resulting from a change in the use of consumable supplies is given by

$$\frac{\partial Y}{\partial CS} = \gamma A L^{\alpha} K^{\beta} CS^{\gamma - 1}$$

$$= \gamma \frac{Y}{CS}$$

Consider now a numeric example of these partial derivatives. Suppose we are interested in a situation in which our institution currently employs 200 units of labor, 100 units of capital, and 600 units of consumable supplies. Also suppose that, after applying regression analysis to historical data, we find that $\alpha = .10$, $\beta = .45$, and $\gamma = .30$. If the constant A is equal to 10, we find that the volume of service is given by

$$Y = AL^{.10} K^{.45} CS^{.20}$$

$$= 10(200)^{.10}(100)^{.45}(600)^{.30}$$

$$= 919 \text{ units}$$

Recalling that $\partial Y/\partial L = \alpha Y/L$, $\partial Y/\partial K = \beta Y/K$, and $\partial Y/\partial CS = \gamma Y/CS$, the change in volume of service resulting from a change in the use of each resource is given by

$$\partial Y/\partial L = .10(919/200)$$

$$\cong .46 \text{ units}$$

$$\partial Y/\partial K = .45(919/100)$$

$$\cong 4.14 \text{ units and}$$

$$\partial Y/\partial CS = .30(919/600)$$

$$\cong .46 \text{ units}$$

At this point, the reader should verify that $\partial Y/\partial L = \alpha AL^{\alpha-1}K^{\beta}CS$, $\partial Y/\partial K = \beta AL^{\alpha}K^{\beta-1}CS$, and $\partial Y/\partial CS = \gamma AL^{\alpha}K^{\beta}CS^{\gamma-1}$ yield equivalent results. These findings, coupled with the estimated volume of service resulting from the use of a given complement of resources, are of rather obvious value to management when discharging the functions of planning, monitoring, and controlling the operational activity of the facility.

Problems for Solution

1. Suppose that the output of a given facility depends on two variable inputs, A and B, and that the dependence is given by the surface $y = f(a, b)$. Specifically, assume that the production function is of the form

$$y = 2Mab - Aa^2 - Bb^2$$

which implies that

$$\frac{y}{a} = 2Mb - Aa - \frac{Bb^2}{a}$$

represents the average product of A. Show that for the fixed amount b_1 of Factor B_1 the average product of A is a maximum when $a = \sqrt{(B/A)}\, b_1$. Hint: the average product is maximized when $\partial y/\partial A = Y/a$.

2. Suppose that the relation between output and the factor inputs L and CS is given by

$$Y = 100L^{.50}CS^{.30}$$

Find the marginal product (i.e., $\Delta Y/\Delta L$) for $L = 20$, assuming CS is held constant and equal to 30.

3. Assume that the relation between the use of service and the age, education, and income of the patient is given by

$$U = .001AE^2 + .0001AI + .003IE^2$$

Find the marginal rate of change in U at $A = 20$ while holding E and I constant and equal to seven years and \$4,000, respectively.

4. Find the first- and second-order partial derivatives for

$$Z = x^3 + y^3 - 2xy$$

where

Z = use of service
x = incidence of disease
y = measure of case severity

What interpretation do you attach to these partial derivatives?

5. Suppose that $50a$ staff hours of labor are combined with b hours of capital equipment and

$$y = 2(24ab - 3a^2 - 7b^2)$$

units of output are obtained. Find the marginal product (i.e. $\partial y/\partial a$) of $a = 6$ if only b hours of capital equipment are available.

6. Suppose that

$$Y = 20L^{.2}K^{.6}$$

is the amount of output obtained when L staff hours are combined with K hours of capital equipment. Find the marginal product (i.e. $\partial y/\partial a$) of 50 hours of labor if (a) 10 hours and (b) 100 hours of capital equipment are available.

7. Suppose that the production function is

$$Y = (2Hab - Aa^2 - Bb^{21/2}$$

What is the marginal product of $Y(\partial y/\partial a)$ at $a = a_1$ while holding B constant and equal to b_1?

8. Find the first- and second-order partial derivatives of

$$Y = \frac{x^2}{x - y}$$

Assuming that $x = y$, what interpretation do you attach to this partial derivative?

9. How would you interpret the partial derivatives of

$$y = x^2 - xw + 2w^2$$

Verify that

$$x\frac{\partial y}{\partial x} + w\frac{\partial y}{\partial w} = 2y$$

10. Suppose the relation between the admission rate R and outpatient price P, revenue per admission A, room charge C, occupancy rate O, and income of the patient I is given by

$$R = P^{-.98}A^{.39}C^{.69}O^{-.20}I^{.30}$$

Find

$$\frac{\partial R}{\partial P}, \frac{\partial R}{\partial A}, \frac{\partial R}{\partial C}, \frac{\partial R}{\partial O}, \frac{\partial R}{\partial I}$$

What interpretation do you attach to these partial derivatives?

11. Suppose that the relation between the number of outpatient visits to a set of hospitals, OV, and the outpatient price P, the revenue per patient day R, the room charge C, and the income of the patient I, is given by

$$OV = P^{-.97}R^{.39}C^{1.27}I^{.32}$$

Find

$$\frac{\partial OV}{\partial P}, \frac{\partial OV}{\partial R}, \frac{\partial OV}{\partial C} \quad \text{and} \quad \frac{\partial OV}{\partial I}$$

What interpretation do you attach to these partial derivatives?

12. Suppose that the net income of an institution is given by

$$Y(x, z) = 2000 + 20x - x^2 + 75z - z^2$$

where x represents the cost of labor and z is the cost of consumable supplies. Find the values of x and z that maximize net income.

13. Find any relative maximum and minimum values for the following functions

$$f(x, y) = 60 + 4x - 2y + 4x^2 + 2y^2 + 4xy$$

$$f(x, y) = 2x^2 + 4y^2 - 5x$$

$$f(x, y) = x^2 - 4xy + 2y^2 + x - 7$$

$$f(x, y) = x^2 + 3xy + 4y^2 - 7x + 5y$$

$$f(x, y) = -60 + x + 2y + x^2 + 2y^2 + 3xy$$

Chapter 12

Integral Calculus

Objectives

After completing this chapter you should be able to:

1. Understand and perform antidifferentiation;
2. Distinguish between definite and indefinite integrals;
3. Use the process of integration to find areas under a curve;
4. Use integration to perform the operation of summation.

Chapter Map

The sections comprising this chapter may be summarized as follows:

Section Number	Required Reading	Optional Reading	Generic Development	Application to Management	Fundamental Principles	Complex Material
	(R)	(O)	(G)	(A)	(F)	(C)
12.1	x		x		x	
12.2	x		x			x
12.3		x	x	x		x
12.3.1		x		x		x
12.3.2		x	x			x
12.4	x		x			x
12.4.1	x		x			x
12.4.2	x		x			x
12.5	x			x		x
12.6		x		x		x
12.7		x		x	x	

12.1 INTRODUCTION (R,G,F)

As mentioned earlier, the concept of integration has essentially two characteristics and two distinct applications. First, the integral may be viewed as the inverse of differentiation and the major objective of the process of antidifferentiation is to obtain a function, $F(x)$, that has as its derivative the function $f(x)$. Thus, when $F'(x) = f(x)$, the function $F(x)$ is said to be an antiderivative of $f(x)$. When viewed from this perspective, the function $F(x)$, if it exists, is called the *indefinite integral* of the function $f(x)$. We also may regard the integral as the limiting value of a specific summation expression that, in diagrammatic terms, corresponds to the area under the curve defined by $f(x)$. When viewed from this perspective, the integral is referred to as a *definite integral.*

It is in this sense that the concept of integration is of crucial importance to the problem of determining the probability that a continuous random variable will assume a value on a specified interval. These and related matters are considered in Chapter 15. Since an examination of the definite integral and the area under the curve defined by $f(x)$ requires an understanding of antidifferentiation, we begin our discussion with an analysis of this process.

12.2 THE ANTIDERIVATIVE (R,G,C)

As the name implies, antidifferentiation is the reverse of differentiation. Given that $F(x)$ is a function whose derivative, $F'(x)$, is $f(x)$, the function $F(x)$ is called the antiderivative of $f(x)$. As seen earlier, the derivative of each term in a polynomial is given by

$$\frac{d}{dx} x^n = nx^{n-1} \tag{12.1}$$

This procedure is simply reversed when we find the antiderivative. For example, if

$$f(x) = x^n$$

the antiderivative is found by simply raising the exponent of x by one and dividing x^{n+1} by the new exponent. For example, the expression $1/4\ x^4$ is an antiderivative of x^3 just as $x^5 - 3x^2 + 3x$ is an antiderivative of $5x^4 - 6x + 3$.

In the following discussion, we assume that $f(x)$ is a single valued function that is continuous for all values in the interval $x = a$ to $x = b$. Recalling that the derivative of some constant C is zero, suppose we are given

$$F(x) = 3x^2 + 2x + 6 \quad \textbf{(12.2.1)}$$

The derivative of this function is found to be

$$F'(x) = f(x) = 6x + 2 \quad \textbf{(12.2.2)}$$

Now, suppose that the function $f(x)$, as defined by Equation 12.2.2, is the only expression available to us and that we want to find the antiderivative $F(x)$. In this case, we find that the antiderivative of $f(x)$ is given by $3x^2 + 2x$, to which we must add an arbitrary constant C in order to compensate for the fact that the constant term (i.e., 6) in Equation 12.2.1 was "lost" when we determined $F'(x) = f(x)$.

This example illustrates the following rule:

> *To determine the antiderivative of some function $f(x)$, it is necessary to find any function* **F(x)** *such that* **F′(x) = f(x)** *and then add the constant of integration that is represented by C.*

The addition of the arbitrary constant C is required since the derivative of any constant value is zero.

This discussion also suggests that an antiderivative is not unique. For example, the antiderivative of

$$f(x) = 1 + x^3 \quad \textbf{(12.3.1)}$$

is given by

$$F(x) = x + 1/4\,x^4 + C \quad \textbf{(12.3.2)}$$

Given that the constant C may assume values that range from $+\infty$ to $-\infty$, there are an infinite number of expressions satisfying the condition $F'(x) = f(x)$. Thus, if a function has an antiderivative, we may assert that it has many antiderivatives.

Frequently, it is necessary to find the antiderivative of the function $f(x)$ and to assign a specific value to the constant C that satisfies a prescribed condition. For example, suppose that

$$f(x) = -x \quad \textbf{(12.4.1)}$$

Applying the rules developed above, we find that the antiderivative of $f(x)$ is given by

$$F(x) = -1/2\,x^2 + C \qquad \textbf{(12.4.2)}$$

Observe that Equation 12.4.2 defines a family of curves whose slope satisfies the equation $y = -1/2\,x^2$. However, only one curve in the family of curves passes through the point (2,3). To obtain the curve that satisfies this condition, we simply substitute $x = 2$ and $y = 3$ in Equation 12.4.2 and solve for C. Thus, we find that

$$3 = -1/2\,(2)^2 + C$$

As a result, we find that the curve that passes through the point (2,3) is given by

$$y = -1/2\,x^2 + 5$$

where C must assume the value of 5.

12.3 THE INDEFINITE INTEGRAL(R,G,A,C)

Because of the fundamental relation between antidifferentiation and integration, antiderivatives frequently are referred to as indefinite integrals while the systematic derivation of antiderivatives is referred to as the technique of integration. Suppose we are given $f(x)$ and we want to find the antiderivative, or integral, that satisfies the condition $F'(x) = f(x)$. The indefinite integral of $f(x)$, which is denoted by $\int f(x)dx$, is given by

$$\int f(x)dx = F(x) + C \qquad \textbf{(12.5)}$$

where $F(x)$ is any function that satisfies the condition $F'(x) = f(x)$ and C is the constant of integration to which we referred earlier.

12.3.1 An Example(O,A,C)

To illustrate the use of Equation 12.5, we define marginal cost as the change in total cost resulting from changing the output of the health care institution by one unit. Employing the analysis of Chapter 9, we might posit that total cost function is given by

$$y = TC = g(x)$$

where x represents the output of the institution, as measured, say, in terms of days of care provided, and TC represents total cost as plotted on the y axis. Given that marginal costs may be represented by $\Delta TC/\Delta x$, we find that dTC/dx yields the marginal cost function. Letting

$$\frac{dTC}{dx} = f(x)$$

we find that

$$\int f(x)dx$$

yields the total cost function. More specifically, suppose that

$$f(x) = 3x \tag{12.6.1}$$

represents the marginal cost function of the institution. Applying the rules introduced earlier we find that

$$\int 3xdx = \frac{3}{2}x^2 + C \tag{12.6.2}$$

is an antiderivative of $f(x)$. Now, suppose that the total cost curve must pass through the point (30,000, 10), which implies, that total costs of \$30,000 are incurred when ten days of care are provided. In this case, the value assumed by C is given by

$$30{,}000 = \frac{3}{2}(10)^2 + C$$

and, as a result, the total cost function is found to be

$$TC = \$29{,}850 + \frac{3}{2}x^2 = \$29{,}850 + \$1.50\,x^2 \tag{12.6.3}$$

Obviously, Equation 12.6.3 may be used to determine the total costs of providing differing amounts of care.

12.3.2 Standard Forms of Integration (O,G,C)

Since $d/dx(x^{n+1}/n + 1) = x^n$ for $x \neq -1$, while $d/dx(\log x) = 1/x$ and $d/dx(e^x) = e^x$, it can be easily verified that

$$\int x^n dx = \frac{x^{n+1}}{n + 1} + C \qquad \textbf{(12.7)}$$

$$\int (1/x)\, dx = \log x + C \qquad \textbf{(12.8)}$$

$$\int e^x dx = e^x + C \qquad \textbf{(12.9)}$$

Equations 12.7, 12.8, and 12.9 define the major standard forms employed in the process of integration.

In addition to these standard forms, it also is possible to derive several rules for the integration of functions. If $f(x)$ and $g(x)$ are single valued functions that are continuous over the relevant range, then

$$\int cf(x)dx = c\int f(x)dx \qquad \textbf{(12.10)}$$

$$\int [f(x) + g(x)]dx = \int f(x)dx + \int g(x)dx \qquad \textbf{(12.11)}$$

where the term c appearing in Equation 12.10 is some constant. Extending and combining Equations 12.10 and 12.11, we find that

$$\int [c_1 f(x) + c_2 g(x)]dx = c_1 \int f(x)dx + c_2 \int g(x)dx \qquad \textbf{(12.12)}$$

We encounter no difficulties when integrating an expression consisting of sums or differences of simple functions. However, it is not possible to formulate rules for the integration of products, quotients, or functions of functions that have general applicability. Rather, the practical approach to integrating equations involving these expressions requires a "trial and error" process, the results of which are validated by ensuring that $F'(x) = f(x)$.

12.4 THE DEFINITE INTEGRAL AND THE AREA UNDER A CURVE (R,G,C)

This section simply extends our previous work. More specifically, our primary objective is to illustrate the usefulness of integration in determining specified areas under a curve that is defined by the function $y = f(x)$ and to introduce the concept of the definite integral.

12.4.1 Areas of Integration (R,G,C)

In the following presentation, we assume that the function $y = f(x)$ is continuous and that the values assumed by y are greater than or equal to zero. On the basis of these assumptions, our concern is to determine the area $A(x)$ that is bounded by: the curve defined by $y = f(x)$, the x axis, the fixed ordinate $x = a$, and a variable ordinate $x = x_i$.

Suppose we are given the function $y = f(x)$ that defines the curve in Figure 12-1. We now let ΔA represent the increase in $A(x)$ that results when the value of x is increased from x to $x + \Delta x$. Observe that ΔA is represented by the area MORN in the figure. Also note that the area of the rectangle MOPN is less than or equal to ΔA while the area of the rectangle MQRN is greater than or equal to ΔA. As a result, we may assert that

$$\text{MOPN} \leq \Delta A \leq \text{MQRN} \qquad \textbf{(12.13)}$$

Since the area of a rectangle is given by the product of width and length, Equation 12.13 may be written in the form

$$y\Delta x \leq \Delta A \leq (y + \Delta y)\,\Delta x \qquad \textbf{(12.14)}$$

Figure 12-1 An Example of the Definite Integral

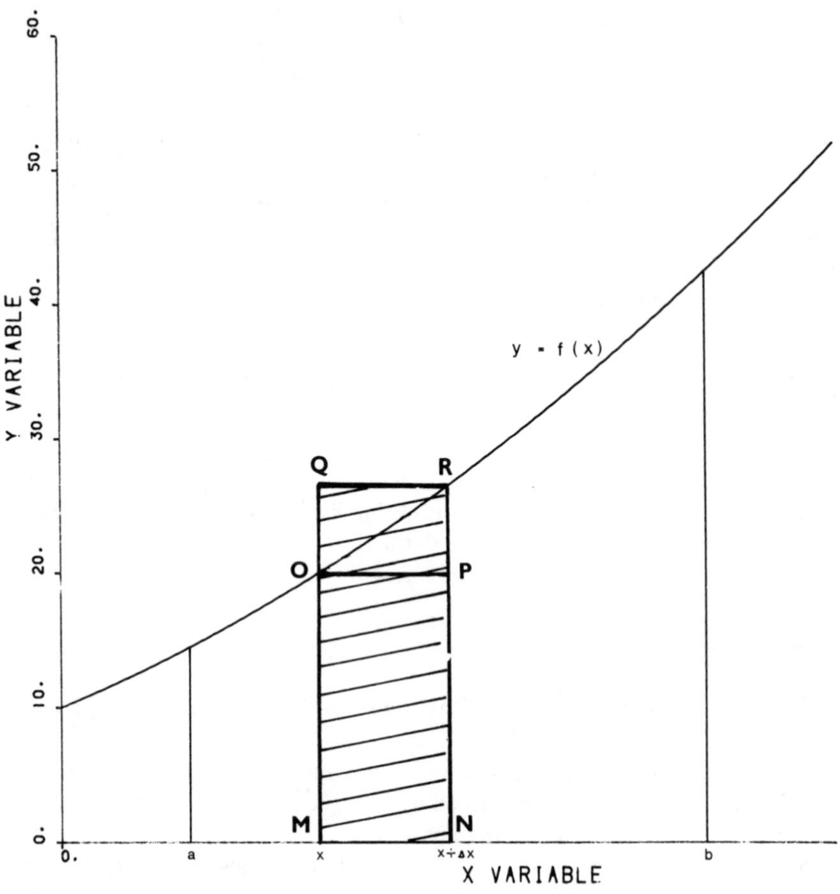

In this equation, Δx represents the width of both rectangles while y and $y + \Delta y$ represent the heights of the two rectangles. Similarly, the terms y and $y + \Delta y$ are given by the vertical distances MO and NR respectively. If we divide each term in Equation 12.14 by Δx we obtain

$$y \leq \frac{\Delta A}{\Delta x} \leq y + \Delta y$$

As Δx approaches zero, Δy also approaches zero and, as a consequence, we find that

$$\frac{dA}{dx} = y = f(x)$$

Applying the analysis of the previous section, we find that

$$\begin{aligned} A(x) &= \int f(x)dx \\ &= F(x) + C \end{aligned} \tag{12.15.1}$$

where $F'(x) = f(x)$. Concerning the value assumed by the constant of integration, we first determine $A(x)$ for $x = a$ which is given by

$$A(a) = F(a) + C \tag{12.15.2}$$

When $x = a = x_i$, the bounding ordinates coincide, which implies that there is no area under the curve (i.e., $A(a) = 0$). As a result, Equation 12.15.2 becomes

$$0 = F(a) + C$$

which implies that $C = -F(a)$. Hence, for $x_i \neq a$, the area under the curve is given by

$$A(x_i) = F(x_i) - F(a)$$

Referring to Figure 12-1, suppose we wish to find the area defined by the function $y = f(x)$, the x axis, and the ordinate $x = a$ and $x = b$, where $b > a$. In this case the desired area is given by

$$A(b) = F(b) - F(a)$$

By way of illustration, suppose we want to find the area bounded by the function $y = 4x^3$, the x axis, and the ordinates $x = 2$ and $x = 5$. In this case, $dA/dx = 4x^3$, which implies that

$$\begin{aligned} A(x) &= \int 4x^3dx \\ &= x^4 + C \end{aligned}$$

When $x = 2$, $A(x) = 0$ and, as a consequence, the constant of integration assumes the value -16. As a result, the area $A(x)$ is given by

$$A(x) = x^4 - 16 \tag{12.16}$$

and substituting $x = 5$ into Equation 12.16 yields

$$\begin{aligned} A(5) &= (5)^4 - 16 \\ &= 609 \text{ square units} \end{aligned}$$

12.4.2 The Definite Integral (R,G,C)

We may now simplify the procedures discussed in the previous section by using definite integrals. As seen earlier, we found that $F(x) = x^4$ is an integral of $4x^3$. Observe that the area bounded by $y = 4x^3$, the x axis, and the ordinates $x = 2$ and $x = 5$ may be also found by

$$\begin{aligned} A &= F(5) - F(2) \\ &= (5)^4 - (2)^4 \\ &= 609 \text{ square units} \end{aligned}$$

which is identical to the results we obtained earlier. We now define $\int_a^b f(x)dx$, which is read the definite integral of $f(x)$ between $x = a$ and $x = b$, as

$$\int_a^b f(x)dx = F(x)\Big|_a^b = F(b) - F(a) \tag{12.17}$$

Employing the concept of the definite integral, then, we may find the area defined by the function $y = f(x) > 0$, the x axis, and the ordinates $x = a$ and $x = b$ by

$$\int_a^b f(x)dx \tag{12.18}$$

As an example of Equation 12.18, suppose we want to find the area bounded by the line $y = 6x$, the x axis, and the ordinates $x = 2$ and $x = 7$. Applying Equation 12.18, we find that

$$A = \int_{2}^{7} 6x dx = 3x^2 \Big|_{2}^{7}$$
$$= 3(7)^2 - 3(2)^2$$
$$= 135 \text{ square units}$$

12.5 THE DEFINITE INTEGRAL AND SUMMATION (R,A,C)

The objective of this section is to examine the relationship between the definite integral and summation. By way of illustration, let us refer to the problem of examining the cost structure of a health care institution whose output may be measured in terms of the number of days of care provided during a specified time period. As before, we define total costs as the sum of the institution's total fixed costs (*TFC*) and the total variable costs (*TVC*) associated with a given rate of operation. We may express the relation between these cost components by

$$TC = TFC + TVC$$

As the name implies, fixed costs remain constant for *all* levels of output while variable costs increase (decrease) as the rate of activity increases (decreases). As before, marginal cost is defined as the change in total cost that results when output is increased or decreased by one unit. Also note that, since fixed costs remain invariant with respect to output, marginal costs also may be found by observing the change in total variable costs that results when output is increased or decreased by one unit.

Suppose that the functional relation between output, x, and total variable costs is given by

$$TVC = f(x) = 4x + .10x^2$$

Table 12-1 presents marginal and total variable costs that correspond to 1, · · ·, 6 days of care. In this case, it will be observed that

1. The total variable costs of \$8.40 for day two may be obtained by summing the marginal costs of \$4.10 and \$4.30.
2. The total variable costs of \$12.90 for day three may be obtained by summing the marginal costs of \$4.10, \$4.30, and \$4.50.

Table 12-1 Marginal and Total Variable Costs for Specified Rates of Output

Days of Care	Total Variable Costs		Marginal Costs
0	—		
(.5)		>	\$4.10
1	\$4.10		
(1.5)		>	\$4.30
2	\$8.40		
(2.5)		>	\$4.50
3	\$12.90		
(3.5)		>	\$4.70
4	\$17.60		
(4.5)		>	\$4.90
5	\$22.50		
(5.5)		>	\$5.10
6	\$27.60		

3. The total variable costs of \$17.60 may be obtained by summing the marginal costs of \$4.10, \$4.30, \$4.50, and \$4.70.
4. The total variable costs of \$22.50 may be obtained by summing the marginal costs of \$4.10, \$4.30, \$4.50, \$4.70, and \$4.90.
5. The total variable costs of \$27.60 may be obtained by summing the marginal costs of \$4.10, \$4.30, \$4.50, \$4.70, \$4.90, and \$5.10.

In short, the total variable costs associated with a given level of output may be found by summing over an appropriate set of marginal costs.

On the basis of this finding, it can be seen that the area under the marginal cost curve defined by the function $f(x) = 4 + .20x$ represents total variable costs. Figure 12-2 is a graphic representation of the marginal cost function confronting our institution. The function $f(x) = 4 + .20x$ is a linear marginal cost curve that has an intercept of \$4 and a slope of \$.20. Now, suppose we want to determine the total variable costs associated with providing b days of care. In this case, it is necessary to determine the area bounded by the marginal cost curve ($f(x) = 4 + .20x$), the x axis, and the ordinates $x = 0$ and $x = b$. As seen in Figure 12-2, the interval $0 < x < b$ is divided into n equal parts of length Δx. At each point of subdivision we then construct a line perpendicular to the x axis. This process divides the desired area into n strips. Even though the areas of the strips are unknown, we may approximate each strip with a rectangle whose area is known. Figure 12-3 is a representative strip and the corresponding rectangle that approximates its area. This

Figure 12-2 The Marginal Cost Function

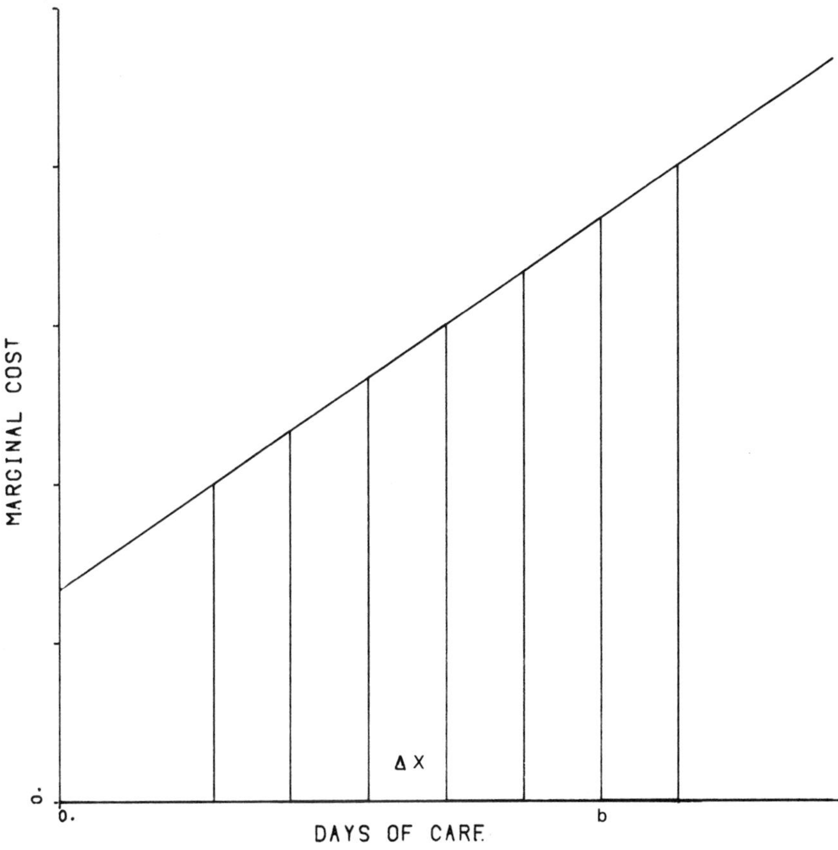

figure reveals that the area *abdfg* is common to the rectangle *acfg* and the strip *abdeg*. By inspection, we may conclude that the area of the triangle *bcd* is approximately equal to the area of triangle *def*, which implies that the area of the strip is approximated by the area of the rectangle.

Now, suppose that the representative strip is strip *i* when counting from the left. Letting $x = x_i$ represent the *midpoint* of the base of the strip, we find that the value of y at the point d is given by $y_i = f(x_i)$. As seen in Table 12-1, the values assumed by x_i are .5, 1.5, $\cdots$, 5.5 days of care. As can be verified, the area of the rectangle is found to be $y_i \Delta x$, which approximates the area of

Figure 12-3 A Representative Strip and Rectangle

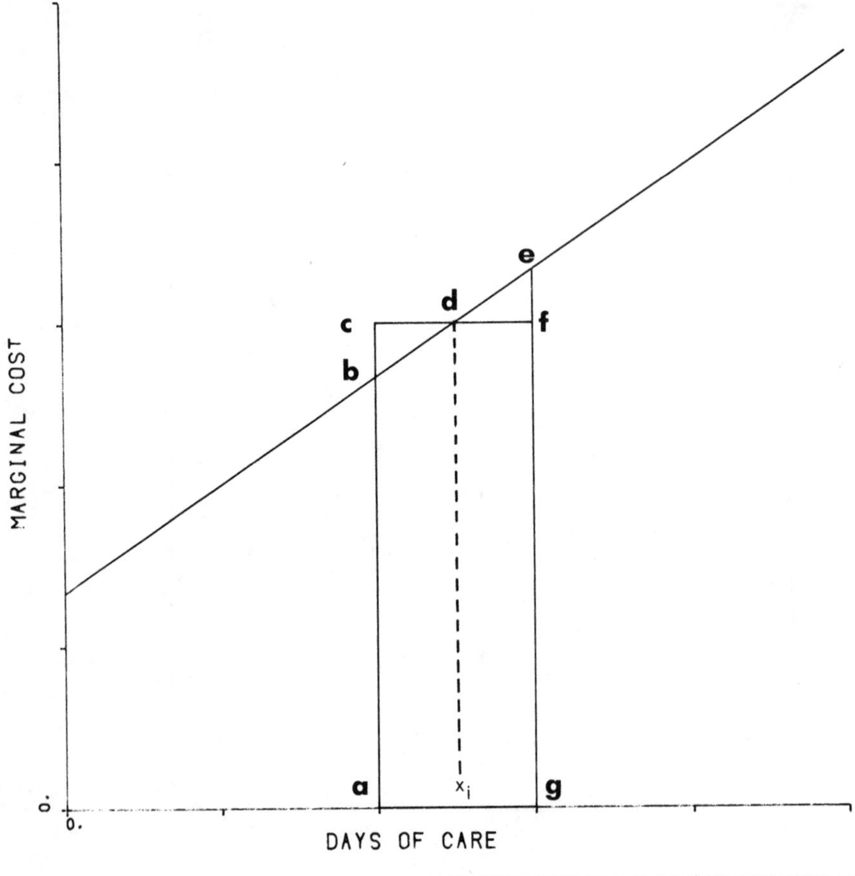

the strip. When each strip is treated similarly, the desired area may be approximated by

$$y_1\Delta x + y_2\Delta x + \cdots + y_i\Delta x + \cdots + y_n\Delta x$$

which may be written in the form

$$\sum_{i=1}^{n} y_i\Delta x$$

Suppose that the number of strips and the number of approximating rectangles are increased indefinitely so that $\Delta x \rightarrow 0$. It is evident that, by increasing the number of approximating rectangles, the desired area is given by

$$A = \lim_{n \rightarrow \infty} \sum_{i=1}^{n} y_i \Delta x = \int_0^b y dx = \int_0^b f(x) dx$$

Returning to our example, it will be recalled that $f(x) = 4 + .20x$ such that

$$\int_0^b 4 + .20x dx = 4x + .10x^2 + C$$

It will be noted that the term $4x + .10x^2$ corresponds to the total variable cost component while the constant of integration represents the fixed cost component. Now suppose that $\int_0^b 4 + .20x dx$ must pass through the point (370, 5). As before, the value assumed by the constant of integration is found by substituting $y = 370$ and $x = 5$ into $\int 4 + 20x dx$ and solving for C. Thus, we find that

$$370 = 4(5) + 10(5)^2 + C$$

and, as a result, the fixed costs of our institution amount to \$100.

To further our understanding, we define marginal revenue as the change in total revenue that results when output is increased or decreased by one unit. Thus, the marginal revenue function is simply the first derivative of the total revenue function. Assuming that total revenue is given by

$$TR = 200x$$

marginal revenue is constant and equal to \$200. In this case, we assume that the average daily charge of the institution is constant and equal to marginal revenue and that revenue is zero when no care is provided. We may now combine the marginal revenue and marginal cost functions to determine the net income or the net loss associated with specific rates of activity. Figure 12-4 presents the marginal revenue and marginal cost curves for our institution. As should be verified, marginal revenue is equal to marginal cost when 980 days of care are provided. Also note that $\int_0^{980} 200dx - \int_0^{980} (4 + .20x)dx$

yields the net loss or the net income that is generated when 980 days of care are provided. In this example, we find that

$$\int_0^{980} 200dx - \int_0^{980} (4 + .20x)dx = 200x - [4x + .10x^2 + 100] \Big|_0^{980}$$

$$= 196x - .10x^2 - 100 \Big|_0^{980}$$

$$= [196(980) - .10(980)^2 - 100]$$

$$- [196(0) - .10(0)^2 - 100]$$

$$= \$95,940 + \$100$$

$$= \$96,040$$

These calculations suggest that a net income of $96,040.00 will be earned by providing 980 days of care. When fewer than 980 days of care are provided, marginal revenue is greater than marginal costs. On the other hand, when more than 980 days of care are provided, marginal revenue is less than marginal costs. As a consequence, the provision of 980 days of care maximizes the net income of the institution.

To illustrate this point, suppose that 900 days of care are provided. In this case, the institution would earn a net income of $95,400 which is less than the net earnings associated with the provision of 980 days of care. Consider next, the provision of 1,960 days of care. As should be verified, the revenues and costs associated with this rate of activity are equal and the institution is said to break even. Finally, notice that when 2,500 days of care are provided, the institution incurs a net loss amounting to $135,000. If on an ex ante basis, the institution is expected to provide 2,500 days of care, this analysis suggests that management should investigate the possibility of: (1) increasing the average daily charge, (2) reducing variable costs, (3) reducing fixed costs, or (4) implementing a combination of these alternatives.

12.6 APPLICATIONS TO HEALTH CARE MANAGEMENT[(O,A,C)]

It is clear that the concept of integral calculus may be used to determine the net revenue or loss associated with various rates of activity. In this section we explore the applicability of integral calculus to other areas of managerial responsibility.

Figure 12-4 Marginal Revenue and Cost Curves

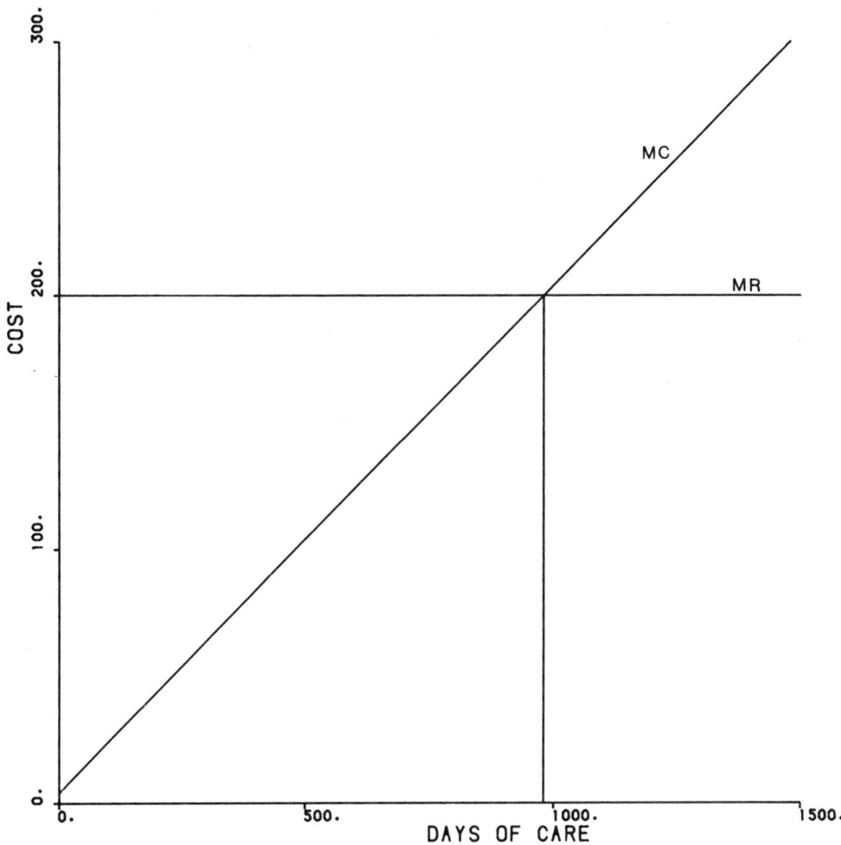

Evaluating Costs and Benefits

Suppose we are interested in evaluating the benefits and costs associated with proposals A and C relative to those generated by the status quo which we represent by the letter B. Suppose further that the annual benefits generated by the proposals and the status quo are different and may be approximated by

$$B_{at} = f(t)$$

$$B_{bt} = h(t)$$

$$B_{ct} = g(t)$$

where

B_{at} = the benefits derived from program A in year t

B_{bt} = the benefits derived from program B in year t

B_{ct} = the benefits derived from program C in year t

Suppose further that the programs will be operated for j years and that the annual benefits of these programs are measured by the monetary value of the number of lives saved, as expressed in present value equivalents. Finally, assume that the functions $f(t)$, $g(t)$ and $h(t)$ may be portrayed as in Figure 12-5. From this figure, we may conclude that the difference between the total benefits of programs A and C relative to program B are given by the lined areas of Figures 12-5(a) and 12-5(b) respectively. In this case,

$$\int_0^j f(t)dt$$

yields the total benefits derived from program A during the period 0 to j; while

$$\int_0^j h(t)dt$$

yields the total benefits derived from the status quo during the same period. Similarly,

$$\int_0^j g(t)dt$$

yields the total benefits of program C. Figure 12-5 shows us that

$$\int_0^j f(t)dt - \int_0^j h(t)dt$$

and

$$\int_0^j g(t)dt - \int_0^j h(t)dt$$

give the difference in the total benefits of programs A and C, respectively, relative to the status quo.

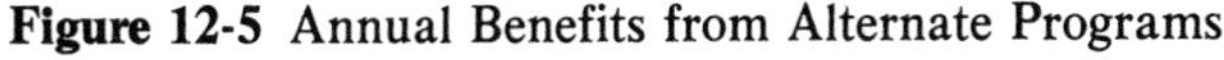

Figure 12-5 Annual Benefits from Alternate Programs

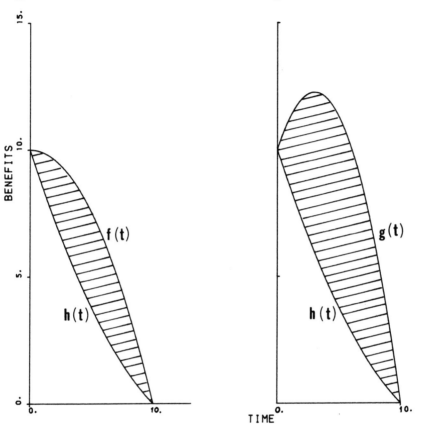

Now, suppose that the annual costs of programs A, B, and C are given by

$$TC_{at} = \theta(t)$$
$$TC_{bt} = \gamma(t)$$
$$TC_{ct} = \alpha(t)$$

where

TC_{at} = the costs of program A in year t,

TC_{bt} = the annual costs of program B in year t,

TC_{ct} = the annual costs of program C in year t.

Also assume that we may display the cost functions of these programs as seen in Figure 12-6. In this case, we find that

$$\int_0^j \theta(t)dt$$

$$\int_0^j \gamma(t)dt$$

and

$$\int_0^j \alpha(t)dt$$

represent the total cost of operating programs A, B, and C, respectively. Further

$$\int_0^j \theta(t)dt - \int_0^j \gamma(t)dt$$

and

$$\int_0^j \alpha(t)dt - \int_0^j \gamma(t)dt$$

represent the difference in the total cost of operating programs A and C, respectively, relative to B.

We may now combine these two sets of results and construct the ratios

$$R_a = \frac{\int_0^j f(t)dt - \int_0^j h(t)dt}{\int_0^j \theta(t)dt - \int_0^j \gamma(t)dt}$$

$$R_c = \frac{\int_0^j g(t)dt - \int_0^j h(t)dt}{\int_0^j \alpha(t)dt - \int_0^j \gamma(t)dt}$$

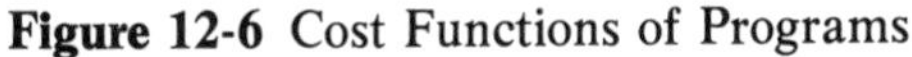

Figure 12-6 Cost Functions of Programs

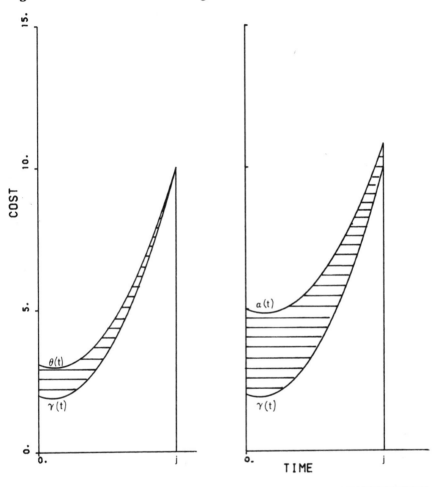

Subject to the condition that none of the programs is dominated by the others, and that the costs of each program are less than or equal to budget constraint, an ordinal ranking of these ratios provides the basis for selecting the project that maximizes the net benefits obtainable from the set of available resources.

Estimating the Future Use of Service

Consider a slightly different situation in which we are interested in estimating the number of admissions in period $t + 1$. Suppose further that the admission rate, A_t, during the period $t = 0$ to $t = t$ is available and provides the basis for estimating the function

$$A_t = f(t)$$

which is displayed in Figure 12-7. In this case the estimated number of admissions for the period $t + 1$ is found to be

$$\int_0^{t+1} f(t)dt - \int_0^t f(t)dt$$

Figure 12-7 Estimation of Number of Admissions

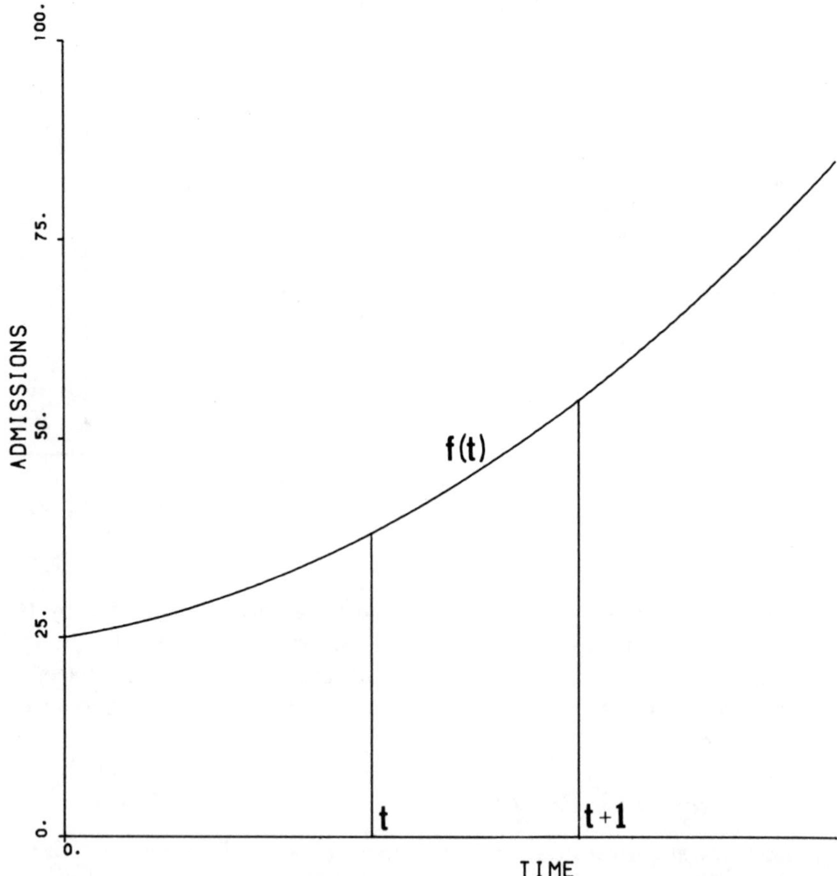

Also note that we might let

$$N_t = g(t) \quad \text{and} \quad S_k = h(t)$$

represent the functional relation between the annual use of stay-specific services (N) and ancillary services (S_k) with respect to time. Estimates concerning the values assumed by these components of care for a specified period may also be obtained using these techniques.

12.7 SUMMARY (O,A,F)

As seen in Chapter 10, the derivative is a mathematical technique that allows management to determine the responsiveness of one variable to changes in another. Similarly, the partial derivative allows us to determine the responsiveness of the dependent variable to changes in one of several independent variables while holding other factors constant. It is in this sense, then, that differential calculus may be regarded as the study of change.

In this chapter, we examined the indefinite integral as a technique of reversing the process of differentiation. In turn, the definite integral was used as a mathematical tool for determining the area under a curve defined by the function $f(x)$. As seen above, integral calculus is a mathematical technique that may be employed in deriving information on which managerial decisions might be based. In Chapter 15 we employ the definite integral when we describe the concept of probability as applied to a continuous variable.

Problems for Solution

1. Suppose that the marginal revenue (dTR/dx) is given by

$$MR = 100$$

Find the total revenue earned if 600 units of service are provided.

2. Suppose that the marginal cost function (dTC/dx) is given by

$$MC = .2x + 5$$

If this function must pass through the point (60, 860), find the total cost of providing:

a. 800 units of care,
b. 2,000 units of care,
c. 600 units of care.

3. Referring to Problems 1 and 2, what is the net income (or loss) associated with:

a. 1,800 units of care,
b. 200 units of care,
c. 1,200 units of care.

4. Find the antiderivative of

$$f(x) = x^3 - 7x^2 + 3x - 10$$

$$f(x) = \frac{2}{3}x^4 + 3x^2 - 4$$

$$f(x) = \frac{3}{2}x - 6x^2 + 4$$

5. Suppose that

$$f(x) = x^2 - 9x + 10$$

and

$$q(x) = 2x^2 - x + 25$$

represent the marginal rate of change in the use of care relative to age for females and males respectively. Find the area between the two curves. What interpretation do you attach to this finding?

6. Suppose we want to introduce proposal A that yields annual savings given by

$$S(t) = 200 - t^2$$

where t is the number of years of operation; the annual costs of proposal A are given by

$$C(t) = t^2 + 11/4\ t$$

a. for how many years should we adopt the proposal?
b. what are the net savings for the first year?
c. what are the total net savings?

7. If a machine provides x thousand units of service, the rate of repair costs as expressed in terms of dollars per service is given by

$$R(x) = .03x^2$$

Find the total repair cost if we produce

a. 200,000 units of service,
b. 700,000 units of service,
c. 100,000 units of service.

8. Our hospital is considering a new method of treatment. Suppose we know that the rate of savings $S(t)$ is given by

$$S(t) = 500t + 4$$

where t is the number of years we use the new method. What is the total savings during the first

a. 2 years?
b. 10 years?
c. 12 years?

9. Suppose the number of admissions per year $A(t)$ is given by

$$A(t) = 200 + .02t^{1/2}$$

How many admissions should we expect in years 1, 2, and 3?

10. Referring to Problem 8, suppose that required capital equipment costs amount to $24,000. Approximately when will the new method result in savings that are equal to these capital costs?

11. The annual benefits derived from proposal A are given by

$$B(t) = 200 + .01t^{3/2}$$

What are the total benefits of

a. the first two years of operation?
b. the first six years of operation?
c. the first eight years of operation?

12. The annual costs of operating the proposal of Problem 11 are given by

$$C(t) = 100 + .02t^{1/2}$$

What are the net benefits after the first four years of operation?

Part IV

Probability

Chapter 13

Sets and Operations on Sets

Objectives

After completing this chapter you should be able to:

1. Define the concepts of:
    - set, subset, element, universe
    - finite and infinite sets
    - set equality
    - disjoint sets
    - complement;
2. Use the proper symbols for defining sets;
3. Draw Venn diagrams representing set operations including:
    - union, intersection
    - disjoint sets
    - complements;
4. Define DeMorgan's Laws and represent them symbolically;
5. Use a tabular presentation of data to identify subsets, unions, and intersections and explain their meaning;
6. Give examples of the use of set concepts in health care management, planning, or research.

Chapter Map

The sections comprising this chapter may be summarized as follows:

Section Number	Required Reading	Optional Reading	Generic Development	Application to Management	Fundamental Principles	Complex Material
	(R)	(O)	(G)	(A)	(F)	(C)
13.1	x		x	x	x	
13.2	x		x		x	
13.3	x		x	x	x	
13.4	x		x		x	
13.5		x		x	x	
13.6		x	x		x	
13.6.1		x	x		x	
13.6.2		x	x		x	
13.6.3		x	x		x	
13.6.4		x	x		x	
13.6.5		x	x		x	
13.7		x		x	x	

13.1 INTRODUCTION(R,A,G,F)

In this part, the focus of analysis shifts from a discussion of algebra and the formalities of calculus to an examination of probability as applied to discrete and continuous variables. It is well known that most managerial decisions are reached in an environment of uncertainty. For example, before deciding to accept a proposal to construct a new wing, health care management must assess the probability of maintaining an adequate occupancy rate. Similarly, when assessing the desirability of investing funds in a given marketable security, management must assess the probability of earning a net gain or incurring a net loss. In each of these situations the concept of probability may be employed to capture an uncertain environment quantitatively.

Since an understanding of probability is enhanced by first examining sets and operations on sets, we begin with this area of mathematics. In turn, the basic principles described in this chapter provide the basis for our analysis of discrete probability that is presented in Chapter 14. Finally, this part concludes with a discussion of continuous probability in which the definite integral described earlier is of central importance.

13.2 DEFINITIONS(R,G,F)

A set is defined as a collection of objects or elements that represent the phenomenon of interest such as patients, beds, or the employees of the institution. Frequently, it is convenient to divide sets into identifiable components, the component parts of which are called subsets. The set of hospital beds consists of subsets of medical, surgical, gynecological, . . . beds. An *element* of any set is one of the objects in the collection. A *finite set* is one in which the objects can be counted, while an *infinite set* is one where the number of objects is beyond counting.

In the real world, then, sets are all finite, but some are so large that they might just as well be considered infinite. The set of natural numbers, called N, is an example of an infinite set. The set of pieces of capital equipment in a hospital is a finite set. A *null set* (or *empty set*) has no members. Some sets are null because they are impossible, such as the set of female fathers. Others are null because there are no objects that meet the conditions for membership in the set. For example, there might be no patients under age 30 in hospital with cancer of the colon. As another example, a hospital might have no employees with more than 40 years of experience.

A set can be defined by listing all of its members, such as the inventory of capital equipment. This is referred to as the *roster* or *enumeration* method of set *definition*. A set also can be defined by specifying a rule or several rules for inclusion that members of the set must satisfy. This is called the *defining method.*

13.3 SYMBOLS USED IN REPRESENTING SETS(R,G,A,F)

Certain symbols are used as shorthand for the treatment of sets in mathematics. Usually, braces of the form { } are used to indicate a set. For example, the set of admitting clerks may be represented by

{Joan K, John L, Mabel B, Arthur D}

This is a specific example of the enumeration method that in general is expressed as follows:

{list of members}

For the defining method a variable and a condition statement, enclosed in braces, are separated by a vertical bar called a *solidus*, |, that is read as *such that*. In general, such a statement assumes the form:

$\{x \mid \text{condition statement}\}$

For example, consider all individuals in a geographical area at a given moment in time. In Figure 13-1 it should be noted that each individual is either a patient or a nonpatient and we might employ this distinction to divide the population into two groups or sets. Similarly, all patients are either male or female and we may use gender to divide the patient population into two groups or subsets. In this case, we might represent the set of male patients symbolically by

$$\{x \mid x \in \text{patients}, x \in \text{males}\}$$

where $\in$ is the Greek letter epsilon and means "is an element of." The foregoing statement is read as "the set of all x such that x is an element of the set of patients and x is an element of the set of males." Employing similar notation, we represent the null set by the symbol $\emptyset$ or empty braces, { }.

Certain common sets are given letter names. The natural numbers are sometimes called N, the integers I, and the reals R. The set of natural numbers less than or equal to 7 can be represented by:

$$A = \{a \mid a \in N, a \leq 7\}$$
$$= \{1, 2, 3, 4, 5, 6, 7\}$$

Sets may be equal to each other, and $A = B$ if and only if every element of A is also an element of B and vice versa. This condition defines set equality. Thus if,

$$A = \{1, 2, 3, 4, 5, 6, 7\}$$

while

$$B = \{5, 4, 3, 1, 6, 7, 2\}$$

we find that

$$A = B$$

On the other hand, if

$$C = \{1, 2, 3\}$$

our earlier discussion allows us to assert that

$$A \neq C \quad \text{and} \quad B \neq C$$

The symbol $\subset$ means *subset* and is read as *"is included in."* When we also want to recognize the possibility of set equality, the symbol $\subseteq$ is employed. Referring to the examples above, C is a *proper subset* of A and is written

$$C \subset A$$

However, B also is equal to A as well as being a subset of A, which implies that the possibility of equality must be recognized as follows:

$$B \subseteq A$$

These symbols are interpreted in a fashion similar to the interpretation of the greater than ($>$) and less than ($<$) symbols described earlier. For example, if $P \subset W$, at least one element of W cannot be found in P.

The set of all potential objects that might be considered for study is called the *universe* and is represented by the symbol U. For example, the set of all patients in a hospital might be regarded as "the universe" for a study. Then the set of male patients is a subset of the universe.

The complement of a subset D of the universe U consists of all members of U that are *not* members of D. The complement of D is written as D', and can be read either as *D complement, D not* or as *not D*. Symbolically this is written as

$$D' = \{x \mid x \in U, x \notin D\}$$

Figure 13-1 An Example of Sets and Subsets

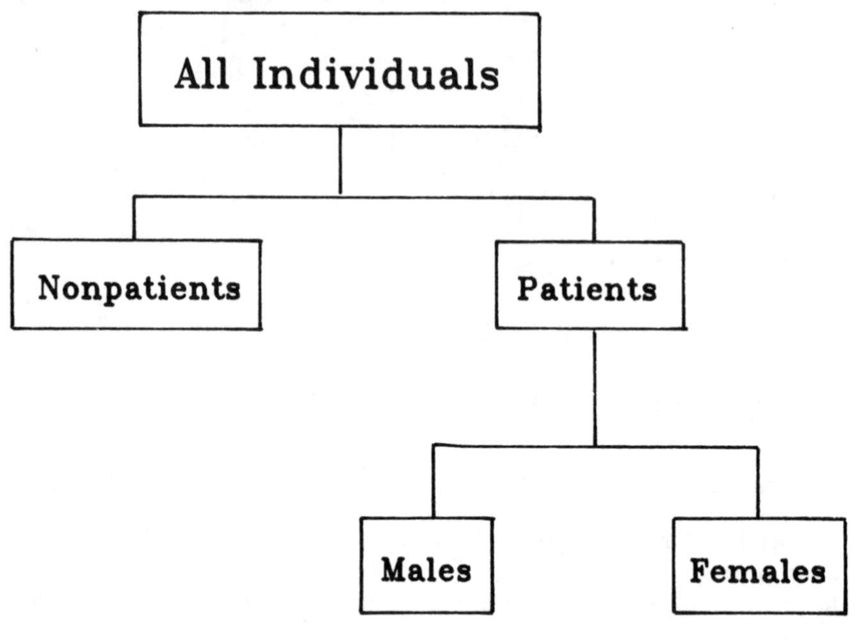

For instance, if U is the set of employees and D is the subset of *male* employees, then D' is the subset of *female* employees. Notice that if:

$$A = \{1, 2, 3, 4, 5, 6, 7\} = U$$

and $$C = \{1, 2, 3\}$$

then $$C' = \{4, 5, 6, 7\}$$

The *union* of two sets is the set that consists of elements from *either or both* sets and is represented by the symbol $\cup$. Thus, referring to the examples previously, we find that

$$A \cup B = \{1, 2, 3, 4, 5, 6, 7\} \cup \{5, 4, 3, 1, 6, 7, 2\}$$

$$= \{1, 2, 3, 4, 5, 6, 7\}$$

$$A \cup C = \{1, 2, 3, 4, 5, 6, 7\}$$

If $$X = \{2, 4, 6, 8, 10\}$$

then $$A \cup X = \{1, 2, 3, 4, 5, 6, 7, 8, 10\}$$

and $$X \cup C = \{1, 2, 3, 4, 6, 8, 10\}$$

Notice that the union of complements is the universe

$$C \cup C' = U = A$$

and Males ∪ Females = All People

The union operation is sometimes called the *logical OR*.

The *intersection* of two sets is the set that consists of elements from both sets. It is represented by the symbol ∩, and is called the *logical AND*. Thus, referring to the sets A and B just introduced, we find that

$$A \cap B = A = B$$

since the two sets are equal. Referring next to the sets A, C, and X, we find that the intersections A and X, A and C, and C and X are given by

$$A \cap X = \{2, 4, 6\}$$

$$A \cap C = \{1, 2, 3\}$$

$$C \cap X = \{2\}$$

respectively. Also notice that the intersection of complements must be a null set. For example,

$$C \cap C' = \emptyset$$

while

Males ∩ Females = ∅

Similarly, the physiological differences between males and females allow us to assert that

Gynecological patients ∩ prostate gland patients = ∅

which, of course, contains no members.

Finally, the union represents the sum of the elements in the sets that are united and each element is counted only once. Conversely, the intersection represents the *overlap* of sets and contains elements that are common to all sets in the intersection.

13.4 VENN DIAGRAMS(R,G,F)

Sets often are represented by *Venn* or circle *diagrams* to portray the Universe, Unions, Intersections, and Complements. In Figure 13-2(a) the rectangle represents the universe, while the circle represents the set of interest, D. The lined area of this rectangle represents D complement. Similarly, the lined area of Figure 13-2(b) represents the union of A and B while the lined area in Figure 13-2(c) indicates the intersection of A and B. Finally, in Figure 13-2(d), the intersection of R and S is null, and they are called *disjoint* or *mutually exclusive*.

13.5 APPLICATION OF SET THEORY(O,A,F)

The current or potential employees of any health care organization usually have different skills, areas of expertise, amounts of experience, etc. When selecting individuals to occupy new and existing positions, reaching decisions concerning promotions, or forming task forces, it frequently is necessary to identify individuals who possess specific characteristics or attributes.

For example, suppose that management is seeking a department head who possesses specific and well-defined qualifications. More specifically, suppose that the individual sought must:

1. be a female,
2. have at least five years of managerial experience,
3. be between 40 and 45 years of age, and
4. possess at least a graduate degree in administration or related discipline.

Suppose further that the information in Table 13-1 is available to management. Using the applicant identification numbers presented in column (1) of the table, we find that the individuals who satisfy our requirements may be represented by the set

$$S = \{03, 10, 19\}$$

In this case, notice that the use of sets has reduced the list of potential applicants from 25 to 3 individuals who possess the required qualifications. Hence, we now may focus on these three and, after obtaining any additional information that might be required, we may reach a well-informed decision concerning the individual who ought to occupy the vacant position.

Figure 13-2 Venn Diagrams of Sets

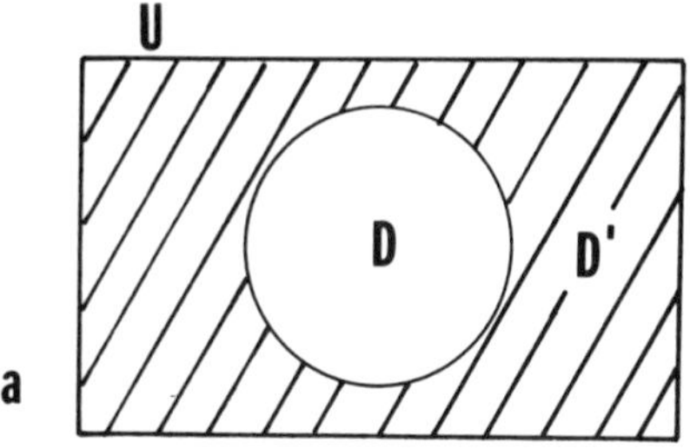

The Universe, *A* Set, and Its Complement

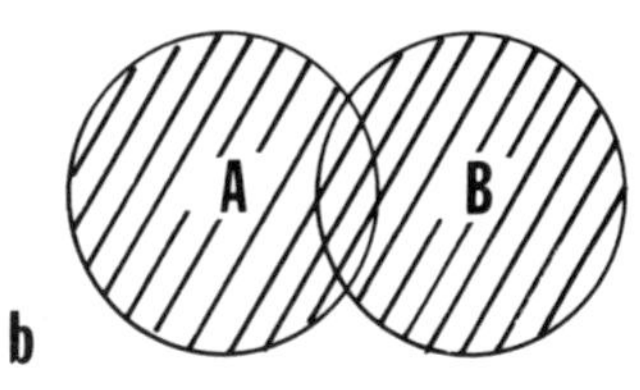

The Union of *A* and *B*

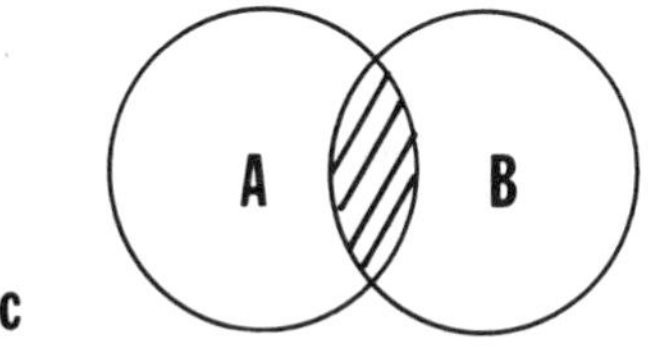

The Intersection of *A* and *B*

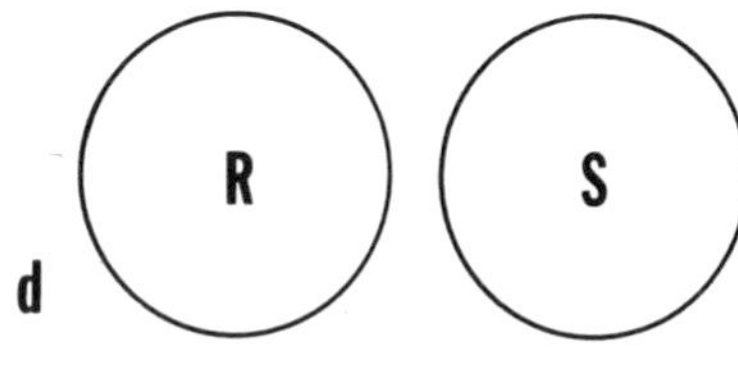

Disjoint Sets

13.6 LAWS OF SET OPERATION(O,G,F)

The primary purpose of this section is to consider the basic laws of set operation. In general, we shall find that the basic laws of algebra described earlier also are applicable to the set operations of union and intersection. For purposes of future illustration, we may define the following sets by letting

$$F = \{1, 2, 3, 4, 5\}$$

$$G = \{3, 4, 5, 6, 7, 8\}$$

$$H = \{2, 4, 6, 8, 10\}$$

13.6.1 The Commutative Law(O,G,F)

Similar to our discussion in Chapter 4, the commutative law permits us to reverse the order of unions and intersections. For example, when considering the union of *A* and *B*, we find that

$$A \cup B = B \cup A$$

Referring to our numeric example, it is easy to verify that

$$F \cup G = \{1, 2, 3, 4, 5, 6, 7, 8\} = G \cup F$$

Similarly, when considering the intersection of A and B, we obtain

$$A \cap B = B \cap A$$

Referring to our numeric examples, an application of the commutative law yields

$$F \cap G = \{3, 4, 5\} = G \cap F$$

Table 13-1 Attributes of Potential Applicants

Applicant Identification Number	Sex (M = male F = female)	Years of Related Experience	Age (in years)	Degree in Administration or Related Discipline
01	M	2	26	M.B.A.
02	M	7	37	B.A.
03	F	12	41	M.H.A.
04	M	1	24	B.A.
05	M	3	28	B.A.
06	F	6	30	B.A.
07	M	7	27	M.B.A.
08	F	4	26	B.A.
09	F	3	32	M.H.A.
10	F	11	43	D.B.A.
11	M	14	38	M.H.A.
12	M	22	43	B.A.
13	M	18	52	M.B.A.
14	M	10	41	M.B.A.
15	F	3	29	B.A.
16	F	2	31	B.A.
17	M	14	47	B.A.
18	M	6	35	M.A.
19	F	9	43	M.A.
20	M	8	42	B.A.
21	F	2	29	B.A.
22	M	3	27	M.A.
23	M	1	35	Ph.D.
24	M	2	23	M.A.
25	M	1	26	M.B.A.

13.6.2 The Associative Law(O,G,F)

The associative law allows us to group pairs of sets by parentheses. For example, when considering the *union* of P, Q, and R, we find that

$$P \cup Q \cup R = P \cup (Q \cup R) = (P \cup Q) \cup R$$

Referring to our example, notice that an application of the associative law yields

$$F \cup G \cup H = (F \cup G) \cup H = F \cup (G \cup H) = \{1, 2, 3, 4, 5, 6, 7, 8, 10\}$$

The associative law may be applied also to the intersection of P, Q, and R. In general, we find that

$$P \cap Q \cap R = P \cap (Q \cap R) = (P \cap Q) \cap R$$

Referring to our numeric example, an application of the associative law yields

$$\begin{aligned}(F \cap G) \cap H &= \{3, 4, 5\} \cap \{2, 4, 6, 8, 10\} \\ &= \{4\}\end{aligned}$$

Also notice that

$$\begin{aligned}F \cap (G \cap H) &= \{1, 2, 3, 4, 5\} \cap \{4, 6, 8\} \\ &= \{4\}\end{aligned}$$

which, of course, agrees with the results already obtained. At this point, the reader should verify that the associative law may be applied also to unions.

13.6.3 The Distributive Law(O,G,F)

The distributive law allows us to distribute unions across intersections and intersections across unions. For example, an application of the distributive law yields

$$A \cup (B \cap C) = (A \cup B) \cap (A \cup C)$$

while

$$A \cap (B \cup C) = (A \cap B) \cup (A \cap C)$$

Referring to our numeric example, notice that

$$F \cap (G \cup H) = \{1, 2, 3, 4, 5\} \cap \{2, 3, 4, 5, 6, 7, 8, 10\}$$
$$= \{2, 3, 4, 5\}$$

Applying the distributive law, we obtain

$$(F \cap G) \cup (F \cap H) = \{3, 4, 5\} \cup \{2, 4\}$$
$$= \{2, 3, 4, 5\}$$

which, as before, agrees with our earlier results.

13.6.4 DeMorgan's Laws(O,G,F)

Thus far, we have applied the commutative law, the associative law, and the distributive law to the set operations of union and intersection. We now examine two additional ones that are called DeMorgan's Laws. The first of these states that

$$(A \cap B)' = A' \cup B'$$

while the second states that

$$(A \cup B)' = A' \cap B'$$

Recalling that

$$F = \{1, 2, 3, 4, 5\}$$

while

$$G = \{3, 4, 5, 6, 7, 8\}$$

we now define the universe of interest by

$$U = \{x \mid x \in N, x \leq 10\}$$

Applying the first of DeMorgan's Laws, we obtain

$$(F \cap G)' = \{3, 4, 5\}'$$
$$= \{1, 2, 6, 7, 8, 9, 10\}$$

Similarly, we find that

$$F' \cup G' = \{6, 7, 8, 9, 10\} \cup \{1, 2, 9, 10\}$$
$$= \{1, 2, 6, 7, 8, 9, 10\}$$

Consequently, these results imply that

$$(F \cap G)' = F' \cup G' = \{1, 2, 6, 7, 8, 9, 10\}$$

At this point, the reader should verify the second of the two laws.

13.6.5 The Laws of Complementation and the Idempotent Laws(O,G,F)

The laws of complementation state that

$$A \cup A' = U$$

and that

$$A \cap A' = \emptyset$$

which are intuitively obvious. On the other hand, the idempotent (eee.dem.pō.tent) laws assert that

$$A \cup A = A$$

while

$$A \cap A = A$$

which also are intuitively plausible.

13.7 AN ANALYSIS OF EMERGENCY ROOM USE(O,G,F)

Suppose that in developing estimates of the resources required to provide ambulatory care, our health care institution is interested in examining the use of the services that are provided by our emergency room. To demonstrate the usefulness of the concepts developed in this chapter, refer to Figure 13-3, where it is assumed that the universe for our analysis consists of the population of the city, say 500,000. The set A consists of the 100,000 patients who visited the emergency room. Similarly, we let B represent the 8,000 patients who presented true emergencies and C correspond to the 20,000 who were admitted to our hospital. Further, assume that 6,000 patients presented true emergencies and visited the emergency room. Of these 6,000, we assume that 1,000 were admitted to the hospital. Assume that an additional 1,500 patients presented valid emergencies and were admitted directly to the hospital without visiting the emergency room. We also assume that patients are admitted to the hospital from our emergency room only if they present a true emergency.

In Figure 13-3, notice that the three circles represent the sets A, B, and C defined earlier. In this case we find that

$$(A \cap B')$$

represents the number of patients who visited the emergency room and did not present a true emergency. Consequently, the use of service by these patients might have been provided in an alternate and perhaps less costly setting.

Similarly, notice that

$$A \cap B \cap C$$

represents the number of patients who

1. presented a true emergency,
2. visited the emergency room, and
3. were admitted to the hospital.

Similarly, the intersection of A and B represents the number of patients who

1. presented a true emergency,
2. visited the emergency room.

In both cases, we might regard the use of service by these patients as appropriate and required when viewed from a clinical or medical perspective. Finally, notice that

$$(A \cup B \cup C)'$$

represents the number of patients who did not encounter an emergency condition and did not visit the emergency room. At this point in the analysis, the reader should be able to express the meaning of areas W, X, Y, and Z, both in words and in terms of the symbols A, B, and C.

Figure 13-3 Venn Diagram of Emergency Department Visits and Hospital Admissions

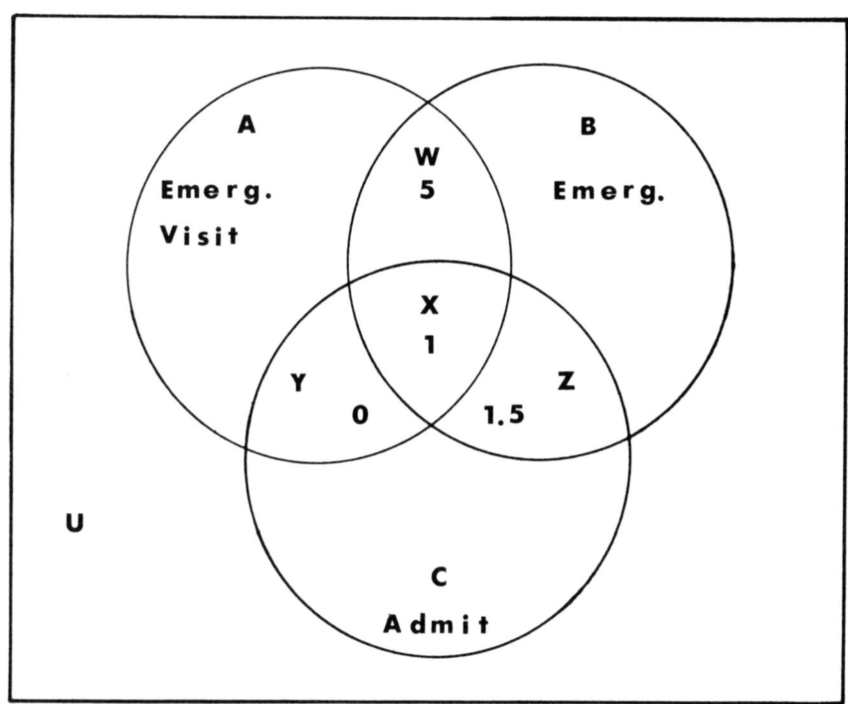

KEY:

Universe—All persons in the city	500,000
A—Persons Visiting Emergency Department	100,000
B—Persons with Emergencies	8,000
C—Persons Admitted to Hospital	20,000

The information in our example also may be displayed in tabular form, as in Table 13-2. Such a table is called a contingency table. Only certain pieces of information are essential and have been included in the table. The number of patients appearing in each of the remaining cells of the table may be obtained by subtraction and addition. At this point, the reader should complete the table so as to ensure an understanding of this approach. Also notice that a contingency table is an additional method of displaying relationships among variables or phenomena of interest.

Table 13-2 Emergency Department Utilization and Hospital Admission

	Emergencies (B)			Nonemergencies (B')			
	Admitted (C))	Not Admitted (C')	Sub Total	Admitted (C)	Not Admitted (C')	Sub Total	TOTAL
Emergency Visit (A)	1	5		0			100
No Emergency Visit (A')	1.5						
TOTAL			8	17.5			500

Problems for Solution

1. Reproduce Figure 13-3 and write in the symbolic definition of each segment of the diagram.
2. Determine the numbers that belong in each segment.
3. Complete Table 13-2.
4. Write the symbolic identification of each position of the table.
5. What is the symbolic identification of the left half of the table? Of the bottom half? Of the top right quarter?
6. Rearrange the table to highlight the Admitted versus Not Admitted dichotomy.
7. Let $U = \{x \mid x \in \text{alphabet}\}$
   Let $A = \{$a, b, c, d, e, f, g, h, x, y, z$\}$
   Let $B = \{$a, c, e, g, m, p, q$\}$
   Let $C = A'$
   Let $D = B'$

   Find the following:

   a. $A \cap B$
   b. $A \cup B$
   c. C
   d. D
   e. $B \cap C$
   f. $A \cap D$
   g. $A \cup C$
   h. $A \cup (B \cap C)$
   i. $(A \cup B)'$
   j. $(A \cup C)'$
   k. $B \cap D$
   l. $(B \cap D)'$
   m. $A' \cap B'$
   n. $D' \cup B'$
   o. $(A \cap B)'$
   p. $A \cap (B \cup C)$

8. Personnel selection categories have been established as indicated in the following list.

   A: Sex: M, F
   B: Age: yrs
   C: Education: 1 = Grade, 2 = High, 3 = Trade, 4 = University, 5 = Advanced
   D: Nursing: 1 = Aide, 2 = Diploma, 3 = Degree, 4 = R.N., 5 = Master's

E: Position: 0 = Nonsupervisory, 1, 2, 3 = Supervisory,
4, 5, 6 = Management
F: Special Nursing: 1 = CPR, 2 = Surgical, 3 = ICU, 4 = Dialysis
G: Other Special: 1 = Management, 2 = Computer,
3 = Management engineering

Some of the persons who have been categorized for the personnel selection system are:

CATEGORIES

INITIALS	A	B	C	D	E	F				G		
						1,	2,	3,	4	1,	2,	3
A. B. C.	F	25	2	2	0	x	—	—	—	—	x	—
J. B. R.	F	30	3	0	1	—	—	—	—	x	x	—
C. D. J.	M	31	4	3	2	—	x	—	x	—	—	—
R. J. M.	F	27	4	4	3	x	x	—	—	x	—	—
M. A. F.	F	33	2	4	4	x	x	x	x	x	—	—
P. E. R.	M	22	4	0	0	—	—	—	—	—	x	—
D. A. J.	M	39	5	4	4	—	—	x	x	x	x	—
M. E. M.	F	41	5	3	4	—	x	x	—	—	x	x

Find the following sets:

a. All persons with management training.
b. All persons with management and computer training, or with management engineering, in a supervisory level position.
c. R.N. with CPR training.
d. Nonnursing background.
e. Nurses with diploma qualifications or better, with management or computer training.
f. Supervisory level personnel.

Chapter 14

Discrete Probability and Mathematical Expectation

Objectives

After completing this chapter, you should be able to:

1. Define probability;
2. Define random experiments, basic outcomes, sample spaces, and an event;
3. Use Venn diagrams to represent unions, complements, intersections;
4. Use the rules of addition and multiplication to calculate desired probabilities;
5. Calculate and interpret conditional probabilities;
6. Use the principles of mathematical expectation in reaching decisions.

Chapter Map

The sections comprising this chapter may be summarized as follows:

Section Number	Required Reading	Optional Reading	Generic Development	Application to Management	Fundamental Principles	Complex Material
	(R)	(O)	(G)	(A)	(F)	(C)
14.1	x		x	x	x	
14.2	x		x		x	
14.2.1	x		x		x	
14.2.2	x		x		x	
14.2.3	x		x		x	
14.2.4	x		x		x	
14.2.5	x		x		x	
14.3	x		x	x	x	
14.3.1	x		x		x	
14.3.2	x		x	x	x	
14.3.2.1	x		x	x	x	
14.3.2.2	x		x	x	x	
14.3.2.3	x		x	x	x	
14.3.3	x		x	x	x	
14.3.4	x		x	x	x	
14.4	x		x	x	x	
14.5		x		x	x	
14.5.1		x		x	x	

In this chapter we employ the notation of set operations and the fundamentals of set theory in an introductory discussion of probability. Specifically, this chapter is devoted to a discussion of probability as applied to discrete random variables and to problems involving statements of mathematical expectation. We consider first the meaning of probability as applied to discrete random variables and then discuss the rules of probability. This chapter concludes with an examination of the usefulness of mathematical expectation in reaching managerial decisions.

14.1 THE MEANING OF PROBABILITY[R,G,A,F]

Probabilities may be interpreted as the percentage or the proportion of the time that an event will occur in a repeated series of trials or in the long run. Applying this interpretation, we may define probability in terms of relative frequencies or, to be more precise, in terms of limits of relative frequencies. When dealing with finite populations of size n, the relative frequency of an event that occurs x times out of n is simply x/n. Obviously, if we multiply the relative frequency by 100 we obtain the percentage of the time that a given event will occur. Both the relative frequency x/n and the percentage $(x/n)100$ may be viewed as probability statements concerning the likelihood that a given event will occur. Although it is possible to define probability in subjective terms, we limit our discussion in this chapter and the next to the objective definition of probability as described above.

14.2 BASIC CONCEPTS OF PROBABILITY[R,G,F]

That probability may be defined in terms of relative frequencies or percentages should be clear from the foregoing discussion. In this section we extend our understanding of probability by examining the fundamental concepts that provide the basis for the development of probability theory. Specifically, our objective is to describe random experiments, basic outcomes, sample spaces, and events.

14.2.1 Random Experiment[R,G,F]

A random experiment may be defined as any process that leads to one of several possible results. For example, the next patient we admit to our hospital will be either a male or female of some specific age. As another example, an outstanding receivable may be transformed into a cash receipt during the first, second, third, or fourth month after care has been provided. In these examples, the objective of the experiment usually dictates the characteristic or phenomenon of interest.

14.2.2 Basic Outcomes(R,G,F)

The results obtained from a random experiment are referred to as basic outcomes. For example, suppose we are interested in the insurance status of a given set of patients. Assume that one of the questions patients were asked was "Do you have any form of health insurance?" In this case, there would be two possible outcomes: yes, and no.

14.2.3 Sample Space(R,G,F)

The set of all possible outcomes in a random experiment is called the sample space. The basic outcomes comprising a given sample space often are characterized as being mutually exclusive and collectively exhaustive. As a consequence, the points comprising the sample space represent all of the possible results of the experiment, and the outcome of a random experiment is characterized by one and only one of the basic outcomes.

14.2.4 Events(R,G,F)

In many problems, it is necessary to focus on results that are broader than the basic outcomes in the sample space. We may define an event as consisting of one or more of the basic outcomes that comprise the sample space. All of the events with which probability theory is concerned are subsets or portions of sample spaces.

14.2.5 Unions, Intersections, and Complements(R,G,F)

Recalling that events consist of one or more basic outcomes comprising the sample space, we turn to the union of events, the intersection of events, and complementary events.

If E_i and E_j are two events, we define their intersection, which is symbolized by $E_i \cap E_j$, as the event that consists of all basic outcomes common to *both* E_i and E_j. This relation is shown in the Venn diagram in Figure 14-1 where the shaded area corresponds to $E_i \cap E_j$.

As before, assume that we are interested in the two events E_i and E_j. We may define the union between these two events as all basic outcomes contained in *either* E_i or E_j or in both E_i and E_j. The union between E_i and E_j is symbolized by $E_i \cup E_j$ and represented by the lined area of Figure 14-2.

Finally, we define the complement of the event E_i as the subset that consists of all basic outcomes in the sample space that are not contained in E_i. Figure 14-3 presents the event E_i and its complement E_i'. The lined area represents the complement of the event E_i.

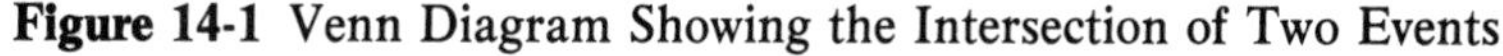

Figure 14-1 Venn Diagram Showing the Intersection of Two Events

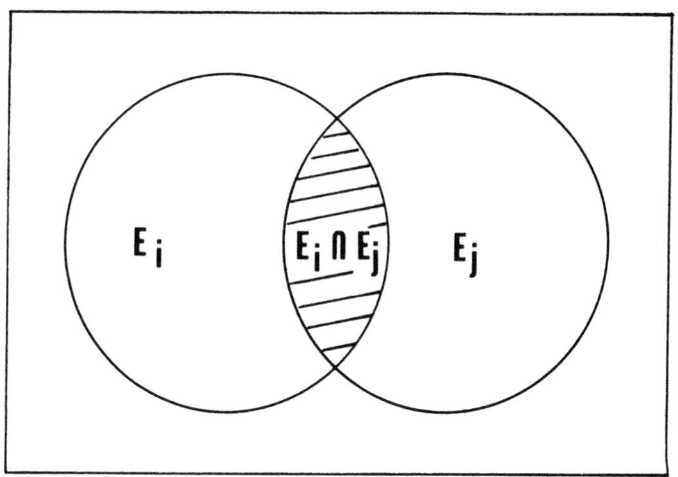

Figure 14-2 Venn Diagram Showing the Union of Two Events

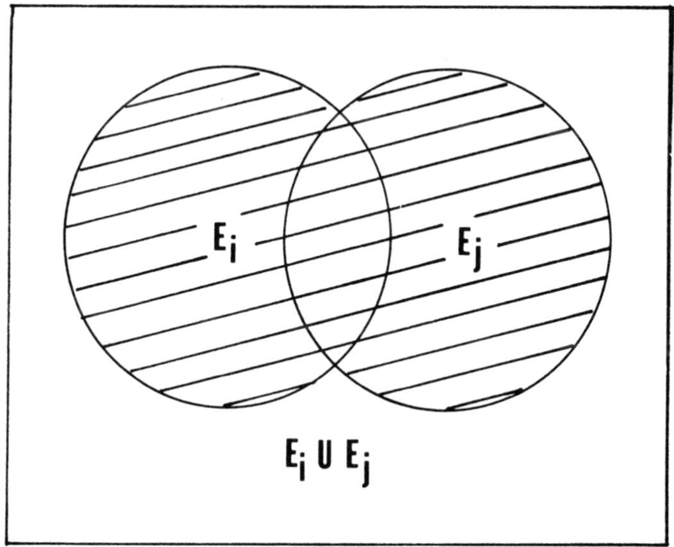

Figure 14-3 Venn Diagram of an Event and Its Complement

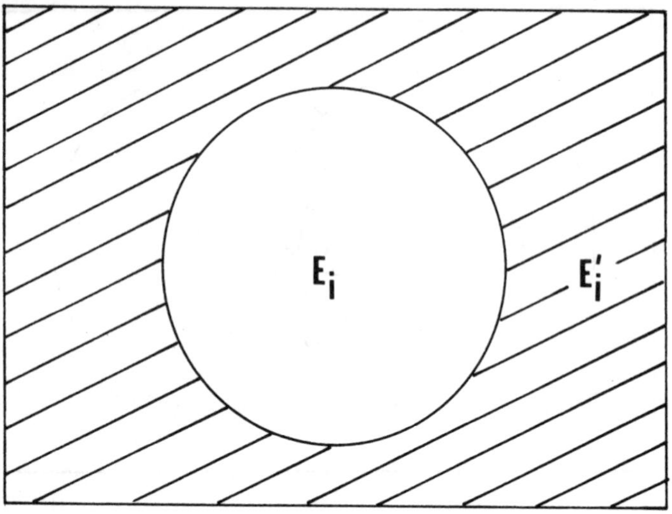

14.3 RULES AND POSTULATES OF PROBABILITY(R,G,A,F)

Having described the meaning and the basic concepts of probability, we turn to the fundamental rules and postulates of probability theory. To facilitate this analysis, we denote the basic outcomes of a random experiment by $O_1, \cdots, O_i, \cdots, O_k$. Further, the probability that O_i will be the outcome of the experiment is represented by $P(O_i)$. Finally, we denote the appropriate sample space by the notation SS.

14.3.1 The Postulates of Probability(R,G,F)

In general, probability theory is based on three fundamental postulates. This section describes these postulates so as to provide the foundation for our examination of probability theory.

The first postulate asserts that each probability $P(O_i)$ must satisfy the relation

$$O \leq P(O_i) \leq 1 \qquad \textbf{(14.1)}$$

which implies that the probability of obtaining the basic outcome O_i cannot be less than zero or greater than one. When viewed from the perspective of relative frequency, the probability associated with outcome O_i cannot be less than 0 or greater than 1. Stated another way, this postulate asserts that the outcome O_i cannot happen less than zero percent or more than 100 percent of the time.

Assuming that there are k basic outcomes in the sample space, the second postulate of probability asserts that

$$\sum_{i=1}^{i=k} P(O_i) = 1 \qquad \textbf{(14.2)}$$

Since the sample space SS has been characterized as being collectively exhaustive and consisting of mutually exclusive outcomes, the second postulate asserts that the probability that one of the possibilities contained in the sample space will occur is equal to 1. Stated differently, this postulate asserts that one of the basic outcomes in the sample space will occur.

The third basic postulate of probability asserts that, for any two basic outcomes O_i and O_j,

$$P(O_i \text{ or } O_j) = P(O_i) + P(O_j) \qquad \textbf{(14.3)}$$

In this formulation $P(O_i \text{ or } O_j)$ refers to the probability that either outcome O_i or O_j will be the result of the random experiment. As an example of this postulate, assume that $P(O_i) = .10$ and $P(O_j) = .45$. Thus, if outcome O_i occurs 10 percent of the time, and O_j occurs 45 percent of the time, one or the other will occur 10% + 45% or 55% of the time. (Remember that *only one* basic outcome can occur at any one time.)

14.3.2 The Rules of Addition(R,G,A,F)

These postulates provide the basis for developing rules of addition that permit us to determine the probability of an event as well as to develop the general and the special rules of addition and multiplication.

For purposes of future illustration, suppose we are interested in evaluating the performance of department A, which is rated as unacceptable, acceptable, or superior. We might regard these ratings as the basic outcomes of a random experiment that might be represented by the three-point scale

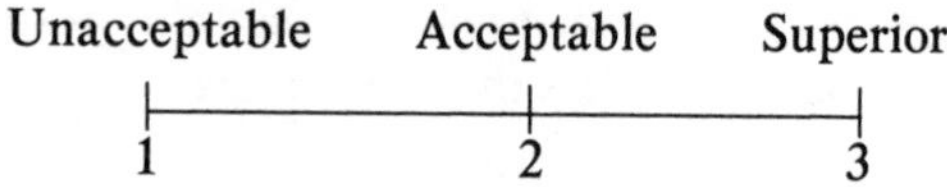

in which we assign numeric values to the outcomes O_1, O_2, and O_3 as follows

$$O_1: \text{unacceptable} = 1$$
$$O_2: \text{acceptable} = 2$$
$$O_3: \text{superior} = 3$$

Hence, the evaluation process may be viewed as a random experiment having the basic outcomes O_1, O_2, and O_3.

Now suppose we are interested in evaluating the performance of two departments, say A and B. In this case we may represent the set of basic outcomes in terms of the points in Figure 14-4. In terms of our earlier discussion, notice that Figure 14-4 corresponds to the sample space of the random experiment.

Figure 14-4 Evaluation of Two Departments

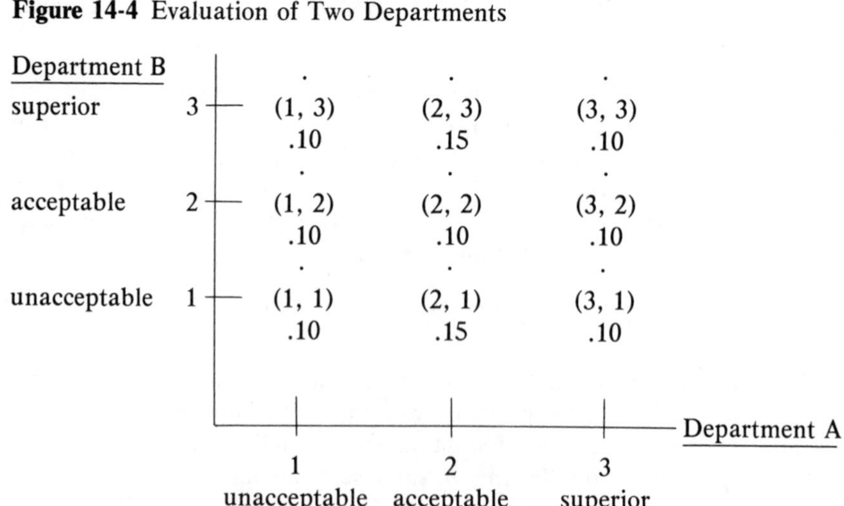

Essentially two sets of data are associated with each outcome. The first, which appears in parentheses, represents our evaluation of the performance of the two units. For example, the coordinate (2, 1) refers to the basic outcome in which the performance of Department A is acceptable and the performance of Department B is unacceptable. The other values in parentheses are interpreted in a similar fashion. The second value associated with each point and located below the information appearing in parentheses refers to the probability $P(O_i)$. Hence, the figure reveals that the probability of both departments receiving an acceptable rating is .10.

14.3.2.1 The Probability of an Event(R,G,A,F)

Consider first the probability of the event E_i, which consists of the basic outcomes $O_1, \cdots, O_j, \cdots, O_n$. The probability of event E_i is given by

$$P(E_i) = P(O_1) + \cdots + P(O_j) + \cdots + P(O_n) \tag{14.4}$$

$$= \sum_{j=1}^{n} P(O_j)$$

In general, then, the probability of any event is equal to the sum of the probabilities of the basic outcomes comprising the event.

To illustrate, assume that E_1 corresponds to the event that the two departments receive the same rating. Referring to Figure 14-4, we see that event E_1 is composed of the points (1, 1), (2, 2), and (3, 3). Applying Equation 14.4, we find that

$$P(E_1) = .10 + .10 + .10$$

$$= .30$$

which implies that the two departments will receive the same rating 30 percent of the time. Similarly, let E_2 correspond to the event that one of the departments receives an unacceptable rating and the other receives a superior rating. Referring to Figure 14-4, we find that event E_2 is composed of the points (1, 3) and (3, 1). Hence, an application of Equation 14.4 yields

$$P(E_2) = .10 + .10$$

$$= .20$$

These calculations imply that event E_2 will occur 20 percent of the time.

14.3.2.2 The Special Rule of Addition (R,G,A,F)

Consider next the *special rule of addition.* If two events are mutually exclusive (i.e., one but not both may occur at the same time) the probability that one or the other will occur is given by the sum of their probabilities. Thus, letting E_i and E_j represent mutually exclusive events, we find that

$$P(E_i \cup E_j) = P(E_i) + P(E_j) \qquad \textbf{(14.5)}$$

Referring to our example, observe that events E_1 and E_2 are mutually exclusive. Since $P(E_1) = .30$ and $P(E_2) = .20$, we find that an application of Equation 14.5 yields

$$P(E_1 \cup E_2) = .30 + .20$$
$$= .50$$

These results imply that event E_1 or E_2 will occur 50 percent of the time.

As expressed previously, the special rule of addition applies only to two mutually exclusive events. However, we may expand Equation 14.5 to accommodate more than two mutually exclusive events. If $E_1, E_2, \cdots, E_k$ are mutually exclusive events, the probability that one of them will occur is given by

$$P(E_1 \cup E_2 \cup \cdots \cup E_k) = P(E_1) + P(E_2) + \cdots + P(E_k)$$
$$= \sum_{i=1}^{k} P(E_i) \qquad \textbf{(14.6)}$$

14.3.2.3 The General Rule of Addition (R,G,A,F)

Since this analysis applies only to mutually exclusive events, we cannot employ Equation 14.6 to determine $P(E_i \cup E_j)$ where E_i and E_j are not mutually exclusive. In deriving a formula that can be used to determine $P(E_i \cup E_j)$ regardless of whether or not E_i and E_j are mutually exclusive, consider Figure 14-5. This figure reveals that

$$P(E_i) = .20 + .10 = .30$$

and

$$P(E_j) = .10 + .30 = .40$$

If we were to apply Equation 14.5, we would conclude erroneously that

$$P(E_i \cup E_j) = .30 + .40$$
$$= .70$$

Here, it will be observed that these calculations result in a value that overstates the correct probability by 10 percent since the area corresponding to $(E_i \cap E_j)$ was counted twice. To avoid overstating the probability $P(E_i \cup E_j)$, we need only subtract $P(E_i \cap E_j)$ from $P(E_i) + P(E_j)$. This procedure results in the *general* rule of addition expressed by

$$P(E_i \cup E_j) = P(E_i) + P(E_j) - P(E_i \cap E_j) \quad \textbf{(14.7)}$$

Referring to Figure 14-5 and using Equation 14.7 we find that the value of the probability $P(E_i \cup E_j)$ is given by

$$P(E_i \cup E_j) = .30 + .40 - .10$$
$$= .60$$

which is, of course, the correct probability.

Note that when E_i and E_j are mutually exclusive events, $(E_i \cap E_j)$ contains no points, which implies that $P(E_i \cap E_j)$ is equal to zero. Thus, for the mutually exclusive events E_i and E_j, we find that

$$P(E_i \cup E_j) = P(E_i) + P(E_j) - 0$$

which, of course, is identical to Equation 14.5.

As in our earlier example, let E_3 correspond to the event that at least one of the departments receives an acceptable rating. Figure 14-4 shows that event E_3 consists of the points (2, 1), (2, 2), (2, 3), (1, 2), and (3, 2). As a consequence, we find that an application of Equation 14.4 yields

$$P(E_3) = .15 + .10 + .15 + .10 + .10$$
$$= .60$$

Figure 14-5 Venn Diagram for Probabilities of Intersecting Events

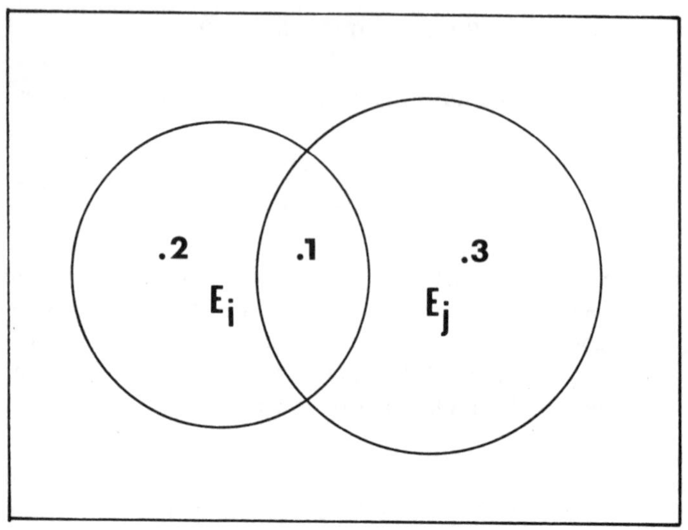

Now suppose we want to find $P(E_1 \cup E_3)$ where the event E_1 consists of the points (1, 1), (2, 2), and (3, 3). Notice that the point (2, 2) is common to both events E_1 and E_3. As a consequence, the two events are not mutually exclusive and the use of Equation 14.7 is required to determine the probability $P(E_1 \cup E_3)$. In this case, we find that

$$P(E_1 \cup E_3) = .30 + .60 - .10$$
$$= .80$$

where the value .10 corresponds to the probability $P(E_1 \cap E_3)$.

14.3.3 Conditional Probability and the Rules of Multiplication(R,G,A,F)

Occasionally we require a statement that indicates the probability that event E_i will be the result of a random experiment given that event E_j is known or assumed to have occurred. Such an expression is referred to as a statement of *conditional probability*. In general, if we have any two events E_i and E_j based on the same sample space, the conditional probability of event E_i given E_j is written as $P(E_i|E_j)$.

To illustrate the meaning of statements of conditional probability, suppose that the operation of our institution is limited to the medical management of the conditions M_1, M_2, and M_3. Suppose further that we are interested in the probability of a patient's requiring some service k, given that the person was admitted with one of the three conditions. From the medical records of patients who were treated previously for the conditions M_1, M_2, and M_3 we obtain the information presented in Table 14-1. Here, we let E_1, E_2, and E_3 refer to the events that a patient presents condition M_1, M_2, and M_3 respectively, while B_1 and B_2 represent the events that the patient required and did not require service S_k, respectively. This table demonstrates that

$$P(E_1) = .20 \qquad P(B_1) \cong .567$$

$$P(E_2) = .30 \qquad P(B_2) \cong .433$$

$$P(E_3) = .50$$

In this case, it will be observed that $P(E_1)$, $P(E_2)$, $P(E_3)$, $P(B_1)$, and $P(B_3)$ are each less than or equal to 1 and greater than or equal to zero. Also note that

$$P(E_1) + P(E_2) + P(E_3) = 1 = P(B_1) + P(B_2) \tag{14.8}$$

We now turn to the calculation of the probabilities concerning the intersection of these events. For example, $P(E_1 \cap B_1)$ refers to the probability of a patient's being admitted with condition M_1 and requiring S_k. As seen in the table, 20 of the 300 patients were admitted with this condition and required service S_k. As a consequence, we find that

$$P(E_1 \cap B_1) = \frac{20}{300} \cong .067$$

Table 14-1 Distribution of Patients by Diagnosis and Service Requirements

Condition	Required S_k (B_1)	Did not Require S_k (B_2)	Total
M_1 (E_1)	20	40	60
M_2 (E_2)	30	60	90
M_3 (E_3)	120	30	150
Total	170	130	300

Similarly, we find that 30 of the 300 patients were admitted with condition M_2 and required service S_k. Thus, we find that

$$P(E_2 \cap B_1) = \frac{30}{300} = .10$$

As can be verified, we may use a similar technique to compute

$$P(E_3 \cap B_1) = \frac{120}{300} = .40$$

$$P(E_1 \cap B_2) = \frac{40}{300} = .133$$

$$P(E_2 \cap B_2) = \frac{60}{300} = .20$$

$$P(E_3 \cap B_2) = \frac{30}{300} = .10$$

We may summarize these calculations by developing the bivariate probability distribution in Table 14-2. Note that a value appearing on the bottom line of this table is simply the sum of the elements appearing in the corresponding column; that is:

$$.567 = .067 + .10 + .40$$

$$.433 = .133 + .20 + .10$$

Similarly, a value appearing in the last column of the table is simply the sum of the elements in the corresponding row. As a consequence, we find that

$$.20 = .067 + .133$$

$$.30 = .10 + .20$$

$$.50 = .40 + .10$$

These data provide the basis for an examination of conditional probability. Suppose that we limit our analysis to only those patients who were admitted with condition M_1. In this case, we find that $P(B_1|E_1)$ is given by

$$\frac{.067}{.20} = .335$$

Table 14-2 Bivariate Probability Distribution

Condition	Requires S_k (B_1)	Does Not Require S_k (B_2)	Total
M_1 (E_1)	.067	.133	.20
M_2 (E_2)	.10	.20	.30
M_3 (E_3)	.40	.10	.50
Total	.567	.433	1.00

This probability statement implies that 33.5 percent of the patients admitted with condition M_1 require service S_k. Stated differently, $P(B_1|E_1)$ is the conditional probability of selecting a patient who requires service S_k, given that the person was admitted with condition M_1. This conditional probability also may be written in the form

$$P(B_1 \mid E_1) = \frac{P(B_1 \cap E_1)}{P(E_1)}$$

In similar fashion, we could find the conditional probability of selecting a patient who required service S_k, given that the person was admitted with condition M_2. This conditional probability is given by

$$P(B_1|E_2) = \frac{P(B_1 \cap E_2)}{P(E_2)}$$

$$= \frac{.10}{.30} = .333$$

In general, if we have two events, E_i and E_j, that are based on the same sample space, the conditional probability of E_i given E_j is defined by

$$P(E_i|E_j) = \frac{P(E_i \cap E_j)}{P(E_j)} \qquad \textbf{(14.9)}$$

Once the set of conditional probabilities has been calculated using Equation 14.9, it is possible to construct a conditional probability distribution of service requirements. For example, the conditional probability distribution of service requirements, given diagnostic condition M_1, is found to be

$$P(B_1|E_1) = \frac{.067}{.20} = .335$$

$$P(B_2|E_1) = \frac{.133}{.20} = \frac{.665}{1.000}$$

Similarly, the conditional probability distributions of service requirements, given conditions M_2 and M_3, are found to be

$$P(B_1|E_2) = \frac{.10}{.30} = .33$$

$$P(B_2|E_2) = \frac{.20}{.30} = \frac{.67}{1.00}$$

and

$$P(B_1|E_3) = \frac{.40}{.50} = .80$$

$$P(B_2|E_3) = \frac{.10}{.50} = \frac{.20}{1.00}$$

respectively. In each of these distributions, the sum of the probabilities is one and the value of each of the probabilities is between zero and one.

14.3.4 The Rules of Multiplication(R,G,A,F)

Thus far in our analysis, we have used the expression

$$P(E_i|E_j) = \frac{P(E_i \cap E_j)}{P(E_j)}$$

to calculate conditional probabilities. If we now multiply both sides of this equation by $P(E_j)$ we obtain

$$P(E_j)P(E_i|E_j) = P(E_i \cap E_j) \qquad \textbf{(14.10.1)}$$

which, when rearranged slightly, yields

$$P(E_i \cap E_j) = P(E_j) \cdot P(E_i | E_j) \qquad \textbf{(14.10.2)}$$

Equation 14.10.2 is referred to as the *general rule of multiplication* that allows us to calculate the probability that two events will occur. In this case, the probability that the events E_i and E_j will both occur is given by $P(E_j)P(E_i|E_j)$. To illustrate, let us return to our example and compute $P(E_1 \cap B_1)$, which is given by

$$\begin{aligned} P(E_1 \cap B_1) &= (.20)(.335) \\ &= .067 \end{aligned}$$

This result, of course, agrees with the findings in Table 14-2.

Equation 14.10.2 also provides the basis for deriving the special rule of multiplication. To derive this rule, it is important to note that, when

(1) $P(E_i|E_j) > P(E_i)$, the occurrence of event E_j increases the probability of E_i occurring, which implies that E_i and E_j are positively associated;
(2) $P(E_i|E_j) < P(E_i)$, the occurrence of event E_j reduces the probability of event E_i occurring, which implies that E_i and E_j are negatively associated;
(3) $P(E_i|E_j) = P(E_i)$, the probability of event E_i occurring does not depend on the occurrence or nonoccurrence of event E_j.

Concerning the last relation, when $P(E_i|E_j) = P(E_i)$ we say that event E_i is independent of E_j and this means that the occurrence or nonoccurrence of event E_i is not influenced by the occurrence or nonoccurrence of event E_j. Substituting $P(E_i)$ for $P(E_i|E_j)$ in Equation 14.10.2 yields

$$P(E_i \cap E_j) = P(E_i)P(E_j) \qquad \textbf{(14.11)}$$

This equation expresses the *special rule of multiplication* that may be used to calculate $P(E_i \cap E_j)$ whenever E_i and E_j are independent events as defined by $P(E_i|E_j) = P(E_i)$.

We might generalize Equation 14.11 so as to accommodate k independent events. For example, if $E_1, E_2, \cdots, E_k$ are independent events, then

$$P(E_1 \cap E_2 \cap \cdots \cap E_k) = P(E_1) \cdot P(E_2) \cdot \cdots \cdot P(E_k) \tag{14.12}$$

For k equiprobable independent events, it can be verified that Equation 14.12 reduces to

$$P(E_1 \cap E_2 \cap \cdots \cap E_k) = [P(E)]^k \tag{14.13}$$

14.4 MATHEMATICAL EXPECTATION AND DECISION MAKING(R,G,A,F)

Having examined the basic meaning of probability as well as the rules and postulates of probability theory, we turn now to an examination of the role played by probability in the process by which decisions are reached in health care facilities. It should be noted, however, that decisions seldom are reached on the basis of probability alone. Rather, when faced with uncertainty, the manager bases decisions on probabilities as well as on the penalties and rewards associated with the different courses of action being evaluated.

As an example, suppose that we are considering the construction of a new wing for our hospital. Further, assume that on the basis of our analysis, we find there is a probability of .1 that we will earn a net income of $250,000 if the occupancy rate of the new wing is in excess of 80 percent and a probability of .70 that we will incur a net loss of $1,000 if the occupancy rate is below 80 percent. In this case, the probability of incurring a net loss is relatively high when compared with the likelihood of earning revenue in excess of costs. On the other hand, when compared with the net losses, the net revenue that might be earned is quite high. This example serves to indicate the necessity of examining the interrelation between the payoffs and the probabilities associated with the outcomes of various courses of action that are considered by management.

In estimating the use of the new wing introduced above, statements such as

a. patients aged 0–14 are expected to use 210.1 bed days per pediatric bed year,

b. patients between the ages of 15 and 44 are expected to use 362.5 bed days per surgical bed year, and
c. patients between 45 and 85 years of age are expected to use 196.2 bed days per intensive care bed year;

would be of considerable value. Here, however, the word "expected" should be interpreted as an average and these statements may be viewed as expressions of mathematical expectation.

To illustrate the concept of mathematical expectation assume that, as of December 31, our institution reports outstanding receivables of $10,000. Suppose further that 10 percent of the institution's outstanding receivables result in bad debts. In this case we would expect that receivables amounting to $.1 \times 10{,}000$ or $1,000 will be written off as a bad debt. In general, if the probability of outcome K occurring is $P(K)$ and the value associated with outcome K is $V(K)$, our mathematical expectation is given by

$$\hat{E} = P(K)\,V(K) \tag{14.14}$$

Thus far in the analysis, we have examined the mathematical expectation concerning the payoff associated with a single outcome. However, Equation 14.14 may be modified to accommodate more than one payoff. To illustrate, let us expand the example and assume that the following data are available to management. In this example, we assume that, as an outstanding receivable ages, the probability of writing off the account as a bad debt also increases. Expanding Equation 14.4, the expected bad debt expense associated with these accounts is given by

$$\begin{aligned}\hat{E} &= .10(\$600{,}000) + .30(\$200{,}000) + .4(\$300{,}000) + .5(\$500{,}000)\\ &\quad + .60(\$120{,}000)\\ &= \$562{,}000\end{aligned}$$

Age of Outstanding Receivables (in Days)	Value of Outstanding Receivables (in $1,000)	Probability of Bad Debt
0–30	600	.10
31–60	200	.30
61–90	300	.40
91–120	500	.50
More than 120	120	.60

This suggests the following rule: if the probabilities of obtaining the amounts $a_1, a_2, \cdots a_k$ are $P(a_1), P(a_2), \cdots P(a_k)$, respectively, the corresponding mathematical expectation is given by

$$\hat{E} = P(a_1)a_1 + P(a_2)a_2 + \cdots + P(a_k)a_k$$

$$= \sum_{i=1}^{i=k} P(a_i)a_i \qquad (14.15)$$

We now have measured the value of one or more outcomes in monetary units. However, statements of mathematical expectation are not limited to only those payoffs that are expressed in financial terms. Rather, we might say that a patient is expected to use 3.5 x-rays per hospital episode, to use 7.36 days of care per hospital episode, to require 4.3 dental visits per year, and to visit the physician 1.7 times per year. Similarly, we might assert that of the 1,000 patients who will be treated in the next period, we expect 100 to present condition M_1 and be admitted to department D_1.

As an example, assume we wish to estimate the number of days of care and the amount of labor of type L_1 and L_2 required to manage a patient population of 1,000 presenting conditions M_1, M_2, and M_3. Further, suppose that our institution is composed of nursing departments D_1, D_2, and D_3. Also assume that an examination of the medical records of 500 patients who were hospitalized previously resulted in the data in Table 14-3 (see page 350). Replicating our earlier work, we find that the probabilities corresponding to the event $(M_i \cap D_i)$ for $i = 1, 2, 3$ may be summarized as seen in Table 14-4 (see page 350). Notice that the product of each probability in the table and the size of the patient population (1,000 in this case) results in this distribution:

	Workload by Department (Number of Patients)			
Condition	D_1	D_2	D_3	Total
M_1	150	0	50	200
M_2	0	400	0	400
M_3	50	200	150	400
Total	200	600	200	1,000

Suppose further that

$$E(N_1) = 10.3 \text{ days}$$
$$E(N_2) = 6.4 \text{ days}$$

and

$$E(N_3) = 12.1 \text{ days}$$

represent the expected length of stay associated with diagnostic conditions M_1, M_2, and M_3, respectively. On the basis of these findings, the projected workload associated with these patients is given by

$$\sum_{i=1}^{3} E(M_i)E(N_i) = 200(10.3) + 400(6.4) + 400(12.1)$$
$$= 9{,}460$$

days of care. In this formulation, $E(M_i)$ represents the number of patients who are expected to present condition M_i for $i = 1, 2, 3$. Similarly, the projected workloads of departments D_1, D_2, and D_3 are given by

$$E(W_1) = 150(10.3) + 0(6.4) + 50(12.1)$$
$$= 2{,}150$$
$$E(W_2) = 0(10.3) + 400(6.4) + 200(12.1)$$
$$= 4{,}980$$

and

$$E(W_3) = 50(10.3) + 0(6.4) + 150(12.1)$$
$$= 2{,}330$$

days of care, respectively.

Now, assume that the data in Table 14-5 (page 352) represent the amount of labor of type L_1 and L_2 that management expects to employ in providing a day of care in each of the three nursing departments. In this example management expects to use

$$e(L_1) = 0(2{,}150) + 6.1(4{,}980) + 2.0(2{,}330)$$
$$= 35{,}038$$

and

$$e(L_2) = 4.3(2{,}150) + 0(4{,}980) + 2.1(2{,}330)$$
$$= 14{,}138$$

manhours of labor of type L_1 and L_2, respectively. Here, the values in parentheses correspond to the expected workload of each department while the coefficients to which these projections are applied correspond to the expected manhours of labor required to provide a day of care.

14.5 APPLICATIONS TO HEALTH CARE MANAGEMENT(O,A,F)

It is well recognized that the manager of the health care facility must reach decisions under conditions of uncertainty. In this section, we explore the applicability of the probability concepts introduced earlier to the managerial

Table 14-3 Distribution of Patient Sample by Diagnosis and by Department

	Department			
Diagnosis	D_1	D_2	D_3	Total
M_1	75	0	25	100
M_2	0	200	0	200
M_3	25	100	75	200
Total	100	300	100	500

Table 14-4 Probabilities Corresponding to the Event $(M_i \cap D_i)$ for $i = 1, 2, 3$

	Department		
Diagnosis	D_1	D_2	D_3
M_1	.15	0	.05
M_2	0	.40	0
M_3	.05	.20	.15

responsibility of planning, monitoring, and controlling the operations of the health care facility.

14.5.1 Laboratory Test Accuracy(O,A,F)

Suppose that we are interested in monitoring and controlling the operational activity of the laboratory department that employs technicians A_1 and A_2 in providing a given test. Suppose further that the data in Table 14-6 (page 352) are available to management. We now may use these data to evaluate the performance of the two technicians by calculating

$$P(A_1) = 120/250 = .48; \qquad P(A_1 \cap B_1) = 100/250 = .40$$
$$P(A_2) = 130/250 = .52; \qquad P(A_1 \cap B_2) = 20/250 = .08$$
$$P(B_1) = 150/250 = .60; \qquad P(A_2 \cap B_1) = 50/250 = .20$$
$$P(B_2) = 100/250 = .40; \qquad P(A_2 \cap B_2) = 80/250 = .32$$

On the basis of these findings we may now calculate $P(B_2|A_1)$ and $P(B_2|A_2)$ as follows:

$$P(B_2|A_1) = \frac{P(A_1 \cap B_2)}{P(A_1)} = \frac{.08}{.48} \cong .167$$

$$P(B_2|A_2) = \frac{P(A_2 \cap B_2)}{P(A_2)} = \frac{.32}{.52} \cong .62$$

Since $P(B_2|A_1) < P(B_2)$ and $P(B_2|A_2) > P(B_2)$, we conclude that the performance of technician A_1 is superior to the performance of technician A_2.

Now, suppose that an expert tells us that no more than 20 percent of the results from this test should be unsatisfactory. In this case, we find that $P(B_2) > .20$, which implies that the performance of our laboratory is substandard. In particular, a further examination of the data reveals that

$$P(B_2|A_1) < .20$$

and

$$P(B_2|A_2) > .20$$

These findings suggest that the performance of technician A_1 is acceptable when compared with a standard error rate of 20 percent. On the other hand, when compared with the same standard, the performance of technician A_2 is found to be substandard and management should implement steps designed to improve the accuracy of the tests performed by this employee.

Laboratory Staffing Requirements

As another example, suppose we are interested in estimating the staffing requirements of the laboratory for the coming period. We assume that management expects to provide care to 15,000 patients presenting conditions M_1 and M_2. Further, assuming that $P(M_1) = .8$ and $P(M_2) = .2$ we find that

$$15{,}000(.8) = 12{,}000$$

and

$$15{,}000(.2) = 3{,}000$$

patients are expected to present diagnoses M_1 and M_2, respectively. Now suppose that the data in Table 14-7 also are available to management. In

Table 14-5 Expected Labor Requirements by Department and Type of Labor

	Labor	
Department	L_1	L_2
D_1	0	4.3
D_2	6.1	0
D_3	2.0	2.1

Table 14-6 Distribution of Satisfactory and Unsatisfactory Tests by Technician

Technician	Number of Satisfactory Tests (B_1)	Number of Unsatisfactory Tests (B_2)	Total
A_1	100	20	120
A_2	50	80	130
Total	150	100	250

Table 14-7 Expected Use of Lab Service

Number of Tests Per Patient	Probability of Use by Diagnosis	
	Diagnosis M_1	Diagnosis M_2
0	.1	.1
1	.3	.2
2	.4	.4
3	.1	.2
4	.1	.1

this case, we find that the expected use of laboratory service by a patient presenting conditions M_1 and M_2 is given by

$$.1(0) + .3(1) + .4(2) + .1(3) + .1(4) = 1.8 \text{ units}$$

and by

$$.1(0) + .2(1) + .4(2) + .2(3) + .1(4) = 2.0 \text{ units}$$

respectively. Thus, management expects to provide

$$12{,}000(1.8) + 3{,}000(2.0) = 27{,}600$$

units of laboratory care during the coming period.

Now assume that labor of type L_1 and L_2 is used in providing laboratory service and that the data in Table 14-8 are available to management. Applying the principles of mathematical expectation, we find that

$$.6(.5) + .2(1.0) + .1(1.5) + .1(2.0) = .85 \text{ manhours}$$

and

$$.1(.5) + .7(1.0) + .1(1.5) + .1(2.0) = 1.1 \text{ manhours}$$

of type L_1 and L_2 labor will be required per unit of laboratory service during the coming period. The total labor requirement of type of manpower is given by

$$27{,}600(.85) = 23{,}460 \text{ manhours}$$

and by

$$27{,}600(1.1) = 30{,}360 \text{ manhours}$$

These calculations suggest that 23,460 manhours of type L_1 labor and 30,360 manhours of type L_2 labor will be required during the coming period.

Table 14-8 Use of Labor in Lab Service

Manhours per Unit of Service	Probability of Use by Type of Labor	
	L_1	L_2
.5	.6	.1
1.0	.2	.7
1.5	.1	.1
2.0	.1	.1

Problems for Solution

1. Suppose that the probability of a patient having a tooth filled during a dental visit is .30, the probability of having another tooth extracted during the visit is .60, and the probability of having both a tooth extracted and a cavity filled is .20. What is the probability that the patient will have a tooth filled and a tooth extracted during the visit?
2. Suppose that a certain laboratory test can be performed on one of three machines (B_1, B_2, or B_3) and the results are found to be satisfactory (A_1) or unsatisfactory (A_2). Also assume that

$$P(B_1) = .20$$
$$P(A_1) = .40$$
$$P(B_3) = .30$$
$$P(A_1|B_1) = .25$$
$$P(A_2|B_3) = .30$$

Calculate

$$P(A_1|B_2)$$
$$P(A_1|B_3)$$
$$P(A_2|B_2)$$

Are A_1 and B_2 independent events?
Are A_1 and B_3 independent events?
Are A_2 and B_2 independent events?

3. Let A_1 represent the event that a person will receive a medical examination during a given year,
   A_2 represent the event that a person will not receive a medical examination during a given year,
   B_1 represent the event that a person is of high socioeconomic status,
   B_2 represent the event that a person is of low socioeconomic status;

In the population under study suppose we know that

$$P(A_1) = .40$$
$$P(B_2) = .30$$
$$P(A_1|B_2) = .10$$

Find $P(A_2 \cap B_1)$, $P(A_2 \cap B_2)$, $P(A_2 \cup B_2)$, and $P(A_2|B_1)$. Are events A_2 and B_1 independent?

4. Suppose we are provided with the following information concerning the number of days of care used by age of patient:

	Days of Care				
Age	0–3 B_1	4–7 B_2	8–11 B_3	12–15 B_4	Total
0–24 (A_1)	20	70	25	80	200
25–49 (A_2)					
50–74 (A_3)		70	70	30	100
Total	100		200	150	600

Find $P(B_2)$, $P(A_2)$, $P(A_2 \cap B_1)$, $P(A_2 \cap B_2)$, $P(A_2 \cap B_3)$, and $P(A_2 \cap B_4)$. Also calculate the conditional probability distribution of use with respect to age. Are these two variables independent?

5. Let A represent the event that a bill was prepared incorrectly and B represent the event that a third party payer delayed payment.

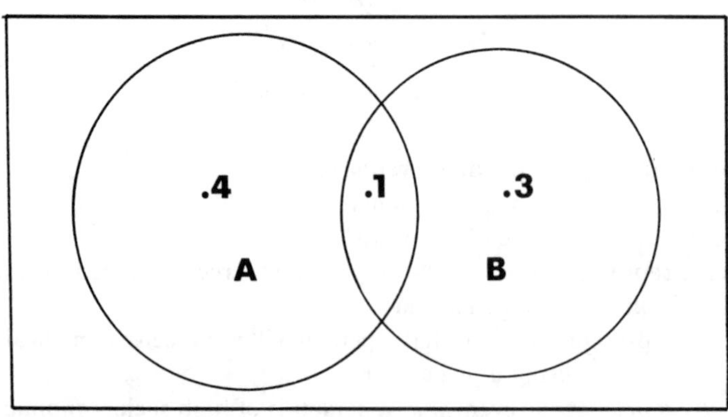

Find $P(A)$, $P(B)$, $P(A \cap B)$, and $P(B|A)$.

6. The following gives the probability of a patient's using 0, 1, 2, 3, 4, and 5 days of care during the hospital stay.

Number of Days	Probability
0	.05
1	.15
2	.20
3	.30
4	.20
5	.10

How many days of care is the patient expected to use?

7. Suppose we want to determine the number of doctor office visits per person. Also assume that the following data are available to us:

Number of Visits	Probability
0	.05
1	.10
2	.15
3	.30
4	.25
5	.10
6	.05

How many visits per person should we expect?

8. Suppose that our institution is provided with the following estimation:

Magnitude of Cash Inflow	Probability
$ 5,000	.40
20,000	.20
40,000	.20
60,000	.10
90,000	.10

What is the expected cash inflow?

9. Suppose further that the institution in Problem 8 also is provided with the following information concerning cash outflow:

Magnitude of Cash Outflow	Probability
$ 5,000	.10
20,000	.10
40,000	.20
60,000	.20
90,000	.40

What is the expected cash outflow?
What is the expected net cash flow?

10. The following information represents the demand for an inventory item during the lead time (i.e., the time between placing and receiving an order). What is the expected usage of this item during the lead time?

Usage	Probability
0	.05
1	.10
2	.15
3	.30
4	.20
5	.10
6	.05
7	.05

Chapter 15

Probability and Continuous Random Variables

Objectives

After completing this chapter, you should be able to:

1. Distinguish between the probability density function and the probability distribution function;
2. Apply integral calculus to find the area under a probability density function;
3. Use the standard normal curve to find desired probabilities.

The sections comprising this chapter may be summarized as follows:

Chapter Map

Section Number	Required Reading	Optional Reading	Generic Development	Application to Management	Fundamental Principles	Complex Material
	(R)	(O)	(G)	(A)	(F)	(C)
15.1	x		x		x	
15.2	x		x	x	x	
15.3	x		x	x	x	
15.4		x		x	x	

15.1 INTRODUCTION(R,G,F)

The purpose of this chapter is to examine the problem of determining the probability that a continuous random variable will assume a value on some specified interval and to illustrate the usefulness of these probability statements in discharging managerial responsibilities in a health care institution. That the definite integral may be used to determine the area under the curve defined by the function $f(x)$ should be clear from Chapter 12. In this chapter, we simply apply the techniques of integration to a continuous sample space.

As seen in the previous chapter, a sample space is the aggregation of all possible outcomes that may result from a random experiment, while any one of these outcomes may be regarded as a sample point. If the sample space consists of the real line or segments of the real line, the sample space is said to be continuous.

15.2 PROBABILITY DENSITY FUNCTIONS AND THE CONTINUOUS PROBABILITY DISTRIBUTION(R,A,G,F)

Assume that we are interested in the duration of operative procedures performed in our institution. To investigate this problem, suppose that we group the duration of 100 procedures performed previously into the five categories that were used to construct Table 15-1. As can be seen in this table, six procedures lasted between 0 and 2.9 hours, 24 procedures lasted between 3.0 and 5.9 hours, etc.

Table 15-1 Frequency Distribution and Relative Frequency Distribution for the Duration of Operative Procedures

Duration (in hours)	Number of Procedures (f_i) (1)	Relative Frequency (f_i/T) (2)
0.0– 2.9	6	.06
3.0– 5.9	24	.24
6.0– 8.9	36	.36
9.0–11.9	26	.26
12.0–14.9	8	.08
Total	100	1.00

On the basis of these data we may develop the relative frequency distribution shown in column (2) of the table. In developing this distribution we simply divide the class frequencies (f_i) by the total number of observations in our sample. For example, consider the relative frequency associated with the first grouping. The relative frequency of .06 was obtained by dividing the class frequency of 6 by 100—the total number of observations in our sample. The other entries in column (2) are derived in similar fashion.

On the basis of the information in Table 15-1 we may now determine the proportion of operative procedures that lasted 2.9 hours or less, 5.9 hours or less, 8.9 hours or less, 11.9 hours or less, and 14.9 hours or less. These calculations are presented in Table 15-2, where columns (1) and (2) represent the relative frequency distribution constructed earlier. The third column represents the number of hours "or less" to which the cumulative relative frequencies refer. In columns (3) and (4) of Table 15-2 is the cumulative "or less" relative frequency distribution for the duration of operative procedures. With respect to the first entry in column (4), observe that 6 percent or .06 of the operative procedures lasted 2.9 hours or less. Hence, the first entry in the column indicates the proportion of procedures in which 2.9 hours or less were required. Similarly, the second entry in column (4) indicates that .06 + .24 or 30 percent of the operations lasted 5.9 hours or less. In this case, the value .30 is the sum of the first two entries in column (2). The last entry in column (4) is the sum of the values in column (2) and suggests that

$$.06 + .24 + .36 + .26 + .08$$

or 100 percent of the operations required 14.9 hours or less to perform.

Table 15-2 Cumulative "Or Less" Relative Frequency Distribution Pertaining to the Duration of Operative Procedures

Duration (in hours) (1)	Relative Frequency (2)	Number of Hours or Less (3)	Cumulative Relative Frequencies (4)
0.0– 2.9	.06	2.9	.06
3.0– 5.9	.24	5.9	.30
6.0– 8.9	.36	8.9	.66
9.0–11.9	.26	11.9	.92
12.0–14.9	.08	14.9	1.00

We now let $F_n(x)$ represent the proportion of the sample values that is less than or equal to a specified time interval, x^*, such that

$$F_n(x) = \frac{\text{Number of } x_i \leq x^*}{n}$$

We may portray $F_n(x)$, for $n = 100$ and $x^* = 2.9, \cdots, 14.9$ hours as seen in Figure 15-1. This figure reveals that $F_n(x)$ is a cumulative distribution that indicates probability as n becomes large. For example, point c in the figure indicates that the likelihood of encountering an operative procedure that lasts 8.9 hours or less is .66.

Figure 15-1 Cumulative "Less Than" Relative Frequency Distribution

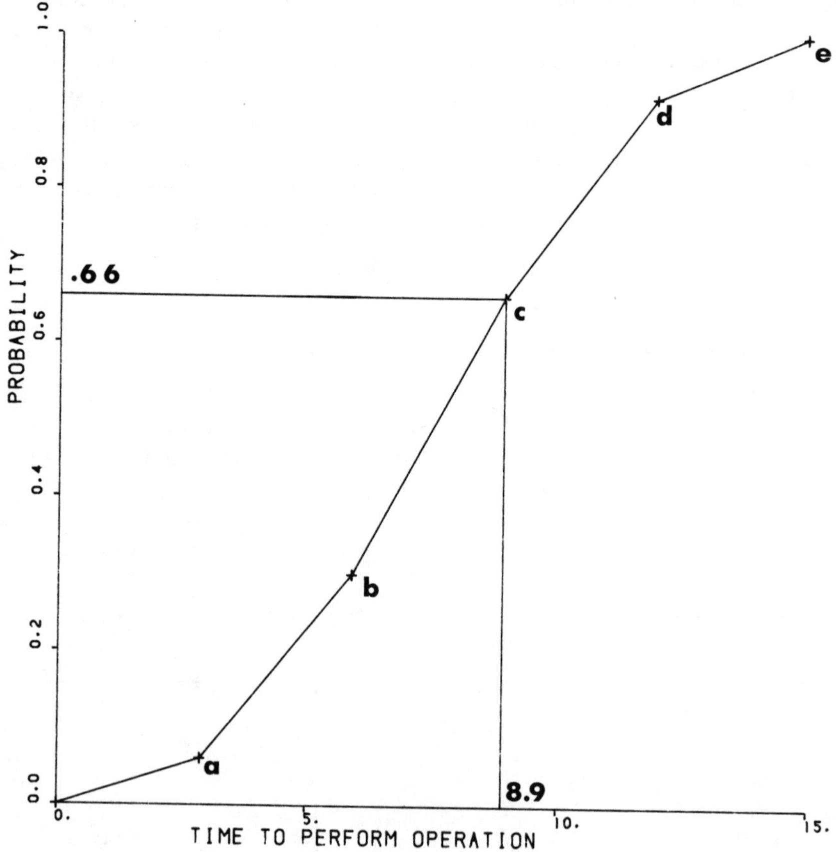

Employing general notation, we now represent $F_n(x)$ by $F(x)$. The function $F(x)$ is referred to as the probability distribution function that gives the probability of obtaining a value of x that is less than or equal to some value we designate as x^*. Symbolically, we may write

$$F(x^*) = P(x \leq x^*)$$

where $P(x \leq x^*)$ indicates the probability of obtaining a value of x that is less than or equal to x^*. Letting n increase indefinitely, we might depict the function $F(x)$ as presented in Figure 15-2.

Figure 15-2 Distribution Function

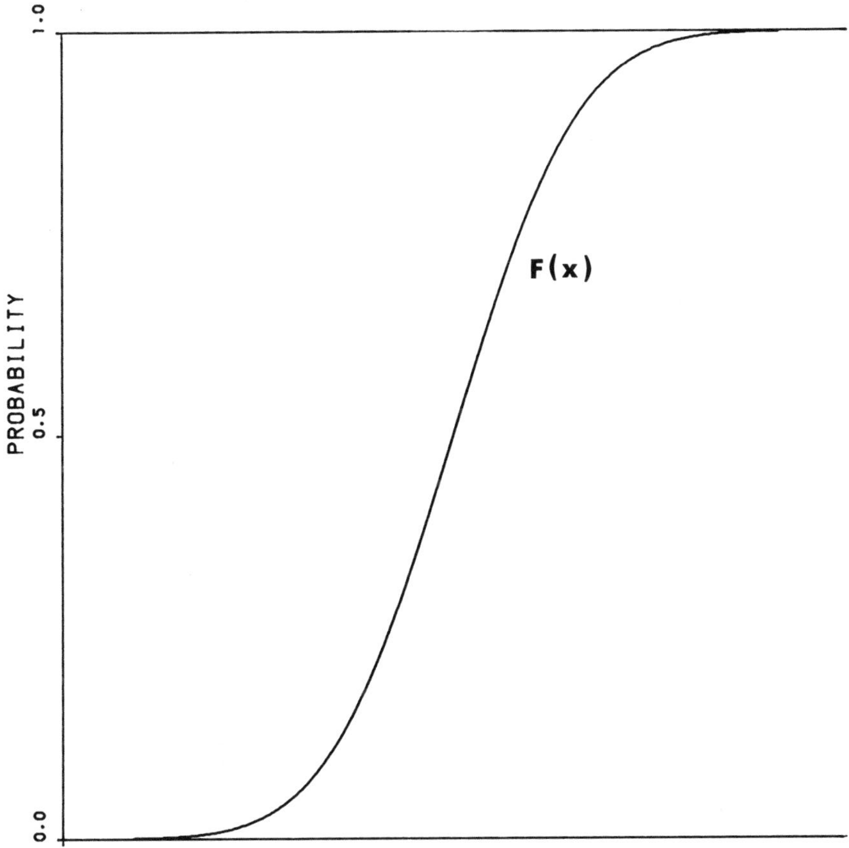

This discussion serves to define a continuous probability distribution for a single variable. In general, we let x represent a random variable that has the real line as its sample space. The function $F(x)$ is referred to as the distribution function and the probability of obtaining a value of x that is less than or equal to x^* is given by

$$F(x^*) = P(x \leq x^*) \tag{15.1}$$

As the value of x increases indefinitely, it can be shown that

$$\lim_{x \to \infty} F(x) = F(+\infty) = 1 \tag{15.2.1}$$

Conversely, if x approaches $-\infty$, it can be shown that

$$\lim_{x \to -\infty} F(x) = F(-\infty) = 0 \tag{15.2.2}$$

Equations 15.2.1 and 15.2.2 suggest that the probability $P(x \leq x^*)$ must lie between zero and one. Also observe that $F(x)$ is a nondecreasing function of x since

$$\begin{aligned} F(b) - F(a) &= P(x \leq b) - P(x \leq a) \\ &= P(a \leq x \leq b) \end{aligned}$$

where $a < b$ and $P(a \leq x \leq b)$ is nonnegative. In the following discussion, we limit our analysis to distribution functions $F(x)$ that are continuous and have a derivative at all points. Letting $F'(x) = f(x)$, our earlier discussion allows us to assert that

$$F(x) - F(\infty) = \int_{-\infty}^{x} f(x)\,dx \tag{15.3.1}$$

and since $\lim_{x \to -\infty} F(x) = 0$, we have

$$F(x) = \int_{-\infty}^{x} f(x)\,dx \tag{15.3.2}$$

Here, the function $f(x)$ is referred to as a probability density function. The reason for this terminology is described below.

The relation between the functions $f(x)$ and $F(x)$ is shown in Figure 15-3 (see page 367). Presented in part (a) of this figure is the function $F(x)$ and the function $f(x)$ in part (b). Applying the discussion in Chapter 12, we find that the probability $P(a \leq x \leq b)$ is given by

$$F(b) - F(a) = \int_{a}^{b} f(x)\,dx$$

which is the area under $f(x)$ between the points a and b.

Consider next the probability of obtaining a value of x that lies on the interval x to $x + \Delta x$. In this case, we find that the desired probability is given by

$$\int_{x}^{x+\Delta x} f(x)\,dx$$

which has the approximate value $f(x)\,\Delta x$ when Δx is small. This example illustrates an important point. As mentioned previously, $f(x)$ simply indicates the height of the curve that implies that $f(x)$ times the length of the interval Δx yields probability. These observations suggest that the function $f(x)$ indicates the density of the probability at a given point and, as a result, $f(x)$ is called a probability density function. Obviously when $\Delta x = 0$, there is no area and, hence, there is no probability.

15.3 THE NORMAL DISTRIBUTION(R,G,A,F)

The normal or the Gaussian distribution is a unimodal and symmetric curve that is asymptotic with respect to the x axis. The normal distribution has two parameters that usually are symbolized by σ (small Greek sigma) and μ (Greek letter mu). The symbol μ refers to the population mean while the symbol σ refers to the standard deviation of the population. The mean of a continuous random variable is defined by

$$\mu = \int_{-\infty}^{+\infty} xf(x)dx \qquad (15.4.1)$$

where $f(x)$ is the probability density function. The variance of a continuous random variable is defined by

$$\sigma^2 = \int_{-\infty}^{+\infty} (x - \mu)^2 f(x)dx \qquad (15.4.2)$$

and the standard deviation, σ, of such a variable is simply the square root of Equation 15.4.2.

Given that μ and σ are defined by Equations 15.4.1 and 15.4.2, respectively, the density function for the normal curve is given by

$$f(x) = \frac{1}{\sigma\sqrt{2\pi}} e^{-\frac{1}{2}\frac{(x-\mu)^2}{\sigma^2}} \qquad (15.5)$$

which implies that the corresponding distribution function is

$$F(x) = \frac{1}{\sigma\sqrt{2\pi}} \int_{-\infty}^{+\infty} e^{-\frac{1}{2}\frac{(x-\mu)^2}{\sigma^2}} dx \qquad (15.6)$$

Assuming that σ must be positive, it can be shown that the area under $f(x)$ is equal to one. The relation between $f(x)$ and $F(x)$ for the normal curve is shown in Figure 15-4, where it is assumed that $\mu = 0$ and $\sigma = 1$.

Figure 15-3 The Relation Between the Functions $f(x)$ and $F(x)$

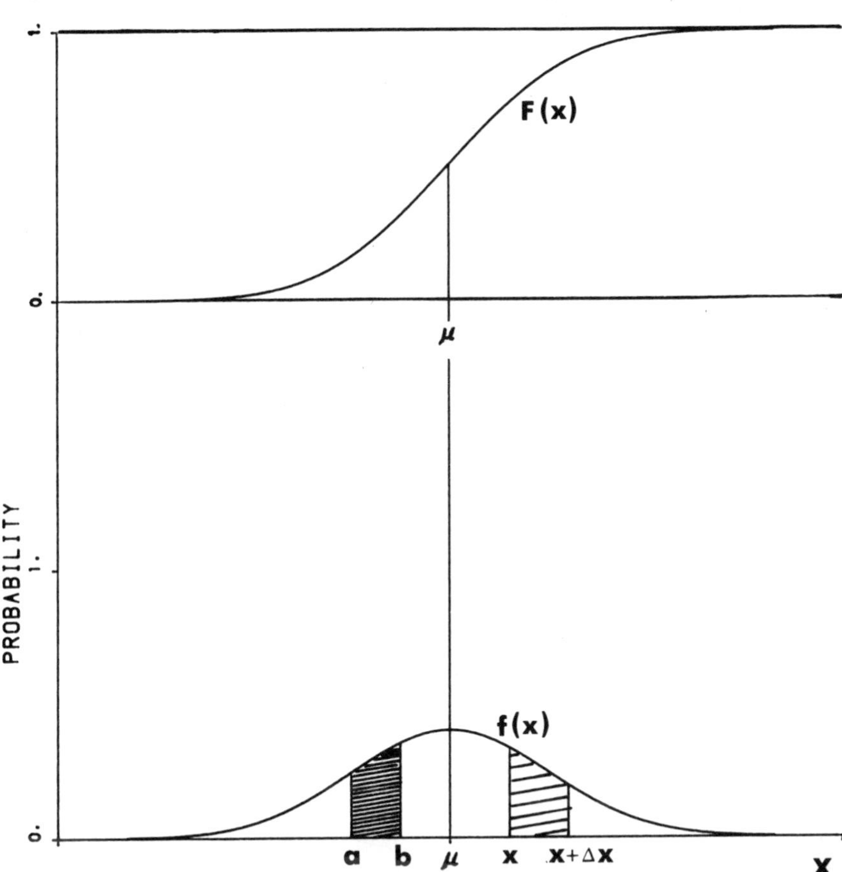

Fortunately, it is not necessary to perform the integration indicated by Equation 15.6 each time we wish to find a desired area under $f(x)$. Rather, when the distribution function of the normal curve is used to calculate probabilities, we need only refer to a special table similar to Table 15-3. This table shows the distribution function for a normal curve that has $\mu = 0$ and $\sigma = 1$. Equation 15.6 reveals that the normal curve depends on the values assumed by μ and σ, which implies that different curves are obtained for differing values of these two parameters. Given that Table 15-3 has been constructed for a normal curve that has $\mu = 0$ and $\sigma = 1$, it is necessary to transform the original scale of measurement into standard normal units in order to obtain the area under $f(x)$ for any normal distribution. Such a transformation is accomplished by the formula

$$Z = \frac{x - \mu}{\sigma} \tag{15.7}$$

Observe that Equation 15.7 simply converts the original scale in which the x's are measured into the corresponding standard normal unit represented by

Figure 15-4 The Normal Probability Density and Distribution Functions

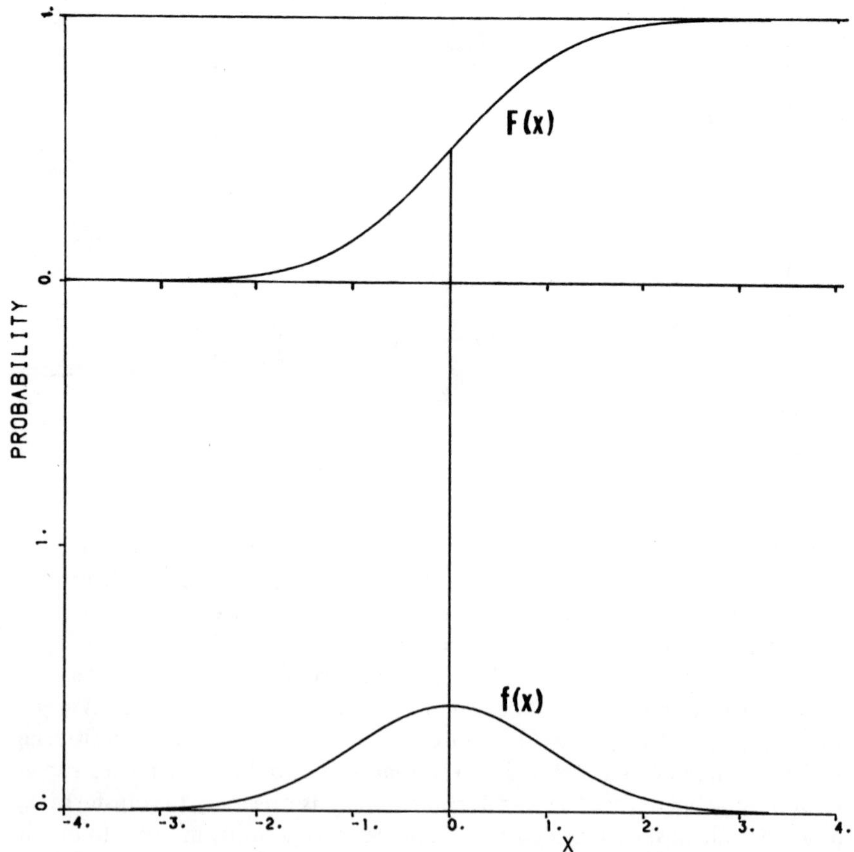

Table 15-3 The Standard Normal Distribution Function, $F(z)$*

z	0.00	0.01	0.02	0.03	0.04	0.05	0.06	0.07	0.08	0.09
−3.4	0.0003	0.0003	0.0003	0.0003	0.0003	0.0003	0.0003	0.0003	0.0003	0.0002
−3.3	0.0005	0.0005	0.0005	0.0004	0.0004	0.0004	0.0004	0.0004	0.0004	0.0003
−3.2	0.0007	0.0007	0.0006	0.0006	0.0006	0.0006	0.0006	0.0005	0.0005	0.0005
−3.1	0.0010	0.0009	0.0009	0.0009	0.0008	0.0008	0.0008	0.0008	0.0007	0.0007
−3.0	0.0013	0.0013	0.0013	0.0012	0.0012	0.0011	0.0011	0.0011	0.0010	0.0010
−2.9	0.0019	0.0018	0.0017	0.0017	0.0016	0.0016	0.0015	0.0015	0.0014	0.0014
−2.8	0.0026	0.0025	0.0024	0.0023	0.0023	0.0022	0.0021	0.0021	0.0020	0.0019
−2.7	0.0035	0.0034	0.0033	0.0032	0.0031	0.0030	0.0029	0.0028	0.0027	0.0026
−2.6	0.0047	0.0045	0.0044	0.0043	0.0041	0.0040	0.0039	0.0038	0.0037	0.0036
−2.5	0.0062	0.0060	0.0059	0.0057	0.0055	0.0054	0.0052	0.0051	0.0049	0.0048
−2.4	0.0082	0.0080	0.0078	0.0075	0.0073	0.0071	0.0069	0.0068	0.0066	0.0064
−2.3	0.0107	0.0104	0.0102	0.0099	0.0096	0.0094	0.0091	0.0089	0.0087	0.0084
−2.2	0.0139	0.0136	0.0132	0.0129	0.0125	0.0122	0.0119	0.0116	0.0113	0.0110
−2.1	0.0179	0.0174	0.0170	0.0166	0.0162	0.0158	0.0154	0.0150	0.0146	0.0143
−2.0	0.0228	0.0222	0.0217	0.0212	0.0207	0.0202	0.0197	0.0192	0.0188	0.0183
−1.9	0.0287	0.0281	0.0274	0.0268	0.0262	0.0256	0.0250	0.0244	0.0239	0.0233
−1.8	0.0359	0.0352	0.0344	0.0336	0.0329	0.0322	0.0314	0.0307	0.0301	0.0294
−1.7	0.0446	0.0436	0.0427	0.0418	0.0409	0.0401	0.0392	0.0384	0.0375	0.0367
−1.6	0.0548	0.0537	0.0526	0.0516	0.0505	0.0495	0.0485	0.0475	0.0465	0.0455
−1.5	0.0668	0.0655	0.0643	0.0630	0.0618	0.0606	0.0594	0.0582	0.0571	0.0559
−1.4	0.0808	0.0793	0.0778	0.0764	0.0749	0.0735	0.0722	0.0708	0.0694	0.0681
−1.3	0.0968	0.0951	0.0934	0.0918	0.0901	0.0885	0.0869	0.0853	0.0838	0.0823
−1.2	0.1151	0.1131	0.1112	0.1093	0.1075	0.1056	0.1038	0.1020	0.1003	0.0985
−1.1	0.1357	0.1335	0.1314	0.1292	0.1271	0.1251	0.1230	0.1210	0.1190	0.1170
−1.0	0.1587	0.1562	0.1539	0.1515	0.1492	0.1469	0.1446	0.1423	0.1401	0.1379
−0.9	0.1841	0.1814	0.1788	0.1762	0.1736	0.1711	0.1685	0.1660	0.1635	0.1611
−0.8	0.2119	0.2090	0.2061	0.2033	0.2005	0.1977	0.1949	0.1922	0.1894	0.1867
−0.7	0.2420	0.2389	0.2358	0.2327	0.2296	0.2266	0.2236	0.2206	0.2177	0.2148
−0.6	0.2743	0.2709	0.2676	0.2643	0.2611	0.2578	0.2546	0.2514	0.2483	0.2451
−0.5	0.3085	0.3050	0.3015	0.2981	0.2946	0.2912	0.2877	0.2843	0.2810	0.2776
−0.4	0.3446	0.3409	0.3372	0.3336	0.3300	0.3264	0.3228	0.3192	0.3156	0.3121
−0.3	0.3821	0.3783	0.3745	0.3707	0.3669	0.3632	0.3594	0.3557	0.3520	0.3483
−0.2	0.4207	0.4168	0.4129	0.4090	0.4052	0.4013	0.3974	0.3936	0.3897	0.3859
−0.1	0.4602	0.4562	0.4522	0.4483	0.4443	0.4404	0.4364	0.4325	0.4286	0.4247
−0.0	0.5000	0.4960	0.4920	0.4880	0.4840	0.4801	0.4761	0.4721	0.4681	0.4641
0.0	0.5000	0.5040	0.5080	0.5120	0.5160	0.5199	0.5239	0.5279	0.5319	0.5359
0.1	0.5398	0.5438	0.5478	0.5517	0.5557	0.5596	0.5636	0.5675	0.5714	0.5753
0.2	0.5793	0.5832	0.5871	0.5910	0.5948	0.5987	0.6026	0.6064	0.6103	0.6141
0.3	0.6179	0.6217	0.6255	0.6293	0.6331	0.6368	0.6406	0.6443	0.6480	0.6517
0.4	0.6554	0.6591	0.6628	0.6664	0.6700	0.6736	0.6772	0.6808	0.6844	0.6879
0.5	0.6915	0.6950	0.6985	0.7019	0.7054	0.7088	0.7123	0.7157	0.7190	0.7224
0.6	0.7257	0.7291	0.7324	0.7357	0.7389	0.7422	0.7454	0.7486	0.7517	0.7549
0.7	0.7580	0.7611	0.7642	0.7673	0.7704	0.7734	0.7764	0.7794	0.7823	0.7852
0.8	0.7881	0.7910	0.7939	0.7967	0.7995	0.8023	0.8051	0.8078	0.8106	0.8133
0.9	0.8159	0.8186	0.8212	0.8238	0.8264	0.8289	0.8315	0.8340	0.8365	0.8389
1.0	0.8413	0.8438	0.8461	0.8485	0.8508	0.8531	0.8554	0.8577	0.8599	0.8621
1.1	0.8643	0.8665	0.8686	0.8708	0.8729	0.8749	0.8770	0.8790	0.8810	0.8830
1.2	0.8849	0.8869	0.8888	0.8907	0.8925	0.8944	0.8962	0.8980	0.8997	0.9015
1.3	0.9032	0.9049	0.9066	0.9082	0.9099	0.9115	0.9131	0.9147	0.9162	0.9177
1.4	0.9192	0.9207	0.9222	0.9236	0.9251	0.9265	0.9278	0.9292	0.9306	0.9319
1.5	0.9332	0.9345	0.9357	0.9370	0.9382	0.9394	0.9406	0.9418	0.9429	0.9441
1.6	0.9452	0.9463	0.9474	0.9484	0.9495	0.9505	0.9515	0.9525	0.9535	0.9545
1.7	0.9554	0.9564	0.9573	0.9582	0.9591	0.9599	0.9608	0.9616	0.9625	0.9633
1.8	0.9641	0.9649	0.9656	0.9664	0.9671	0.9678	0.9686	0.9693	0.9699	0.9706
1.9	0.9713	0.9719	0.9726	0.9732	0.9738	0.9744	0.9750	0.9756	0.9761	0.9767
2.0	0.9772	0.9778	0.9783	0.9788	0.9793	0.9798	0.9803	0.9808	0.9812	0.9817
2.1	0.9821	0.9826	0.9830	0.9834	0.9838	0.9842	0.9846	0.9850	0.9854	0.9857
2.2	0.9861	0.9864	0.9868	0.9871	0.9875	0.9878	0.9881	0.9884	0.9887	0.9890
2.3	0.9893	0.9896	0.9898	0.9901	0.9904	0.9906	0.9909	0.9911	0.9913	0.9916
2.4	0.9918	0.9920	0.9922	0.9925	0.9927	0.9929	0.9931	0.9932	0.9934	0.9936
2.5	0.9938	0.9940	0.9941	0.9943	0.9945	0.9946	0.9948	0.9949	0.9951	0.9952
2.6	0.9953	0.9955	0.9956	0.9957	0.9959	0.9960	0.9961	0.9962	0.9963	0.9964
2.7	0.9965	0.9966	0.9967	0.9968	0.9969	0.9970	0.9971	0.9972	0.9973	0.9974
2.8	0.9974	0.9975	0.9976	0.9977	0.9977	0.9978	0.9979	0.9979	0.9980	0.9981
2.9	0.9981	0.9982	0.9982	0.9983	0.9984	0.9984	0.9985	0.9985	0.9986	0.9986
3.0	0.9987	0.9987	0.9987	0.9988	0.9988	0.9989	0.9989	0.9989	0.9990	0.9990
3.1	0.9990	0.9991	0.9991	0.9991	0.9992	0.9992	0.9992	0.9992	0.9993	0.9993
3.2	0.9993	0.9993	0.9994	0.9994	0.9994	0.9994	0.9994	0.9995	0.9995	0.9995
3.3	0.9995	0.9995	0.9995	0.9996	0.9996	0.9996	0.9996	0.9996	0.9996	0.9997
3.4	0.9997	0.9997	0.9997	0.9997	0.9997	0.9997	0.9997	0.9997	0.9997	0.9998

* This table is based on Table 1 of *Biometrika Tables for Statisticians, Volume I*, 3rd ed., Cambridge: University Press, 1966, by permission of the *Biometrika* trustees.

the letter Z. For example, suppose $x = 20$, $\mu = 10$, and $\sigma = 5$. Substituting into Equation 15.7, we find that

$$Z_{20} = \frac{20 - 10}{5} = 2$$

which implies that $x = 20$ is two standard deviations away from the mean of $\mu = 10$. Once the original scale in which the x's are measured has been transformed into standard normal units, we may use Table 15-3 to calculate desired probabilities.

Duration of Operative Procedures

To illustrate the use of the table, let us return to the problem concerning the duration of operative procedures. Assume that the duration of a given operative procedure is a continuous random variable that is distributed normally with a mean of $\mu = 8$ hours and a standard deviation of $\sigma = 2$ hours. Consider first the probability of encountering an operative procedure that lasts less than or equal to 12 hours. We might portray this problem as in Figure 15-5 where the shaded area represents the desired probability. In this case, the required probability is given by

$$\int_{-\infty}^{12} f(x)\,dx$$

After substituting $x = 12$, $\sigma = 2$, and $\mu = 8$ into Equation 15.7, we find that

$$Z_{12} = \frac{12 - 8}{2} = 2$$

Table 15.3 shows that $F(Z = 2)$ is .9772, which implies that we may be 98 percent certain that an operative procedure will last less than or equal to 12 hours.

When scheduling the use of the operating room, suppose we also are interested in determining the probability that an operative procedure will last less than or equal to six hours. This situation may be portrayed graphically as in Figure 15-6, where, as before, the shaded area represents the required probability. As can be verified the desired area is given by

$$\int_{-\infty}^{6} f(x)\,dx$$

As before, the use of Table 15.3 requires that we transform the original scale of measurement into standard normal units. In this case we have

$$Z_6 = \frac{6 - 8}{2} = -1$$

Table 15.3 shows that the tabular value of $F(Z = -1)$ is .1587 and, on the basis of this finding, we would infer that the likelihood of encountering an operative procedure that lasts less than or equal to six hours is .1587.

Figure 15-5 The Probability of Encountering an Operative Procedure That Lasts 12 Hours or Less

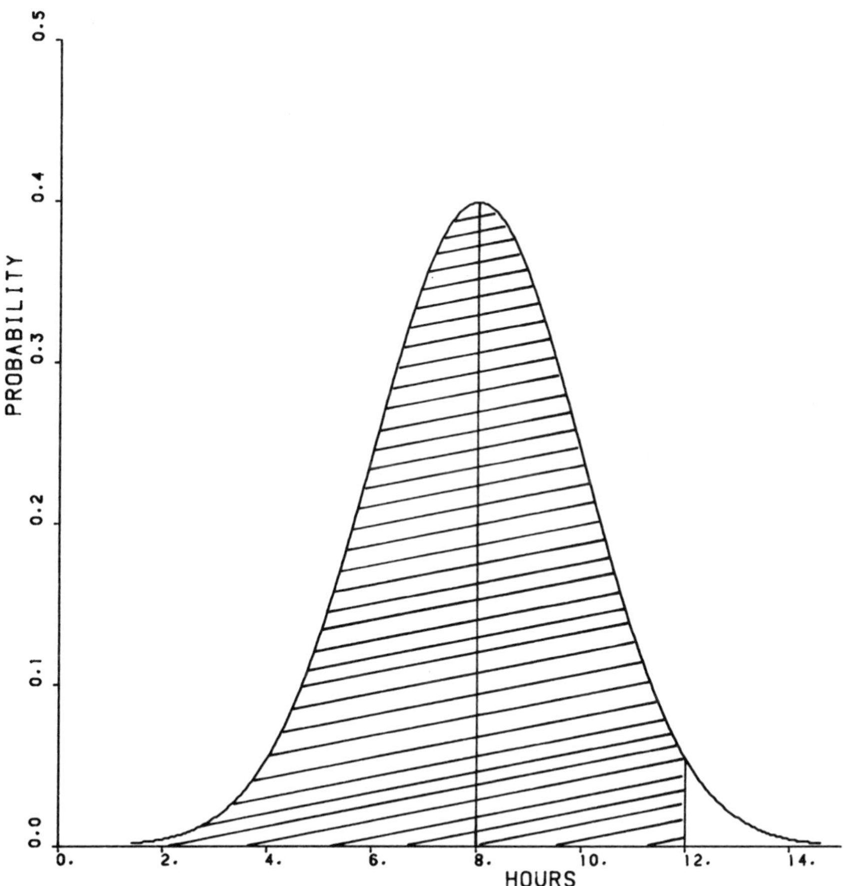

Figure 15-6 The Probability of Encountering an Operative Procedure That Lasts 6 Hours or Less

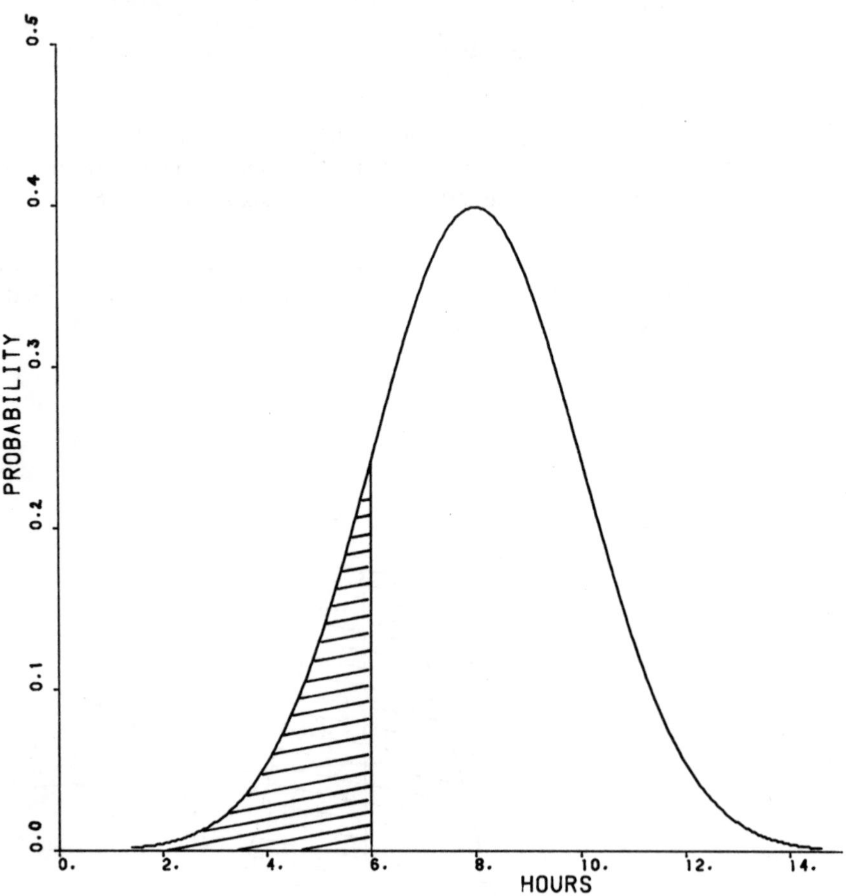

Consider next a slightly different problem in which we want to determine the percentage of operative procedures that have a duration between 9 and 13 hours. In this situation we are interested in the shaded area of Figure 15-7. On the basis of our earlier work, it can be readily seen that the required area is given by

$$\int_{-\infty}^{13} f(x)\,dx - \int_{-\infty}^{9} f(x)\,dx$$

Figure 15-7 The Probability of Encountering an Operative Procedure That Lasts Between 9 and 13 Hours

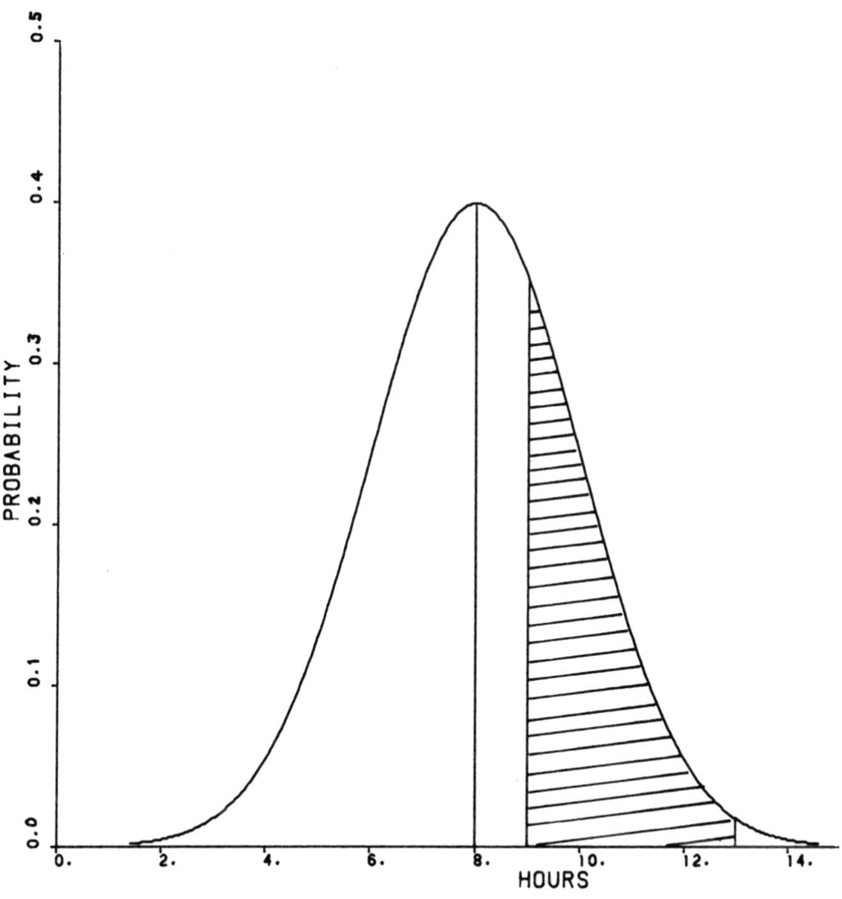

Transforming $x = 9$ and $x = 13$ into standard normal units, we find that

$$Z_9 = \frac{9 - 8}{2} = .5$$

and

$$Z_{13} = \frac{13 - 8}{2} = 2.5$$

Given that $F(Z = .5) = .6915$ and $F(Z = 2.5) = .9938$, we find that the desired probability is given by

$$F(Z = 2.5) - F(Z = .5) = .9938 - .6915$$
$$= .3023$$

In a similar fashion, suppose that we want to find the percentage of operative procedures that last two hours or more. This situation is displayed in Figure 15-8, where the shaded area represents the desired probability.

Figure 15-8 The Probability of Encountering an Operative Procedure That Lasts 2 Hours or More

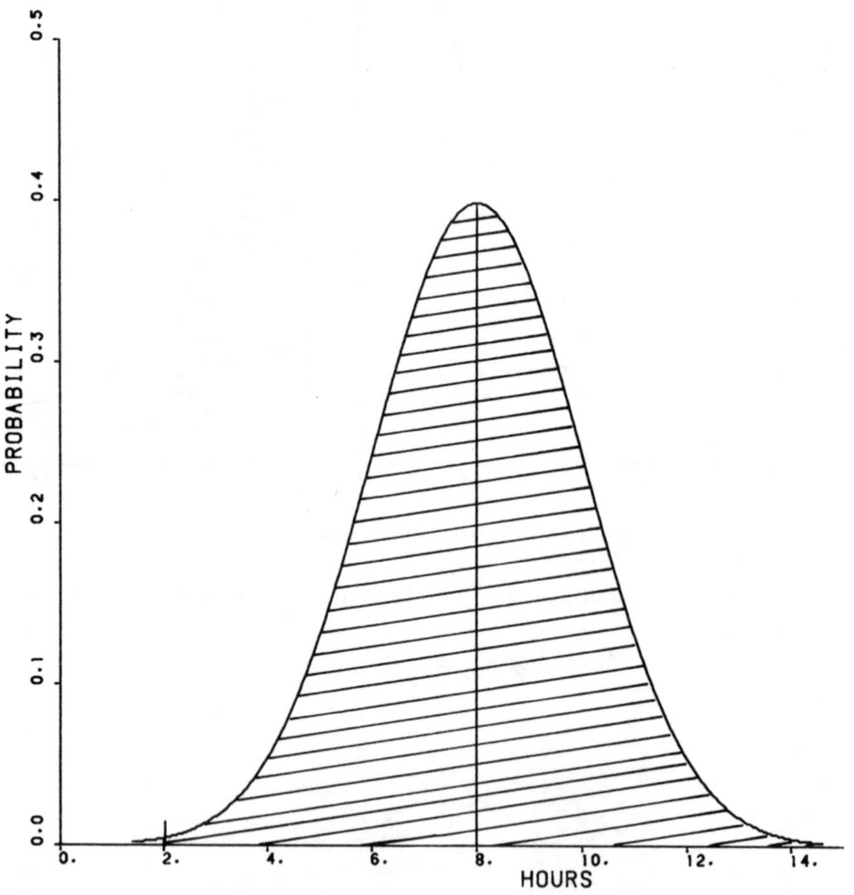

As can be easily verified, the desired probability is given by

$$\int_{-\infty}^{\infty} f(x)\,dx - \int_{-\infty}^{2} f(x)\,dx$$

However, since

$$\int_{-\infty}^{\infty} f(x)\,dx = 1$$

we find that the shaded area of Figure 15-8 is represented by

$$1 - \int_{-\infty}^{2} f(x)\,dx$$

As before, we transform $x = 2$ hours into standard normal units as follows

$$Z_2 = \frac{2 - 8}{2} = -3$$

Referring to Table 15-3, we find that $F(Z = -3) = .0013$, which implies that the probability of encountering an operative procedure that lasts two hours or more is $1 - .0013$ or .9987.

As an additional example, suppose we want to find the probability that an operative procedure will last between 3.5 and 10.2 hours. This situation is displayed in Figure 15-9 where the shaded area represents the required probability. In this case, the shaded area is given by

$$\int_{-\infty}^{10.2} f(x)\,dx - \int_{-\infty}^{3.5} f(x)\,dx$$

Similar to our earlier work, we transform the original scale of measurement into standard normal units as follows

$$Z_{10.2} = \frac{10.2 - 8}{2} = 1.1$$

$$Z_{3.5} = \frac{3.5 - 8}{2} = -2.25$$

Referring to Table 15-3, notice that $F(Z = 1.1)$ is .8643, while $F(Z = -2.25)$ is .0122. Consequently, the probability of encountering an operative procedure that lasts between 3.5 and 10.2 hours is $.8643 - .0122$ or .8521.

Now consider the probability of encountering an operative procedure that lasts 11.5 hours or more. In this situation, the shaded area of Figure 15-10 is represented by

$$\int_{-\infty}^{\infty} f(x)\,dx - \int_{-\infty}^{11.5} f(x)\,dx = 1 - \int_{-\infty}^{11.5} f(x)\,dx$$

Transforming the original scale of measurement into standard normal units we find that

Figure 15-9 The Probability of Encountering an Operative Procedure That Lasts Between 3.5 and 10.2 Hours

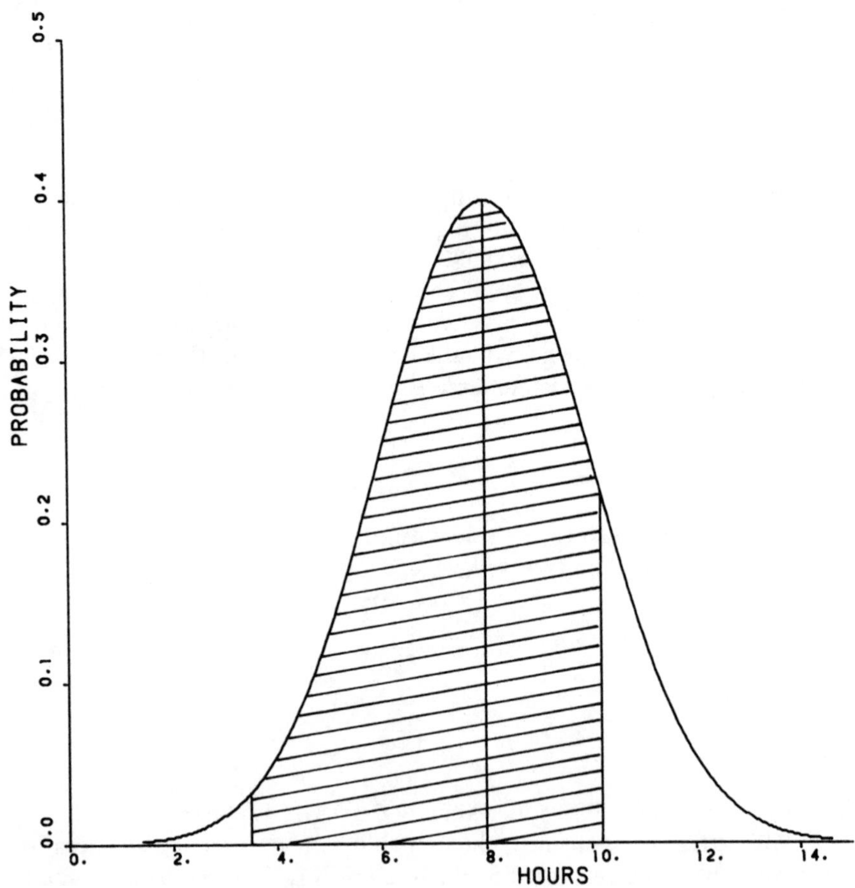

$$Z_{11.5} = \frac{11.5 - 8}{2} = 1.75$$

Referring to Table 15-3, we find the tabular value of $F(Z = 1.75)$ is .9599 and the required probability is given by

$$1 - F(Z = 1.75) = 1 - .9599$$
$$= .0401$$

Figure 15-10 The Probability of Encountering an Operative Procedure That Lasts 11.5 Hours or More

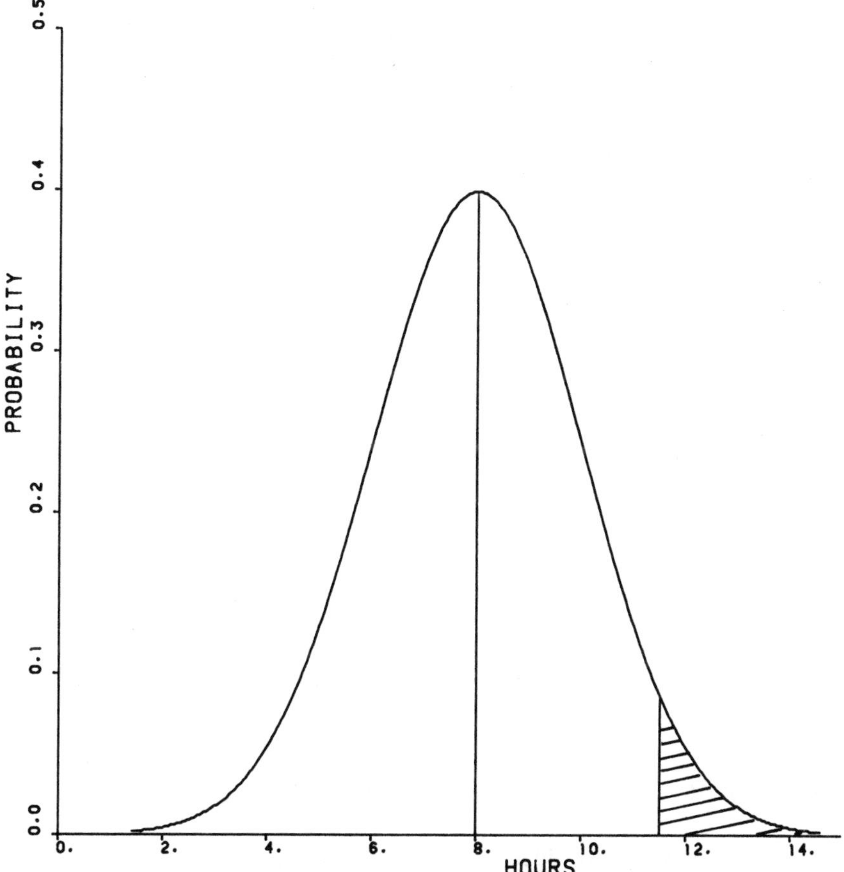

Emergency Room Staffing

The normal curve also may be used to address a slightly different problem. Suppose that the use of the emergency room may be viewed as a continuous random variable with a mean of 12 hours and a standard deviation of 4 hours. Now, suppose we want to be 95 percent confident that our institution is staffed so that required emergency services are provided to the population. In this case, we are given a percentage or a probability rather than a Z or x value. Referring to Figure 15-11, we find that our task is to find the standard normal unit that corresponds to .95. Table 15-3 shows that $F(Z = 1.64) =$ .9495 and, as a result we have

$$1.64 = \frac{x - 12}{4}$$

and

$$\begin{aligned} x &= 4(1.64) + 12 \\ &= 18.56 \text{ hours} \end{aligned}$$

Now, suppose that the additional labor required to provide an additional hour of emergency room service is constant and equal to 2 hours. Under this assumption, 2(18.56) or 37.12 manhours will be required in order to be 95 percent certain that our institution will be capable of providing required emergency room care. On the other hand, suppose that the additional labor, ΔL, required to provide an additional hour of care, Δx, is not constant. In this case, we might argue that

$$\frac{\Delta L}{\Delta x} = f(x)$$

As before, we find that

$$\int_0^{18.56} f(x)\,dx$$

indicates the number of manhours that are required to be 95 percent certain our institution will provide the required amount of emergency care to the community.

15.4 APPLICATIONS TO HEALTH CARE MANAGEMENT(O,A,F)

In this section we consider the use of normal distribution in monitoring, controlling, and planning the operations of the health care institution. Sup-

Figure 15-11 Determining the Number of Required Manhours

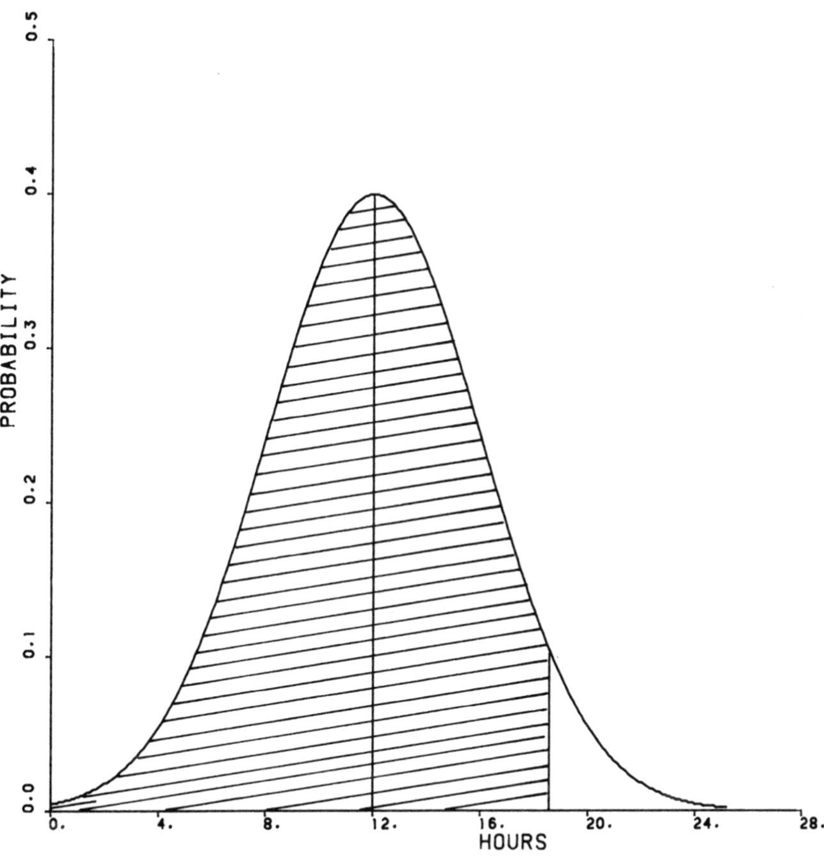

pose, for example, that the quantity of labor required to provide a given service is distributed normally with a mean of ten minutes and a standard deviation of five minutes. This situation may be displayed graphically, as in Figure 15-12. Suppose further that a given technician consistently requires 15 to 20 minutes to perform the service. In evaluating the performance of this employee, suppose management is interested in determining the probability that between 15 and 20 minutes is required to perform the service. As seen in Figure 15-12, the desired probability is represented by the shaded area and is given by

$$\int_{-\infty}^{20} f(x)\,dx - \int_{-\infty}^{15} f(x)\,dx$$

Figure 15-12 The Probability of Requiring Between 15 and 20 Minutes to Provide a Service

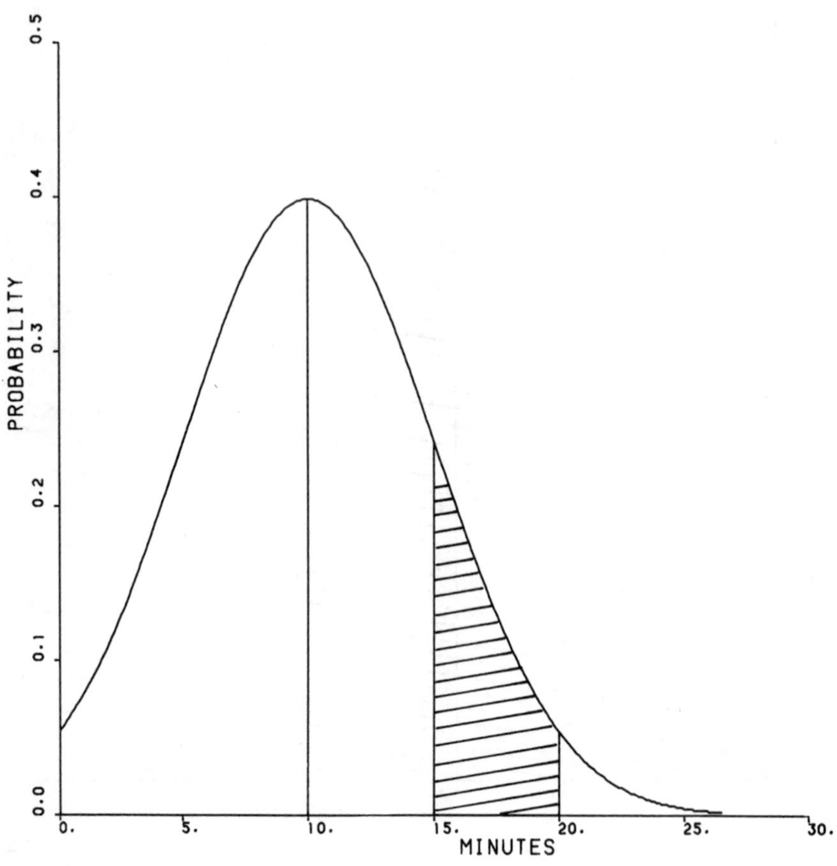

Given that $f(x)$ is assumed to be distributed normally, we need only calculate

$$Z_{15} = \frac{15 - 10}{5} = 1$$

and

$$Z_{20} = \frac{20 - 10}{5} = 2$$

Referring to Table 15-3 we find that $F(Z = 2) = .9772$ and $F(Z = 1) = .8413$. As a result, the required probability is given by

$$F(Z = 2) - F(Z = 1) = .9772 - .8413$$

Thus, there is a probability of .136 that the performance of this procedure will require between 15 and 20 minutes.

In referring to the same problem, suppose we are told that it requires more than 23 minutes to perform this procedure. In evaluating this assertion we need only determine the lined area of the normal curve in Figure 15-13. To find this area, we first calculate Z_{23}

$$Z_{23} = \frac{23 - 10}{5} = 2.6$$

Figure 15-13 The Probability of Requiring 23 Minutes or More to Provide Service

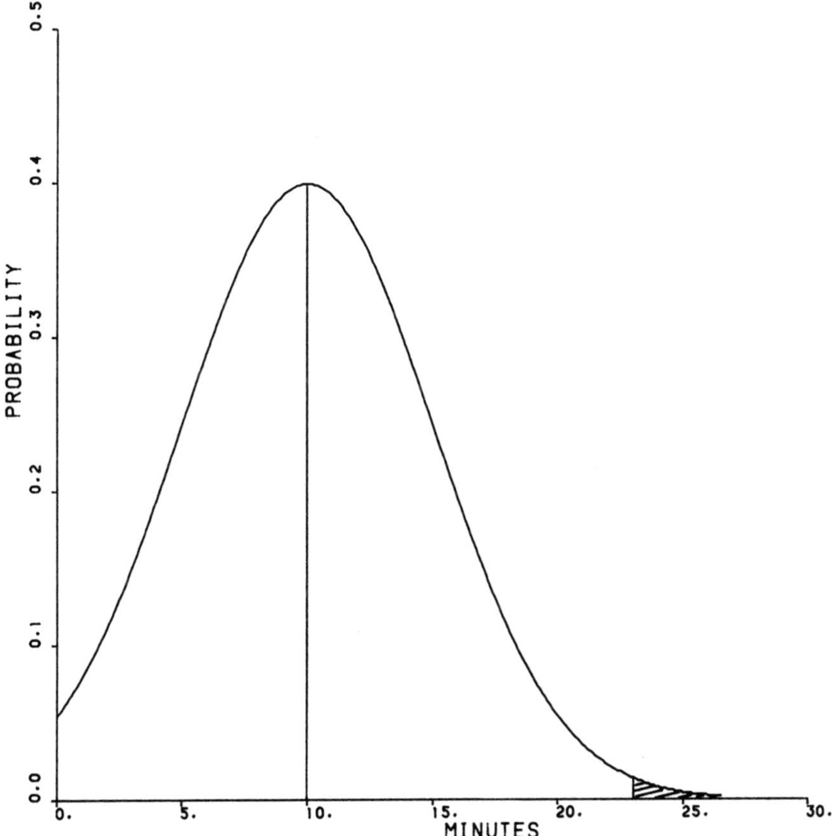

The required probability is given by $1 - F(Z = 2.6)$, which is found to be

$$1 - .9953 = .0047$$

Hence, there is a probability of .0047 that the procedure requires more than 23 minutes to perform.

A slightly different problem might be addressed by determining the amount of time that the procedure will require in 95 percent of the occasions on which the service is provided. In this case we are given relevant percentages and asked to determine the corresponding value as measured in time. Referring to Figure 15-14, we see that 2.5 percent of the data appear in each

Figure 15-14 The Probability of .95 for the Time Required to Provide a Service

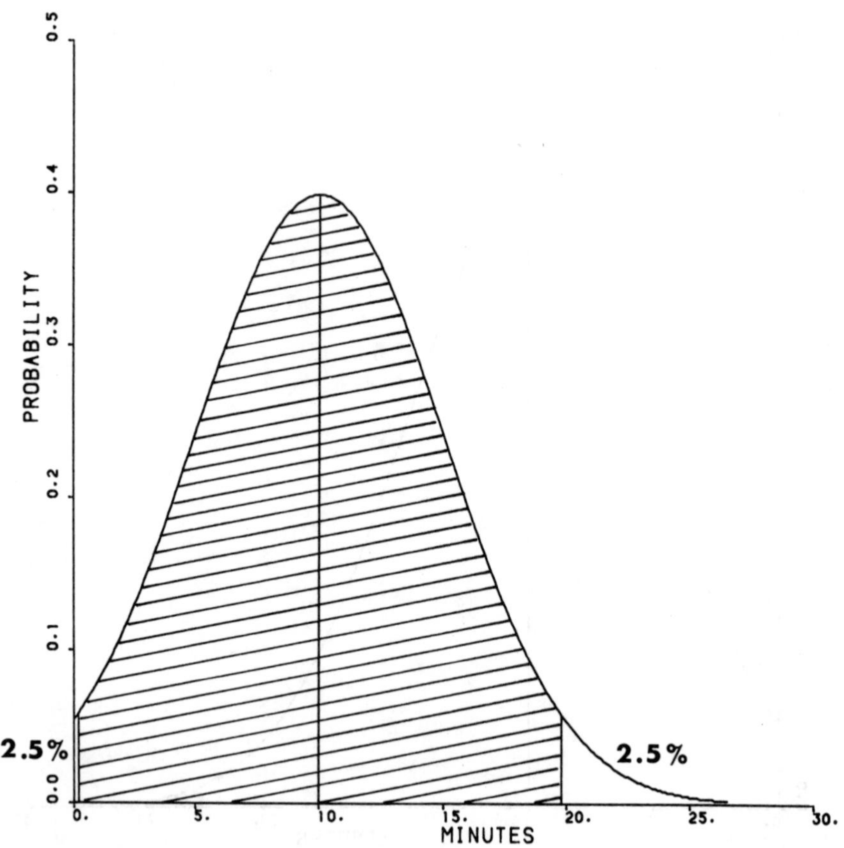

of the two tails of the distribution. Consequently, on 95 percent of the occasions, the service will require between x_1 and x_2 minutes. Referring to Table 15-3, we note that the standard normal unit that corresponds to 0.25 is 1.96. Hence, the values of x_1 and x_2 are found by substituting $Z = \pm 1.96$ into $Z = (x - \mu)/\sigma$ and solving for x. Thus, with respect to x_2 we find that

$$-1.96 = \frac{x_1 - 10}{5}$$

and

$$\begin{aligned} x_2 &= 5(1.96) + 10 \\ &= 19.8 \text{ minutes} \end{aligned}$$

Similarly, the value of x_1 is found by

$$-1.96 = \frac{x_1 - 10}{5}$$

which implies that

$$\begin{aligned} x_1 &= 5(-1.96) + 10 \\ &= .20 \end{aligned}$$

These calculations suggest that on 95 percent of the occasions on which the service is provided, between .20 and 19.8 minutes is required.

Problems for Solution

1. Suppose that the time required to perform a given operative procedure is distributed normally with mean $\overline{x} = 1.7$ hours and a standard deviation of 3.8 hours. Find the probability that a procedure will last:

   a. between 1 and 2.5 hours,
   b. less than 1.0 hour,
   c. more than 2.7 hours,
   d. between 2.0 and 2.4 hours.

2. Suppose that the demand for inventory during the lead time (i.e., the time between the placement and receipt of orders) is distributed normally with a mean of 20 units and a standard deviation of 15 units. Find the probability of using:

   a. between 17 and 26 units,
   b. fewer than 18 units,
   c. more than 22 units,
   d. more than 25 units,
   e. between 23 and 26 units

   during the lead time.
3. Suppose that the net cash flows (i.e., cash inflows − cash outflows) of the institution are distributed normally with a mean of $150,000 per period and a standard deviation of $80,000. Find the probability that the net cash flow is

   a. less than $60,000,
   b. between $10,000 and $70,000,
   c. more than $180,000,
   d. between $160,000 and $175,000.

4. Referring to Problem 1 above, determine the time the procedure will require in 95 percent of the occasions on which it is performed.
5. Referring to Problem 2, determine the largest and smallest demand that will occur during 88 percent of the lead time, 95 percent of the lead time, 99 percent of the lead time.

6. If the ages of the patients admitted to our institution are distributed normally with a mean of 32 years and a standard deviation of 16 years, what percentage of the patients is:

   a. greater than 65 years of age?
   b. between 28 and 45 years of age?
   c. less than 15 years of age?

Part V

Matrix Algebra

Chapter 16

Introduction to Matrix Algebra

Objectives

After completing this chapter, you should be able to:

1. Define a matrix, a vector, and a scalar;
2. Describe the size or the order of a matrix;
3. Construct a matrix from a tabular display of information;
4. Use and interpret basic matrix notation;
5. Use and interpret basic summation notation.

Chapter Map

The sections comprising this chapter may be summarized as follows:

Section Number	Required Reading	Optional Reading	Generic Development	Application to Management	Fundamental Principles	Complex Material
	(R)	(O)	(G)	(A)	(F)	(C)
16.1	x		x	x	x	
16.2	x		x	x	x	
16.2.1	x		x	x	x	
16.2.2	x		x	x	x	
16.3	x		x		x	
16.4		x		x	x	

The management of the health care facility is becoming increasingly quantitative, as has been mentioned. As a consequence, health care managers are confronted with vast amounts of numerical data that are generated by medical records, the accounting department, and agencies such as various levels of government that are external to the institution. In many cases, it is necessary to analyze these data so as to provide information useful to management. Any quantitative analysis involves the use of mathematics, and matrix algebra is one of a wide variety of mathematical techniques that are of value to the manager of the health care facility.

In this section, we examine the basic operations of matrix algebra and apply these techniques to problems that confront the health administrator. It should be noted, however, that the application of matrix algebra is not limited to the problems and illustrations in this book. Indeed, the rapid growth in the development of new theories and the application of matrix algebra to the problems of managing the health care facility preclude an exhaustive examination of the potential areas in which these techniques may be employed. In describing the application of matrix algebra to problems of managing the health care facility, we focus on examples of current practice as well as situations in which it may be used in most health care institutions.

Matrix algebra is a tool that enables us to describe the mathematical operations that must be performed on numeric information irrespective of the size of the set of data or the problems to which the information pertains. Thus, when matrix algebra is employed, the size of the problem or the amount of information does not influence our understanding of the mathematical operations that must be performed in order to obtain the desired results. Rather, these two factors influence only the amount of time required to perform the mathematical calculations. However, the time needed to perform mathematical operations has been reduced significantly in recent years by the increased use of electronic computers, which is another reason for examining the application of matrix algebra to the management of the health care facility. It almost goes without saying that, when a set of mathematical operations is expressed in terms of matrix algebra, computers may then be used to perform required calculations. For this reason, matrix algebra and high speed electronic computers are complementary tools that reduce both the time and energy required to perform mathematical manipulations.

More specifically, this chapter examines the use of matrices to display numeric information as well as the basic notation employed in matrix algebra. Chapter 17 describes the basic operations of matrix addition, subtraction, and multiplication. We then consider the usefulness of matrix algebra in problems requiring a linear transformation of one variable into another as well as several other matrix operations in Chapter 18. Finally, this part of the book concludes with a discussion of the determinant and the inverse of the matrix as applied to problems that confront the health care administrator. We begin our discussion by describing a matrix.

16.1 DESCRIPTION OF A MATRIX(R,G,A,F)

A matrix may be viewed as a device that is useful when displaying the information contained in a two-dimensional table. When dealing with two variables that may be classified into distinct groups, one of the variables defines the rows of the table and the other defines the columns. For example, suppose we are provided with information concerning the number of days of care used by the patient population of our institution during 1979 and that these data have been grouped in accordance with the presenting diagnosis and the age of the patient. We might display this information as shown in Table 16-1, where the age variable defines the columns and the presenting diagnosis defines the rows.

Ignoring the row and column totals, the data here may be extracted and presented in the form

$$\begin{bmatrix} 10 & 89 & 1{,}750 & 2{,}228 \\ 20 & 1{,}890 & 30 & 0 \\ 45 & 262 & 590 & 1{,}800 \\ 62 & 897 & 972 & 1{,}200 \end{bmatrix}$$

where the position of the entry determines its meaning. For example, the entry 262, which appears in the third row and the second column, refers to the number of days of care used by patients in the age range 20–39 who were

Table 16-1 The Number of Days of Care Used in 1979 by Diagnosis and Age

	Age				
Diagnosis	0–19	20–39	40–59	60–69	Total
Heart Disease	10	89	1,750	2,228	4,077
Pregnancy	20	1,890	30	0	1,940
Malignant Neoplasm	45	262	590	1,800	2,697
Benign Neoplasm	62	897	972	1,200	3,131
Total	137	3,138	3,342	5,228	11,845

hospitalized with a malignant neoplasm. Similarly, the entry 1,890 in the second row and second column is the number of days of care used by pregnant females in the range 20–39 years of age. These observations suggest that a row represents the use of care by patients presenting a given condition and the columns the use of care by all patients associated with a given age category.

In this example, we simply extract the utilization data from the two-dimensional table and employ this information to construct a matrix. A matrix, then, is nothing more than a rectangular array of numbers arranged in rows and columns. A matrix usually is identified by enclosing it in square brackets and we follow this procedure in this book.

> Given that matrix algebra is the algebra of matrices, each matrix is represented by a single symbol and treated as an entity. Even though we deal with the elements or individual entries within the arrays, it is the treatment of the matrix as an entity that constitutes matrix algebra.

Before presenting a formal definition of a matrix, it is necessary to review the mathematical notation frequently used in matrix algebra.

16.2 MATHEMATICAL NOTATION(R,G,A,F)

The purpose of the following discussion is to examine essentially two aspects of mathematical notation. The first aspect pertains to the notation commonly employed to denote a matrix and its elements while the second involves a specification and the meaning of summation notation.

16.2.1 Basic Matrix Notation(R,G,A,F)

When displaying matrices, the elements or terms of the matrix usually are denoted by lowercase letters with subscripts attached. Referring to a matrix that consists of four rows and four columns, we might define the matrix **A** by

$$\mathbf{A} = \begin{bmatrix} a_{11} & a_{12} & a_{13} & a_{14} \\ a_{21} & a_{22} & a_{23} & a_{24} \\ a_{31} & a_{32} & a_{33} & a_{34} \\ a_{41} & a_{42} & a_{43} & a_{44} \end{bmatrix}$$

where the elements or terms of the matrix are represented by the a's and the *subscripts* refer to the position of the *element* in the array. For example, the term a_{32} appears in the third row and the second column, just as the term a_{23} appears in the second row and third column. Referring to the matrix formed from Table 16-1, we find that the term a_{41} represents the use of care by patients in the age range 0–19 who were hospitalized with benign neoplasms.

> Thus, the first subscript attached to the term a refers to the row in which the element is located while the second identifies the column.

Consider next the general notation normally used to identify elements of a matrix. Using the variable subscripts i and j suggests that the element a_{ij} appears in row i and column j of the matrix. Accordingly, if we let $i = 3$, the term a_{3j} refers to the third row of the matrix. In terms of matrix **A**, the notation a_{3j}, for $j = 1, \cdots, 4$, refers to the elements a_{31}, a_{32}, a_{33}, and a_{34}. In this case, these elements define the third row of the matrix. Similarly, if we let $j = 3$, the term a_{i3} refers to the third column of the matrix. By way of illustration, consider the term a_{i3} of matrix **A**. Letting $i = 1, \cdots, 4$, it will be observed that the term a_{i3} represents the elements a_{13}, a_{23}, a_{33}, and a_{43}, which of course defines the third column of the matrix. Referring to the matrix formed by the data in Table 16-1, we find that the term a_{i3} for $i = 1$, 2, 3, 4 represents the use of care by patients in the age range 40–59 years while a_{i4} for $i = 1, 2, 3, 4$ represents the use of care by patients in the age range 60–69 years.

16.2.2 Summation Notation$^{(R,G,A,F)}$

Summation is a frequently used mathematical operation. The subscript notation introduced earlier provides the basis for expressing the operation of addition in concise form. For example, if we want to obtain the sum of the numbers represented by x_1, x_2, x_3, and x_4, we observe that

$$x_1 + x_2 + x_3 + x_4$$

provides us with the desired summation. This operation may be expressed more succinctly by

$$\Sigma x_i \qquad \text{for } i = 1, \cdots, 4$$

where the symbol Σ (the Greek letter sigma) indicates that the operation of summation is to be performed on the variable x_i for $i = 1, \cdots, 4$. An alternate method of expressing this summation is given by

$$\sum_{i=1}^{i=4} x_i$$

where the *limits of summation* on x_i are given by the terms "$i = 1$" and "$i = 4$" that appear above and below the Greek letter sigma. This notation indicates that x_i is to be summed for all integer values of i that range from 1 to 4. We might also express this summation by

$$\sum_{i=1}^{4} x_i$$

where the term "$i = 4$" has been replaced by the number 4. Thus,

$$\sum_{i=1}^{4} x_i = x_1 + x_2 + x_3 + x_4$$

In general, when indicating the sum of the numbers $x_1, x_2, \cdots, x_n$, we may write

$$\sum_{i=1}^{i=n} x_i = \sum_{i=1}^{n} x_i = x_1 + x_2 + \cdots + x_n = \Sigma x_i \qquad \text{for } i = 1, \cdots, n$$

Modifying this notation slightly, we may indicate the summation of x_i to include specific values of i while excluding others. For example, given the numbers x_1, x_2, x_3, x_4, x_5, and x_6, we find that

$$\sum_{i=3}^{5} x_i = x_3 + x_4 + x_5$$

includes the values of x_i for $i = 3$, 4, and 5 in the summation while the values of x_i for $i = 1$, 2, and 6 have been excluded. Similarly, we find that

$$\sum_{\substack{i=1 \\ i \neq 4}}^{i=6} x_i = x_1 + x_2 + x_3 + x_5 + x_6$$

includes all values of x_i for $i = 1, 2, 3, 5$, and 6. In this case, the value of x_i for $i = 4$ has been excluded from the summation.

By way of illustration, suppose that the use of hospital service by patients 1, 2, 3, 4, and 5 is measured by x_1, x_2, x_3, x_4, and x_5 days of care, respectively. Also suppose that $x_1 = 4$ days, $x_2 = 7$ days, $x_3 = 10$ days, $x_4 = 46$ days, and $x_5 = 13$ days. In this example, the total use of care by the five patients is given by

$$\sum_{i=1}^{5} x_i = 4 + 7 + 10 + 46 + 13$$

$$= 80 \text{ days}$$

Observe that the use of care by the fourth patient is much greater than the use of care by the other patients and it may be desirable to eliminate this patient from our calculations. In this case, we find that the sum

$$\sum_{\substack{i=1 \\ i \neq 4}}^{i=5} x_i = 4 + 7 + 10 + 13$$

$$= 34 \text{ days}$$

represents the use of service by patients 1, 2, 3, and 5.

> Although we have described summation notation in terms of simple addition, we may also use the Σ sign to refer to the sum of squares, sums of products, and, for that matter, the sum of any series that may be expressed by the use of appropriately attached subscripts.

For example,

$$\sum_{i=1}^{n} a_i^2 = a_1^2 + a_2^2 + \cdots + a_n^2$$

expresses the sum of the squared values of a_i for $i = 1, \cdots, n$. Similarly,

$$\sum_{i=1}^{n} P_i Q_i = P_1 Q_1 + P_2 Q_2 + \cdots + P_n Q_n$$

gives the sum of the products between the two series of numbers represented by P_i and Q_i for $i = 1, \cdots, n$.

On the basis of this discussion, we may now apply summation notation to the elements of a matrix. As before, we let the first subscript that is attached to the term a refer to the row of the matrix and the second subscript specify the column. If we define the matrix **A** as consisting of r rows and c columns, it is apparent that

$$\sum_{j=1}^{c} a_{1j} = a_{11} + a_{12} + \cdots + a_{1c}$$

represents the summation of the elements appearing in the first row of the matrix. As an example, let us refer to the matrix in which data pertaining to the use of service were displayed. As will be recalled, the rows of the matrix were defined by diagnosis and the columns by age, resulting in four rows and four columns. If we are interested in the total number of days of care used by all patients hospitalized with malignant neoplasms, the desired sum is given by

$$\begin{aligned}\sum_{j=1}^{4} a_{3j} &= a_{31} + a_{32} + a_{33} + a_{34} \\ &= 45 + 262 + 590 + 1{,}800 \\ &= 2{,}697\end{aligned}$$

days of care.

Alternatively, we might be interested in column summation. In general, if the matrix **A** consists of r rows and c columns, the expression

$$\sum_{i=1}^{r} a_{i3} = a_{13} + a_{23} + \cdots + a_{r3}$$

yields the summation of the third column of the matrix. As an example, let us refer to the matrix containing the use of service by diagnosis and age. If we are interested in the use of care by patients who were 40–59 years of age, the desired sum is given by

$$\sum_{i=1}^{4} a_{i3} = a_{13} + a_{23} + a_{33} + a_{34}$$

$$= 1{,}750 + 30 + 590 + 972$$
$$= 3{,}342$$

days of care.

We now consider sums of products that are germane to multiplication of two matrices. To anticipate this discussion, suppose we were given the two matrices

$$\mathbf{A} = \begin{bmatrix} a_{11} & a_{12} & a_{13} \\ a_{21} & a_{22} & a_{23} \end{bmatrix} \quad \text{and} \quad \mathbf{B} = \begin{bmatrix} b_{11} & b_{12} \\ b_{21} & b_{22} \\ b_{31} & b_{32} \end{bmatrix}$$

It will be observed that

$$\sum_{j=1}^{3} a_{1j}b_{j1} = a_{11}b_{11} + a_{12}b_{21} + a_{13}b_{31}$$

results in the sum of the products between the elements of the first row of matrix **A** and the elements of the first column of matrix **B**. Similarly,

$$\sum_{j=1}^{3} a_{2j}b_{j2} = a_{21}b_{12} + a_{22}b_{22} + a_{23}b_{32}$$

yields the sum of the products between the elements of the second row of matrix **A** and the elements of the second column of matrix **B**. Observe that, in b_{j1} and b_{j2}, the subscript j has been used to designate the rows of the matrix.

> There is nothing sacred about using i and j to represent rows and columns, respectively. Rather, it must be remembered that the first and second subscripts attached to the element of a matrix refer to the rows and columns respectively.

Finally, let us consider one further property of summation. For example, the summation of a_{ij} for $i = 1$ and $j = 1, \cdots, 4$ is given by

$$\sum_{j=1}^{4} a_{1j} = a_{11} + a_{12} + a_{13} + a_{14}$$

while summing a_{ij} for $j = 1$ and $i = 1, 2, 3$ is found to be

$$\sum_{i=1}^{3} a_{i1} = a_{11} + a_{21} + a_{31}$$

We might express these two summations in general terms by

$$\sum_{j=1}^{4} a_{ij} = a_{i1} + a_{i2} + a_{i3} + a_{i4}$$

and

$$\sum_{i=1}^{3} a_{ij} = a_{1j} + a_{2j} + a_{3j}$$

If we now sum $\Sigma^4{}_{j=1}\, a_{ij}$ with respect to i, we obtain the double summation

$$\begin{aligned}\sum_{i=1}^{3}\left(\sum_{j=1}^{4} a_{ij}\right) &= \sum_{i=1}^{3} (a_{i1} + a_{i2} + a_{i3} + a_{i4}) \\ &= (a_{11} + a_{12} + a_{13} + a_{14}) + (a_{21} + a_{22} + a_{23} + a_{24}) + \\ &\quad (a_{31} + a_{32} + a_{33} + a_{34})\end{aligned}$$

Similarly, we find that

$$\begin{aligned}\sum_{j=1}^{4}\left(\sum_{i=1}^{3} a_{ij}\right) &= \sum_{j=1}^{4} (a_{1j} + a_{2j} + a_{3j}) \\ &= (a_{11} + a_{21} + a_{31}) + (a_{12} + a_{22} + a_{32}) + \\ &\quad (a_{13} + a_{23} + a_{33}) + (a_{14} + a_{24} + a_{34})\end{aligned}$$

which is the same as $\Sigma^3{}_{i=1}\, (\Sigma^4{}_{j=1}\, a_{ij})$. Removing the parentheses, we find that

$$\sum_{i=1}^{3} \sum_{j=1}^{4} a_{ij} = \sum_{j=1}^{4} \sum_{i=1}^{3} a_{ij}$$

Thus, when performing double summation, the order in which the addition is performed does not influence the final result. As can be verified, the expression $\Sigma^r{}_{i=1}\, \Sigma^c{}_{j=1}\, a_{ij}$ yields the sum of the row totals in a matrix consisting of r rows and c columns while $\Sigma^c{}_{j=1}\, \Sigma^r{}_{i=1}\, a_{ij}$ yields the sum of the corresponding column totals. In terms of our earlier example, these expressions both yield a total of 11,845 days of care. This may be seen by observing that the sum of the row totals is given by

$$4{,}077 + 1{,}940 + 2{,}697 + 3{,}131$$

while the sum of the column totals is found to be

$$137 + 3{,}138 + 3{,}342 + 5{,}228$$

Both, of course, sum to a total of 11,845 days of care.

16.3 MATRIX DEFINITIONS(R,G,F)

> A matrix is a rectangular array of numbers that are arranged in rows and columns. Implicit in this definition is the requirement that the rows must be of equal length as must all of the columns.

As before, we let a_{ij} correspond to the element that appears in row i and column j of matrix **A**.

In general, if the matrix **A** has r rows and c columns, we may display it as follows

$$\mathbf{A} = \begin{bmatrix} a_{11} \cdot & \cdot & \cdot a_{1j} \cdot & \cdot & \cdot a_{1c} \\ \cdot & & \cdot & & \cdot \\ \cdot & & \cdot & & \cdot \\ \cdot & & \cdot & & \cdot \\ a_{i1} \cdot & \cdot & \cdot a_{ij} \cdot & \cdot & \cdot a_{ic} \\ \cdot & & \cdot & & \cdot \\ \cdot & & \cdot & & \cdot \\ \cdot & & \cdot & & \cdot \\ a_{r1} \cdot & \cdot & \cdot a_{rj} \cdot & \cdot & \cdot a_{rc} \end{bmatrix}$$

Here, the dots indicate the presence of other elements. For example, the dots in row i indicate that the elements a_{i1}, a_{i2} continue sequentially until the element a_{ij} is reached. Similarly, the dots in the first column indicate that the element a_{11} is followed sequentially by the elements a_{21}, a_{31}, etc., until the element a_{i1} is reached. This notation is common and will be used in later sections.

The *size of the matrix* is measured by the number of rows and columns that appear in the array. Thus, a matrix consisting of r rows and c columns is said to be of *order* (or *dimension*) $r \times c$, which is read "r by c." We may describe the order or size of a matrix by the notation $\mathbf{A}_{r\times c}$ where the r refers to the number of rows and the c the number of columns in the matrix. For example, if we let

$$\mathbf{B} = \begin{bmatrix} -7 & 1/3 & -14 \\ 0 & 17.3 & 2 \\ -6 & 0 & 6 \\ 2 & 8.85 & 7 \end{bmatrix}$$

we would say that the matrix **B** is of order 4×3 since it consists of four rows and three columns. It also will be observed that the elements of a matrix may

1. assume the value zero,
2. be integers,
3. be positive or negative, and
4. be expressed in fractional or decimal form.

In many situations we are interested in a matrix in which the number of rows is equal to the number of columns (i.e., $r = c$). This is referred to as a *square matrix* of order r. When dealing with such a matrix, the elements appearing along its *diagonal* are found to be $a_{11}, a_{22}, \cdots, a_{rr}$. For example, consider the square matrix

$$\mathbf{B} = \begin{bmatrix} b_{11} & 0 & 0 \\ 0 & b_{22} & 0 \\ 0 & 0 & b_{33} \end{bmatrix}$$

where the elements b_{11}, b_{22}, and b_{33} appear along the diagonal. When all of the nondiagonal elements of a square matrix are zero, the matrix is referred to as a *diagonal matrix*. As an example, the matrix

$$\mathbf{C} = \begin{bmatrix} 1 & 0 & 0 \\ 0 & 7 & 0 \\ 0 & 0 & -6 \end{bmatrix}$$

is a diagonal matrix. A slightly different form of the square matrix involves one in which all of the elements above or below the diagonal are zero. As examples, consider

$$\mathbf{C} = \begin{bmatrix} 5 & 1 & 3 \\ 0 & 6 & 7 \\ 0 & 0 & 8 \end{bmatrix} \quad \text{and} \quad \mathbf{F} = \begin{bmatrix} 1 & 0 & 0 \\ 6 & 9 & 0 \\ 8 & 3 & 2 \end{bmatrix}$$

which are called *triangular matrices.* More specifically, the matrix **F** is a *lower triangular* matrix and the matrix **C** is an *upper triangular* matrix.

When a matrix consists of a single row or a single column, it is referred to as a row or column *vector*. For example,

$$\mathbf{x} = \begin{bmatrix} x_1 \\ x_2 \\ x_3 \end{bmatrix}$$

is a *column vector* of order 3 × 1. Similarly a matrix consisting of a single row is called a *row vector*. Employing a prime to represent row vectors, we find that

$$\mathbf{x}' = [x_1 \quad x_2 \quad x_3 \quad x_4 \quad x_5]$$

is a matrix of order 1 × 5.

As a final definition, we refer to a single number such as 5, −3.7, or 1/3 as a *scalar*. Following the procedure employed above, we might view a scalar as a 1 × 1 matrix.

16.4 APPLICATIONS TO HEALTH CARE MANAGEMENT(O,A,F)

Matrices may be used to display data pertaining to a two-dimensional table, as should be clear from the foregoing discussion. In addition, once these data have been presented in matrix form, summation notation may be used to calculate desired row and column totals as well as the sum of all elements appearing in the matrix. It is obvious that the computation of these totals does not require the use of a matrix to display the information. However, as will be seen in later chapters, summarizing data in matrix form permits us to express complex mathematical operations in simplified terms irrespective of the size of the matrix.

Although it is always desirable to demonstrate the applicability of each new topic as it is introduced, the development and operations of matrix algebra require an understanding of several intermediate steps for which few problems in health administration may be used as illustrations. As a result, even though several of the problems at the end of this chapter are related to health care management, the primary purpose of these exercises is to demonstrate the concepts and mathematical procedures presented in this chapter. After these concepts and procedures have been mastered, it is possible to introduce more advanced techniques, which in turn may be used when solving problems confronting the health administrator.

Problems for Solution

1. Suppose we are provided with the following information concerning the number of dental visits during the past year that have been grouped by time of day and the age of the patient.

	Age			
Time of Day	5–14	15–24	25–34	35–44
8–12 a.m.	50	95	62	78
1–5 p.m.	72	107	96	81
6–8 p.m.	12	23	72	186

a. Display these data in the matrix **A** that has the order 3 × 4.
b. Letting a_{ij} represent the elements of **A**, compute

$$\sum_{j=1}^{4} a_{1j};\ \sum_{j=1}^{4} a_{2j};\ \sum_{j=1}^{4} a_{3j};\ \sum_{i=1}^{3} a_{i1};\ \sum_{i=1}^{3} a_{i2};\ \sum_{i=1}^{3} a_{i3};\ \sum_{i=1}^{3} a_{i4};$$

$$\sum_{j=1}^{4} \sum_{i=1}^{3} a_{ij};\ \sum_{i=1}^{3} \sum_{j=1}^{4} a_{ij}$$

c. Describe the meaning of each of these summations.

2. Suppose that the average cost of producing service S_1 is \$10, of S_2 is \$20, and of S_3 is \$30. Also, suppose that the number of units of each of these three services produced during April and May were as follows:

	April	May
S_1	500	600
S_2	50	90
S_3	70	300

Letting

$$\mathbf{a}' = [10 \quad 20 \quad 30] = [a_{11} \quad a_{12} \quad a_{13}]$$

and

$$\mathbf{B} = \begin{bmatrix} 500 & 600 \\ 50 & 90 \\ 70 & 300 \end{bmatrix} = \begin{bmatrix} b_{11} & b_{12} \\ b_{21} & b_{22} \\ b_{31} & b_{32} \end{bmatrix}$$

compute $\Sigma^3{}_{j=1}\, a_{1j}b_{j1}$ and $\Sigma^3{}_{j=1}\, a_{1j}b_{j2}$. What meaning would you attribute to each of these sums?

3. When computing depreciation charges under the sum of the year digits methods, the amount to be depreciated is multiplied by a fraction consisting of a numerator that equals the number of years of life remaining and a denominator that equals the sum of the total years of life. For assets having 5, 6, 7, 8, and 9 years of estimated life, compute the denominator of this fraction (Hint: $\Sigma^n{}_{r=1}\, r$).
4. Given that

$$\mathbf{A} = \begin{bmatrix} 5 & -8 & -2 \\ 1/2 & 12 & 16 \\ -4 & -2 & 0 \end{bmatrix}; \qquad \mathbf{B} = \begin{bmatrix} 1 & 0 & 0 \\ 5 & 1 & 0 \\ 6 & 7 & 1 \end{bmatrix}$$

$$\mathbf{C} = \begin{bmatrix} 15 & 60 & 93 \\ 12 & 30 & -1/2 \\ 0 & 0 & 4 \\ .8 & .1 & .9 \end{bmatrix}; \qquad \mathbf{D} = \begin{bmatrix} 1 & 6 & -9 \\ 0 & 3 & 4 \\ 0 & 0 & 6 \end{bmatrix}$$

$$\mathbf{E} = \begin{bmatrix} 0 & 1 \\ 1 & 0 \end{bmatrix}$$

which are:

1. square matrices?
2. diagonal matrices?
3. upper triangular matrices?
4. lower triangular matrices?

5. Suppose we are provided the following information concerning the number of physicians who have been classified by area of practice and specialization.

Specialization	East	Central	West
	Area		
Surgery	9,000	400	800
General Medicine	6,500	1,200	975
ENT	800	400	326
Dermatology	540	746	930
Urology	627	400	150
Neurology	480	96	75
Other	532	492	82

a. Display these data in matrix form.

b. Letting a_{ij} represent the elements of this matrix, compute

$$\sum_{j=1}^{3} a_{1j};\ \sum_{i=1}^{7} a_{i2};\ \sum_{i=1}^{7} a_{i3};\ \sum_{j=1}^{3} a_{4j};\ \sum_{j=1}^{3} a_{6j};\ \sum_{j=1}^{3} a_{3j};$$

$$\sum_{j=1}^{3} \sum_{i=1}^{7} a_{ij} \quad \text{and} \quad \sum_{i=1}^{7} \sum_{j=1}^{3} a_{ij}$$

c. What is the meaning of each of these summations?

6. Show that

$$\sum_{i=1}^{3} P_i Q_i \neq \left(\sum_{i=1}^{3} P_i\right)\left(\sum_{i=1}^{3} Q_i\right)$$

7. Show that

$$\sum_{i=1}^{r} \sum_{j=1}^{c} 6\, a_{ij} = 6 \sum_{j=1}^{c} \sum_{i=1}^{r} a_{ij}$$

Chapter 17

Basic Matrix Operations

Objectives

After completing this chapter, you should be able to:

1. Perform the matrix operations of addition, subtraction, and multiplication;
2. Define the null and identity matrices;
3. Understand the laws of algebra as applied to matrices;
4. Define the conditions under which matrices are conformable for addition, subtraction, and multiplication.

Chapter Map

The sections comprising this chapter may be summarized as follows:

Section Number	Required Reading	Optional Reading	Generic Development	Application to Management	Fundamental Principles	Complex Material
	(R)	(O)	(G)	(A)	(F)	(C)
17.1	x		x		x	
17.2	x		x	x	x	
17.3	x		x	x	x	
17.4	x		x	x	x	
17.4.1	x		x	x	x	
17.4.2	x		x	x	x	
17.4.3	x		x	x	x	
17.4.4		x	x		x	
17.5	x		x		x	
17.6		x	x		x	
17.7		x		x	x	
Appendix	x		x		x	

17.1 INTRODUCTION(R,G,F)

On the basis of our understanding of the concepts and mathematical procedures presented in the previous chapter, we now turn to the simplest of the operations that may be performed on matrices. In this chapter, we consider the operations of matrix addition, subtraction, and multiplication. In addition, we introduce the matrix algebra counterparts of zero and unity.

17.2 MATRIX ADDITION(R,A,G,F)

Suppose that we measure the output of institutions A_1, A_2, and A_3 in terms of the number of laboratory procedures, surgical procedures, outpatient visits, and days of care provided during a given period. Also, assume that the data in Table 17-1 pertain to the first year of our study period. As before, we may extract the information in this table and display the data in the form

$$\mathbf{A} = \begin{bmatrix} 97 & 102 & 83 \\ 120 & 132 & 156 \\ 47 & 56 & 47 \\ 20 & 14 & 15 \end{bmatrix}$$

which is a matrix of order 4×3. Retaining the same order of presentation, we also may display the provision of these components of care by the three institutions during the second year of our study period in matrix form as follows

$$\mathbf{B} = \begin{bmatrix} 98 & 132 & 76 \\ 196 & 223 & 195 \\ 48 & 53 & 62 \\ 16 & 23 & 34 \end{bmatrix}$$

As can be verified, the number of laboratory procedures performed by institution A_1 during the *two-year* period is given by the sum of the elements appearing in the first row and the first column of matrix **A** and matrix **B**. Thus, we find that institution A_1 provided $97 + 98$ or 195 laboratory procedures during the two-year period. Similarly, the number of outpatient visits handled by institution A_2 in the two-year period is found by summing

Table 17-1 Provision of Care by Type of Institution and Component of Service, 198–

Component of Service	Institution		
	A_1	A_2	A_3
Laboratory Procedures	97	102	83
Days of Care	120	132	156
Outpatient Visits	47	56	47
Operative Procedures	20	14	15

the elements in the third row and the second column of each matrix. In this case, we find that 56 + 53 or 109 outpatient visits were provided by the institution during the period. If we compute all such sums, we find that

$$\mathbf{A} + \mathbf{B} = \begin{bmatrix} 97 + 98 & 102 + 132 & 83 + 76 \\ 120 + 196 & 132 + 223 & 156 + 195 \\ 47 + 48 & 56 + 53 & 47 + 62 \\ 20 + 16 & 14 + 23 & 15 + 34 \end{bmatrix}$$

$$= \begin{bmatrix} 195 & 234 & 159 \\ 316 & 355 & 351 \\ 95 & 109 & 109 \\ 36 & 37 & 49 \end{bmatrix}$$

yields the quantity of each component of care at each of the institutions during the two-year period. We might represent this matrix summation by

$$\mathbf{A} + \mathbf{B} = \mathbf{C}$$

where

$$\mathbf{C} = \begin{bmatrix} c_{11} & c_{12} & c_{13} \\ c_{21} & c_{22} & c_{23} \\ c_{31} & c_{32} & c_{33} \end{bmatrix}$$

Representing the elements of **A** and **B** by a_{ij} and b_{ij}, respectively, we find that the element c_{11} of the matrix **C** is obtained by the sum of a_{11} and b_{11} just as the element c_{32} is obtained by summing the elements a_{32} and b_{32}. In general, then, the element c_{ij} is obtained by summing the elements a_{ij} and b_{ij}, which implies that the matrix **C** is formed by adding the matrices **A** and **B** element by element.

As should be obvious from this illustration,

> matrix addition is permitted only if the two matrices are of the same order. That is, in order to add **A** and **B**, the two matrices must have the same number of rows and columns.

Thus, we say that

$$\mathbf{A} = \begin{bmatrix} 1 & 7 & 6 \\ 4 & 12 & 16 \end{bmatrix} \quad \text{and} \quad \mathbf{B} = \begin{bmatrix} 14 & 23 & 16 \\ 1 & 0 & 7 \end{bmatrix}$$

are *conformable for addition* and their sum is

$$\begin{bmatrix} 15 & 30 & 22 \\ 5 & 12 & 23 \end{bmatrix}$$

However, the matrices

$$\mathbf{A} = \begin{bmatrix} 15 & 5 \\ 30 & 12 \\ 22 & 23 \end{bmatrix} \quad \text{and} \quad \mathbf{B} = \begin{bmatrix} 14 & 23 & 16 \\ 1 & 0 & 7 \end{bmatrix}$$

may not be added since the number of rows and columns in the two matrices are not equal.

A special case of matrix addition is referred to as *scalar multiplication*. For example, if we let

$$\mathbf{A} = \begin{bmatrix} 1 & 2 \\ 3 & 4 \end{bmatrix}$$

we find that

$$\mathbf{A} + \mathbf{A} = \begin{bmatrix} 2 & 4 \\ 6 & 8 \end{bmatrix} = 2\mathbf{A} \qquad \textbf{(17.1)}$$

Extending this finding to a more general case, we have

$$\theta\mathbf{A} = \mathbf{A} + \mathbf{A} + \cdots + \mathbf{A}$$

where there are $\theta\mathbf{A}$'s in the summation. This result, when using any numerical value of θ, serves to define scalar multiplication. Thus,

> when we multiply the matrix **A** by the scalar θ, every element in **A** is multiplied by θ.

Note that a scalar is a single number rather than a matrix or vector. For example, letting

$$\theta = \frac{1}{3} \quad \text{and} \quad \mathbf{A} = \begin{bmatrix} 9 & 6 & 12 \\ 3 & 9 & 15 \\ 21 & 90 & 180 \end{bmatrix}$$

we find that

$$\theta\mathbf{A} = \frac{1}{3}\begin{bmatrix} 9 & 6 & 12 \\ 3 & 9 & 15 \\ 21 & 90 & 180 \end{bmatrix} = \begin{bmatrix} 3 & 2 & 4 \\ 1 & 3 & 5 \\ 7 & 30 & 60 \end{bmatrix}$$

17.3 MATRIX SUBTRACTION(R,G,A,F)

Suppose that the data in Table 17-2 represent the quantity and composition of care provided by four institutions during the course of a given year. As before, these data may be extracted and displayed in the form

$$\mathbf{A} = \begin{bmatrix} 1{,}896 & 2{,}740 & 982 & 4{,}860 \\ 30{,}620 & 8{,}962 & 1{,}723 & 12{,}690 \\ 1{,}926 & 18{,}630 & 14{,}792 & 6{,}973 \\ 1{,}027 & 892 & 763 & 4{,}826 \end{bmatrix}$$

which is a matrix of order 4 × 4. Suppose we also are provided with the cumulative amount of each service at these institutions as of June 30 in the same year. Retaining the same order of presentation, these data may be displayed in matrix form as follows:

$$\mathbf{B} = \begin{bmatrix} 896 & 740 & 82 & 860 \\ 620 & 962 & 723 & 690 \\ 926 & 630 & 792 & 973 \\ 27 & 92 & 63 & 826 \end{bmatrix}$$

In this case, we may determine the quantity of laboratory services provided by institution A_1 between July 1 and December 31 by subtracting the element appearing in the first row and the first column of matrix **B** from the element in the first row and the first column of matrix **A**. Thus, we find that 1,896 − 896 or 1,000 laboratory procedures were performed by this institution during the last half of the year. Similarly, the number of operative procedures performed in institution A_2 is obtained by subtracting the element in the fourth row and the second column of matrix **B** from the element in the fourth row and the second column of matrix **A**. Thus, we find that 892 − 92 or 800 operative procedures were performed in this institution in the last half of the year. Replicating this procedure for all services and all institutions we find that

$$\mathbf{A} - \mathbf{B} = \begin{bmatrix} 1{,}896 - 896 & 2{,}740 - 740 & 982 - 82 & 4{,}860 - 860 \\ 30{,}620 - 620 & 8{,}962 - 962 & 1{,}723 - 723 & 12{,}690 - 690 \\ 1{,}926 - 926 & 18{,}630 - 630 & 14{,}792 - 792 & 6{,}973 - 973 \\ 1{,}027 - 27 & 892 - 92 & 763 - 63 & 4{,}826 - 826 \end{bmatrix}$$

$$= \begin{bmatrix} 1{,}000 & 2{,}000 & 900 & 4{,}000 \\ 30{,}000 & 8{,}000 & 1{,}000 & 12{,}000 \\ 1{,}000 & 18{,}000 & 14{,}000 & 6{,}000 \\ 1{,}000 & 800 & 700 & 4{,}000 \end{bmatrix}$$

Table 17-2 Provision of Care by Type of Institution and Component of Service, 197-

Component of Service	Institution			
	A_1	A_2	A_3	A_4
Laboratory Procedures	1,896	2,740	982	4,860
Days of Care	30,620	8,962	1,723	12,690
Outpatient Visits	1,926	18,630	14,792	6,973
Operative Procedures	1,027	892	763	4,826

represents the quantity and composition of service provided by the four institutions during the last half of the year. In general, if we let a_{ij} represent the elements of **A** and b_{ij} represent the elements of **B**, we find that

$$\mathbf{A} - \mathbf{B} = \mathbf{C} \quad \mathbf{(17.2)}$$

where the element c_{ij} of the matrix **C** is given by $a_{ij} - b_{ij}$. This implies that

> the difference between two matrices is the matrix formed by subtracting matrix **B** from matrix **A**, element by element. As should be obvious, *matrix subtraction* is permitted only when the matrices involved are of the same order. As a result, matrices that are conformable for addition are also comformable for subtraction.

17.4 MATRIX MULTIPLICATION(R,G,A,F)

In this section, we consider the multiplication of matrices. First, however, it is useful to examine the product of vectors and vector-matrix products since these operations provide the basis for understanding the operation of matrix multiplication.

17.4.1 Vector Multiplication(R,G,A,F)

Suppose that institution A_1 provides three services that we denote by S_1, S_2, and S_3. Also assume that during the year, 500 units of S_1, 800 units of S_2, and 1,000 units of S_3 were provided for which the institution charges fees of

\$10, \$20, and \$30 respectively. As can be verified, the total revenue generated by providing these services is

$$\$10(500) + \$20(800) + \$30(1{,}000) = \$51{,}000$$

Now, assume that we express the quantity of these services as the column vector

$$\mathbf{q} = \begin{bmatrix} 500 \\ 800 \\ 1{,}000 \end{bmatrix}$$

and the fees charged for each of these services as a row vector of the form

$$\mathbf{p}' = [\$10 \ \$20 \ \$30]$$

As has been seen, the total revenue generated by providing service is given by the sum of the products of the elements appearing in $\mathbf{p}'$ and the corresponding elements in $\mathbf{q}$. Thus, we find that the product

$$\begin{aligned} \mathbf{p}'\mathbf{q} &= [\$10 \ \$20 \ \$30] \begin{bmatrix} 500 \\ 800 \\ 1{,}000 \end{bmatrix} \\ &= \$10(500) + \$20(800) + \$30(1{,}000) \\ &= \$51{,}000 \end{aligned}$$

yields the total revenue generated by providing these services. The general procedure for obtaining the product $\mathbf{p}'\mathbf{q}$ is as follows. First, multiply each element appearing in $\mathbf{p}'$ by the corresponding element appearing in $\mathbf{q}$ and sum the resulting products. Thus, if

$$\mathbf{p}' = [p_1 \ \ p_2 \ \ p_3 \cdots p_n] \quad \text{and} \quad \mathbf{q} = \begin{bmatrix} q_1 \\ q_2 \\ q_3 \\ \cdot \\ \cdot \\ \cdot \\ q_n \end{bmatrix}$$

the product $\mathbf{p}'\mathbf{q}$ is given by

$$\mathbf{p}'\mathbf{q} = p_1q_1 + p_2q_2 + p_3q_3 + \cdots + p_nq_n \qquad \textbf{(17.3)}$$

Obviously, this definition is valid only when the row and column vectors have the same number of elements.

17.4.2 Vector-Matrix Multiplication[R,G,A,F]

Suppose that we extend this illustration and consider the quantity of services S_1, S_2, and S_3 provided by institutions A_1, A_2, A_3, and A_4. Assume further that the data in Table 17-3 reflect the quantities of S_1, S_2, and S_3 that were provided by the four institutions. Recall that the revenue of hospital A_1 was found to be \$10(500) + \$20(800) + \$30(1,000) = \$51,000. If the institutions A_2, A_3, and A_4 also charge \$10, \$20, and \$30 for each unit of S_1, S_2, and S_3 used, respectively, similar calculations can be made for these institutions, as seen in the following.

Institution	Revenue Earned
A_2	\$10(600) + \$20(900) + \$30(400) = \$36,000
A_3	\$10(300) + \$20(200) + \$30(650) = \$26,500
A_4	\$10(950) + \$20(870) + \$30(1,200) = \$62,900

Table 17-3 Quantity of Services Provided by Institutions A_1, A_2, A_3, and A_4

Service	Institution			
	A_1	A_2	A_3	A_4
S_1	500	600	300	950
S_2	800	900	200	870
S_3	1,000	400	650	1,200

If we now extract the information in Table 17-3 and form the matrix

$$\mathbf{Q} = \begin{bmatrix} 500 & 600 & 300 & 950 \\ 800 & 900 & 200 & 870 \\ 1{,}000 & 400 & 650 & 1{,}200 \end{bmatrix}$$

we find that the total revenue earned by each of these institutions by providing S_1, S_2, and S_3 during the year is given by

$$\begin{aligned}
\mathbf{p}'\mathbf{Q} &= [\$10 \quad \$20 \quad \$30] \begin{bmatrix} 500 & 600 & 300 & 950 \\ 800 & 900 & 200 & 870 \\ 1{,}000 & 400 & 650 & 1{,}200 \end{bmatrix} \\
&= [\$10(500) + \$20(800) + \$30(1{,}000) \\
&\quad \$10(600) + \$20(900) + \$30(400) \\
&\quad \$10(300) + \$20(200) + \$30(650) \\
&\quad \$10(950) + \$20(870) + \$30(1{,}200)] \\
&= [\$51{,}000 \quad \$36{,}000 \quad \$26{,}500 \quad \$62{,}900]
\end{aligned}$$

Each element of the resulting row vector represents the revenue earned by one of the four institutions. The elements of this vector are obtained in the same way that we found $\mathbf{p}'\mathbf{q}$ earlier. Here, however, we use successive columns appearing in $\mathbf{Q}$ as the vector $\mathbf{q}$. Thus, $\mathbf{p}'\mathbf{Q}$ is obtained by repetitions of $\mathbf{p}'\mathbf{q}$ where $\mathbf{q}$ is represented by successive columns of the matrix $\mathbf{Q}$.

General notation for this example is as follows. Letting

$$\mathbf{p}' = [p_1 \quad p_2 \quad p_3]$$

$$\mathbf{Q} = \begin{bmatrix} q_{11} & q_{12} & q_{13} & q_{14} \\ q_{21} & q_{22} & q_{23} & q_{24} \\ q_{31} & q_{32} & q_{33} & q_{34} \end{bmatrix}$$

we find that

$$\begin{aligned}
\mathbf{p}'\mathbf{Q} = [&p_1q_{11} + p_2q_{21} + p_3q_{31} \quad p_1q_{12} + p_2q_{22} + p_3q_{32} \\
&p_1q_{13} + p_2q_{23} + p_3q_{33} \quad p_1q_{14} + p_2q_{24} + p_3q_{34}]
\end{aligned} \qquad \textbf{(17.4)}$$

Here, we see that each term of $\mathbf{p}'\mathbf{Q}$ is the sum of the products of each element in $\mathbf{p}'$ and the corresponding elements appearing in successive columns of $\mathbf{Q}$.

Workload Determination

As another example of vector-matrix products, consider a situation in which we wish to estimate the institution's workload during the coming period. For the sake of simplicity, assume that the operational activity of our institution is limited to the medical management of four diagnostic conditions denoted by M_1, M_2, M_3, and M_4 and that 1,000, 2,000, 4,000, and 6,000 patients are expected to present these diagnostic conditions, respectively. Also assume that the data in Table 17-4 reflect the number of days of care per patient, the number of laboratory procedures per patient, and the number of operative procedures per patient that we expect to provide during the coming period. In this case, we let

$$\mathbf{p}' = [1{,}000 \quad 2{,}000 \quad 4{,}000 \quad 6{,}000]$$

and

$$\mathbf{Q} = \begin{bmatrix} 7.2 & 2.1 & 1.2 \\ 6.1 & 4.0 & .5 \\ 3.0 & 6.8 & 1.3 \\ 1.5 & 9.0 & .6 \end{bmatrix}$$

Table 17-4 Expected Quantity and Composition of Service Per Patient by Diagnosis

Diagnosis	Service Component		
	Days of Care Per Patient	Laboratory Procedures Per Patient	Operative Procedures Per Patient
M_1	7.2	2.1	1.2
M_2	6.1	4.0	.5
M_3	3.0	6.8	1.3
M_4	1.5	9.0	.6

Observe that $\mathbf{p}'$ is a row vector, the elements of which correspond to the number of patients expected to be treated for each of the four conditions, while $\mathbf{Q}$ is a matrix of order 4×3 that consists of the per capita rates at which each of the services is expected to be provided during the coming period. Thus, we find that

$$\mathbf{p}'\mathbf{Q} = [1{,}000 \quad 2{,}000 \quad 4{,}000 \quad 6{,}000] \begin{bmatrix} 7.2 & 2.1 & 1.2 \\ 6.1 & 4.0 & .5 \\ 3.0 & 6.8 & 1.3 \\ 1.5 & 9.0 & .6 \end{bmatrix}$$

$$= [40{,}400 \quad 91{,}300 \quad 11{,}000]$$

yields the quantity and composition of services we expect to provide to the patient population during the coming period. As a result, these calculations suggest that the anticipated workload of the institution consists of 40,400 days of care, 91,300 laboratory examinations, and 11,000 operative procedures. The operation expressed by $\mathbf{p}'\mathbf{Q}$ is appropriate irrespective of the number of diagnostic conditions or the number of services provided by the institution.

17.4.3 Multiplication of Two Matrices(R,A,G,F)

The multiplication of two matrices may now be viewed as an extension of multiplying a matrix by a vector. To illustrate, suppose that the data in Tables 17-5 and 17-6 are available. In this case, Table 17-5 represents the volume of services S_1, S_2, and S_3 that were provided to patients presenting conditions M_1 and M_2. Similarly, the data in Table 17-6 represent the manhours by type of labor required to provide a unit of each of the services. In this case, we may extract the data in Table 17-5 and form the matrix

$$\mathbf{A} = \begin{bmatrix} 2{,}000 & 0 & 6{,}000 \\ 15{,}000 & 28{,}000 & 1{,}000 \end{bmatrix}$$

Similarly, we may employ the data in Table 17-6 to form the matrix

$$\mathbf{B} = \begin{bmatrix} 2.0 & 4.0 \\ 0.0 & 5.0 \\ 3.0 & 0.0 \end{bmatrix}$$

Table 17-5 Service Requirements by Diagnosis

Diagnosis	Volume of Service		
	S_1	S_2	S_3
M_1	2,000	0	6,000
M_2	15,000	28,000	1,000

Table 17-6 Labor Requirements in Manhours, by Type of Service

Service	Labor	
	L_1	L_2
S_1	2.0	4.0
S_2	0.0	5.0
S_3	3.0	0.0

To anticipate the results of our discussion, we find that the product matrix, represented by

$$\mathbf{AB}$$

depicts the amount of labor, by type of labor, required to provide service to patients presenting conditions M_1 and M_2.

Notice that the first row of matrix **A** represents the use of services S_1, S_2, and S_3 by patients presenting M_1. On the other hand, the elements of the first (second) column in matrix **B** represent the amount of labor of type L_1 (L_2) required to produce a unit of service S_1, S_2, and S_3. Hence, when matrix **B** is multiplied by the first row of **A** we obtain

$$[2{,}000 \quad 0 \quad 6{,}000] \begin{bmatrix} 2.0 & 4.0 \\ 0.0 & 5.0 \\ 3.0 & 0.0 \end{bmatrix} = [22{,}000 \quad 8{,}000]$$

These findings imply that the provision of services S_1, S_2, and S_3 to patients presenting diagnostic condition M_1 require 22,000 and 8,000 manhours of labor of type L_1 and L_2, respectively. Similarly, recall that elements of the second row of matrix **A** represent the use of services S_1, S_2, and S_3 by patients presenting diagnosis M_2. Hence, the product of this row and the matrix **B** yields

$$[15{,}000 \quad 28{,}000 \quad 1{,}000] \begin{bmatrix} 2.0 & 4.0 \\ 0.0 & 5.0 \\ 3.0 & 0.0 \end{bmatrix} = [33{,}000 \quad 200{,}000]$$

These calculations imply that the provision of the three services to patients presenting diagnostic condition M_2 require 33,000 and 200,000 staffhours of labor of type L_1 and L_2, respectively.

Consequently, we find that the product matrix **AB** is found by

$$\mathbf{AB} = \begin{bmatrix} 2{,}000 & 0 & 6{,}000 \\ 15{,}000 & 28{,}000 & 1{,}000 \end{bmatrix} \begin{bmatrix} 2.0 & 4.0 \\ 0.0 & 5.0 \\ 3.0 & 0.0 \end{bmatrix}$$

$$= \begin{bmatrix} 22{,}000 & 8{,}000 \\ 33{,}000 & 200{,}000 \end{bmatrix}$$

The general notation for this example is as follows. Letting

$$\mathbf{A} = \begin{bmatrix} a_{11} & a_{12} & a_{13} \\ a_{21} & a_{22} & a_{23} \end{bmatrix} \quad \text{and} \quad \mathbf{B} = \begin{bmatrix} b_{11} & b_{12} \\ b_{21} & b_{22} \\ b_{31} & b_{32} \end{bmatrix}$$

the product of these two matrices is given by

$$\mathbf{AB} = \begin{bmatrix} a_{11}b_{11} + a_{12}b_{21} + a_{13}b_{31} & a_{11}b_{12} + a_{12}b_{22} + a_{13}b_{32} \\ a_{21}b_{11} + a_{22}b_{21} + a_{23}b_{31} & a_{21}b_{12} + a_{22}b_{22} + a_{23}b_{32} \end{bmatrix}$$

$$= \begin{bmatrix} \sum_{j=1}^{3} a_{1j}b_{j1} & \sum_{j=1}^{3} a_{1j}b_{j2} \\ \sum_{j=1}^{3} a_{2j}b_{j1} & \sum_{j=1}^{3} a_{2j}b_{j2} \end{bmatrix} \qquad \textbf{(17.5)}$$

Thus, in order to obtain the element appearing in row i and column j of the product matrix **AB**, we multiply each element of row i appearing in matrix **A** by the corresponding element in column j of **B** and sum the resulting products.

A slightly different method of illustrating the multiplication procedure requires an examination of

$$\mathbf{AB} = \mathbf{C}$$

where, as before,

$$\mathbf{A} = \begin{bmatrix} 2{,}000 & 0 & 6{,}000 \\ 15{,}000 & 28{,}000 & 1{,}000 \end{bmatrix} \quad \text{and} \quad \mathbf{B} = \begin{bmatrix} 2.0 & 4.0 \\ 0.0 & 5.0 \\ 3.0 & 0.0 \end{bmatrix}$$

In this case, the element appearing in the *first* row and the *first* column of the product matrix (i.e., c_{11}) is based on the *first* row of **A** and the *first* column of **B** and is given by

$$2{,}000(2.0) + 0(0.0) + 6{,}000(3.0) = 22{,}000 \text{ manhours}$$

The element in the *first* row and the *second* column of the product matrix (c_{12}) is based on the *first* row of **A** and the *second* column of **B** and is given by

$$2{,}000(4.0) + 0(5.0) + 6{,}000(0.0) = 8{,}000 \text{ manhours}$$

The element in the *second* row and the *first* column of the product matrix (c_{21}) is based on the *second* row of **A** and the *first* column of **B** and is given by

$$15{,}000(2.0) + 28{,}000(0.0) + 1{,}000(3.0) = 33{,}000 \text{ manhours}$$

Finally, the element in the *second* row and the *second* column of the product matrix (c_{22}) is based on the *second* row of **A** and the *second* column of **B** and is given by

$$15{,}000(4.0) + 28{,}000(5.0) + 1{,}000(0.0) = 200{,}000 \text{ manhours}$$

Thus, as before, the product matrix

$$\mathbf{AB} = \begin{bmatrix} 22{,}000 & 8{,}000 \\ 33{,}000 & 200{,}000 \end{bmatrix}$$

represents the use of labor of types L_1 and L_2 in providing service to patients presenting conditions M_1 and M_2. This illustration also demonstrates that the element appearing in row i and column j of the product matrix is given by

$$c_{ij} = \sum_{j=1}^{k} a_{ij} b_{ji} \qquad \text{for } i = 1, \cdots, r$$

17.4.4 The Conformability for Multiplication[O,G,F]

As might be surmised from the foregoing discussion, the product matrix **AB** is defined only when the number of elements appearing in each row of **A** is equal to the number of elements in each column of **B**. Stated another way, the product **AB** is defined only when the number of columns in **A** is equal to the number of rows in **B**. When this condition is satisfied, the matrices **A** and **B** are said to be *conformable for multiplication.* For example, if we let

$$\mathbf{A} = \begin{bmatrix} 1 & 7 & 6 \\ 4 & 3 & 2 \end{bmatrix} \quad \text{and} \quad \mathbf{B} = \begin{bmatrix} 12 & 15 & 16 \\ 10 & 6 & 4 \\ 4 & 1 & 2 \end{bmatrix}$$

it will be observed that the product matrix **AB** is defined, since the number of columns in **A** is equal to the number of rows in **B**. However, the product matrix **BA** is *not defined* since the number of columns in **B** is *not* equal to the number of rows in **A**.

The conformability of two or more matrices for multiplication may be ascertained by examining the order of the matrices. As noted earlier, we may express the order of a matrix **A**, which consists of r rows and k columns, by the notation $\mathbf{A}_{r \times k}$. Thus, the product of **A** and **B** may be expressed by

$$\mathbf{A}_{r \times k} \mathbf{B}_{k \times m} = \mathbf{S}_{r \times m} \tag{17.6}$$

which allows us to determine the conformability of the two matrices for multiplication and to specify the order of the product matrix.

As an example of Equation 17.6, let us refer to our earlier example in which we determined the mix of labor required to provide service to diagnoses M_1 and M_2. In this case, the matrix **A** consisted of two rows and three columns, the matrix **B** of three rows and two columns, and the product matrix **AB** of two rows and two columns. Hence, we find that the product

$$\mathbf{A}_{2 \times 3} \mathbf{B}_{3 \times 2}$$

yields a matrix consisting of two rows and two columns.

Similarly, we may modify Equation 17.6 to accommodate the multiplication of more than two matrices. This is shown below where the product

$$\mathbf{A}_{r\times c}\mathbf{B}_{c\times m}\mathbf{D}_{m\times k} = \mathbf{F}_{r\times k}$$

demonstrates that the matrices **A**, **B**, and **D** are conformable for multiplication and that the resulting product matrix will be of order $r \times k$.

As seen previously, the product matrix **BA** may not exist even though the product matrix **AB** is defined. For example, if **A** is of order $r \times s$ and **B** is of order $s \times k$, we find that

$$\mathbf{A}_{r\times s}\mathbf{B}_{s\times k}$$

is defined and yields a product matrix of order $r \times k$. However,

$$\mathbf{B}_{s\times k}\mathbf{A}_{r\times s}$$

is not defined unless k is equal to r. In general, if the matrix **A** is of order $r \times k$

1. the product matrix **AB** exists only if **B** has k rows
2. the product matrix **BA** exists only if **B** has r columns
3. the product matrices **AB** and **BA** exist only if **B** is of order $k \times r$.

Also observe that $\mathbf{A}^2$ exists only if **A** is square (i.e., $r = k$) and that the product matrices **AB** and **BA** are always defined when **A** and **B** are square and of the same order. As a means of distinguishing between **AB** and **BA**, the first is referred to as **A** postmultiplied by **B** and the second is referred to as **A** premultiplied by **B**.

17.5 THE NULL AND THE IDENTITY MATRICES(R,G,F)

Two matrices are said to be equal when every element in one is equal to every element in the other. For example, if

$$\mathbf{A} = \begin{bmatrix} 1 & 6 & 8 \\ 2 & 1 & 3 \end{bmatrix} \quad \text{and} \quad \mathbf{B} = \begin{bmatrix} 1 & 6 & 8 \\ 2 & 1 & 3 \end{bmatrix}$$

the matrices **A** and **B** are equal since they are identical, element by element. Combining the notion of equality with the operation of matrix subtraction allows us to define the *null* or the *zero matrix*. Here, we observe that if $\mathbf{A} = \mathbf{B}$, the operation

$$\mathbf{A} - \mathbf{B}$$

results in a matrix consisting of zeros. Returning to our example we find that

$$\mathbf{A} - \mathbf{B} = \begin{bmatrix} 1 & 6 & 8 \\ 2 & 1 & 3 \end{bmatrix} - \begin{bmatrix} 1 & 6 & 8 \\ 2 & 1 & 3 \end{bmatrix} = \begin{bmatrix} 0 & 0 & 0 \\ 0 & 0 & 0 \end{bmatrix}$$

Thus, the null matrix is the zero of matrix algebra.

The *identity matrix* is defined as a square matrix in which the diagonal elements are all ones and the off-diagonal elements are all zeros. Thus, an identity matrix of order 4 assumes the form

$$\mathbf{I}_4 = \begin{bmatrix} 1 & 0 & 0 & 0 \\ 0 & 1 & 0 & 0 \\ 0 & 0 & 1 & 0 \\ 0 & 0 & 0 & 1 \end{bmatrix}$$

Observe that if $\mathbf{A}$ is of order 2,

$$\mathbf{A}_2\mathbf{I}_2 = \mathbf{I}_2\mathbf{A}_2 = \mathbf{A}_2$$

As an example, let

$$\mathbf{A} = \begin{bmatrix} 1 & 2 \\ 4 & 5 \end{bmatrix} \quad \text{and} \quad \mathbf{I} = \begin{bmatrix} 1 & 0 \\ 0 & 1 \end{bmatrix}$$

Thus

$$\mathbf{IA} = \begin{bmatrix} 1 & 0 \\ 0 & 1 \end{bmatrix} \begin{bmatrix} 1 & 2 \\ 4 & 5 \end{bmatrix} = \begin{bmatrix} 1 & 2 \\ 4 & 5 \end{bmatrix} = \mathbf{AI}$$

Similarly, when any matrix of order $r \times c$ is multiplied by the null matrix of an appropriate order, the product matrix consists of zeros. For example, if $\mathbf{Z}$ is a null matrix of order $r \times c$ and $\mathbf{A}$ is a matrix of order $c \times m$, then $\mathbf{Z}_{r\times c}\mathbf{A}_{c\times m}$ results in a null matrix of order $r \times m$.

17.6 THE LAWS OF MATRIX ALGEBRA(O,G,F)

Earlier we described the commutative, associative, and distributive laws of basic arithmetic and algebra. In general, provided that matrices are conformable for addition or multiplication, the addition or multiplication of matrices is associative. Similarly, assuming that matrices are appropriately conformable for multiplication and addition, the distributive law also is applicable to matrices. Further, provided that matrices are conformable for addition, the addition of matrices is commutative. However, the multiplication of matrices is *not* generally commutative, which implies that **AB** does not equal **BA**. These observations are discussed in the appendix at the end of this chapter.

17.7 APPLICATIONS TO HEALTH CARE MANAGEMENT(O,A,F)

It should be clear now that the operations of matrix addition, subtraction, and multiplication may simplify the expression of many computational tasks confronting the health administrator. As a further example of the applicability of these techniques, consider the problem of determining the effective rate at which the bed complements of institutions A_1 and A_2 are used in serving the population residing in regions R_1, R_2, and R_3.

Suppose that the desired occupancy rates of institutions A_1 and A_2 are 90 percent and 95 percent respectively. Thus, the number of bed days per bed year associated with institution A_1 is given by

$$\text{Bed days per Bed year } (A_1) = .90(365) = 328.5$$

while the number of bed days per bed year associated with institution A_2 is found by

$$\text{Bed days per Bed year } (A_2) = .95(365) = 346.75$$

Observe that the number of beds in effective use by each of the hospitals is given by

$$\frac{\text{\# of days of care/year}}{\text{\# Bed days/Bed year}}$$

which is equivalent to

$$\text{\# of days of care/year} \times \frac{1}{\text{\# Bed days/Bed year}}$$

Now suppose that the data in Table 17-7 reflect the annual number of days of care provided to residents of regions R_1, R_2, and R_3 by hospitals A_1 and A_2. We may now extract this information and form the matrix

$$\mathbf{A} = \begin{bmatrix} 160{,}000 & 20{,}000 \\ 20{,}000 & 60{,}000 \\ 40{,}000 & 200{,}000 \end{bmatrix}$$

Recalling that the effective use of beds is given by

$$\text{Annual Use of Care} \times \frac{1}{\#\ \text{Bed days/Bed year}}$$

we also form the matrix

$$\mathbf{J} = \begin{bmatrix} \frac{1}{328.5} & 0 \\ 0 & \frac{1}{346.75} \end{bmatrix}$$

Thus, we find that

$$\mathbf{E} = \mathbf{AJ} = \begin{bmatrix} 160{,}000 & 20{,}000 \\ 20{,}000 & 60{,}000 \\ 40{,}000 & 200{,}000 \end{bmatrix} \begin{bmatrix} \frac{1}{328.5} & 0 \\ 0 & \frac{1}{346.75} \end{bmatrix}$$

$$= \begin{bmatrix} 487.06 & 57.68 \\ 60.88 & 173.03 \\ 121.76 & 576.78 \end{bmatrix}$$

This finding suggests that 487.06 beds are "effectively" used in hospital 1 by residents of region 1, while the element a_{12} of the product matrix **AJ** indicates that 57.68 beds in hospital 2 are effectively used by residents of region 1. Similarly, the column total $\Sigma_{i=1}{}^{3}\, e_{i1}$ indicates that 669.71 beds are effectively used in institution A_1 while the column total $\Sigma_{i=1}{}^{3}\, e_{i2}$ indicates that 807.50 beds are effectively used in institution A_2.

We now translate the data in the matrix **E** into the number of beds effectively used per 1,000 population. Here, we assume that the populations of

Table 17-7 Annual Number of Days of Care Provided to Residents of Regions R_1, R_2, and R_3 by Hospitals A_1 and A_2

Region	Hospital	
	A_1	A_2
R_1	160,000	20,000
R_2	20,000	60,000
R_3	40,000	200,000

regions R_1, R_2, and R_3 are 200,000, 100,000, and 50,000 respectively. Thus, the number of effectively used beds per 1,000 is given by

$$\mathbf{P} = \mathbf{ME}$$
$$= \begin{bmatrix} 1/200 & 0 & 0 \\ 0 & 1/100 & 0 \\ 0 & 0 & 1/50 \end{bmatrix} \begin{bmatrix} 487.06 & 57.68 \\ 60.88 & 173.04 \\ 121.77 & 576.78 \end{bmatrix}$$
$$= \begin{bmatrix} 2.44 & .29 \\ .61 & 1.73 \\ 2.44 & 11.54 \end{bmatrix}$$

Assuming that there are no differences in use attributable to the socio-demographic characteristics of the regions, the number of beds in effective use per 1,000 population in hospital A_1 is 2.44 + .61 + 2.44, or 5.49/1,000 population, while the corresponding ratio associated with institution A_2 is .29 + 1.73 + 11.54, or 13.56 beds per 1,000 population. These ratios may then be compared with the desired number of beds per 1,000, which, of course, facilitates an evaluation of the bed distribution between the two hospitals.

Appendix 17A

The Laws of Matrix Algebra (R,G,F)

In this appendix we examine the associative, commutative, and distributive laws as they relate to the operations of matrix addition and multiplication.

THE ASSOCIATIVE LAW

Provided that two matrices are conformable for addition, the addition of matrices is associative. Thus, if **A**, **B**, and **C** are of the same order, we find that

$$(\mathbf{A} + \mathbf{B}) + \mathbf{C} = \mathbf{A} + (\mathbf{B} + \mathbf{C}) = \mathbf{A} + \mathbf{B} + \mathbf{C}$$

Similarly, provided that the matrices are conformable for multiplication, the associative law is applicable to the product of matrices. Thus, if **A** is of order $r \times k$, **B** is of order $k \times m$, and **C** is of order $m \times s$, we find that

$$(\mathbf{AB})\mathbf{C} = \mathbf{A}(\mathbf{BC}) = \mathbf{ABC}$$

As an example, letting

$$\mathbf{A} = \begin{bmatrix} 1 & 3 \\ 5 & 6 \end{bmatrix} \quad \mathbf{B} = \begin{bmatrix} 4 & 1 & 3 \\ 5 & 0 & 2 \end{bmatrix} \quad \text{and} \quad \mathbf{C} = \begin{bmatrix} 1 & 0 \\ 2 & 1 \\ 1 & 0 \end{bmatrix}$$

we find that (**AB**)**C** is found by first calculating the product matrix **AB**

$$\begin{bmatrix} 1 & 3 \\ 5 & 6 \end{bmatrix} \begin{bmatrix} 4 & 1 & 3 \\ 5 & 0 & 2 \end{bmatrix} = \begin{bmatrix} 19 & 1 & 9 \\ 50 & 5 & 27 \end{bmatrix}$$

and (**AB**)**C** is given by

$$\begin{bmatrix} 19 & 1 & 9 \\ 50 & 5 & 27 \end{bmatrix} \begin{bmatrix} 1 & 0 \\ 2 & 1 \\ 1 & 0 \end{bmatrix} = \begin{bmatrix} 30 & 1 \\ 87 & 5 \end{bmatrix}$$

Similarly **A**(**BC**) may be found by first calculating the product matrix **BC** as follows

$$\begin{bmatrix} 4 & 1 & 3 \\ 5 & 0 & 2 \end{bmatrix} \begin{bmatrix} 1 & 0 \\ 2 & 1 \\ 1 & 0 \end{bmatrix} = \begin{bmatrix} 9 & 1 \\ 7 & 0 \end{bmatrix}$$

Thus, **A**(**BC**) is given by

$$\begin{bmatrix} 1 & 3 \\ 5 & 6 \end{bmatrix} \begin{bmatrix} 9 & 1 \\ 7 & 0 \end{bmatrix} = \begin{bmatrix} 30 & 1 \\ 87 & 5 \end{bmatrix}$$

As can be seen, the product **ABC** must be formed by one of these approaches, and they are equivalent.

THE DISTRIBUTIVE LAW

Provided that **B** and **C** are conformable for addition (i.e., of the same order) and that **A** and **B** (as well as **A** and **C**) are conformable for multiplication, the distributive law is true. Thus, if **A** is of order $r \times c$ while **B** and **C** are both of the order $c \times j$, we may write

$$\mathbf{A}(\mathbf{B} + \mathbf{C}) = \mathbf{AB} + \mathbf{AC}$$

For example, letting

$$\mathbf{A} = \begin{bmatrix} 1 & 3 \\ 4 & 5 \end{bmatrix}; \quad \mathbf{B} = \begin{bmatrix} 1 & 6 & 7 \\ 4 & 7 & 2 \end{bmatrix} \quad \text{and} \quad \mathbf{C} = \begin{bmatrix} 4 & 4 & 3 \\ 2 & 3 & 5 \end{bmatrix}$$

we find that

$$\mathbf{A}(\mathbf{B} + \mathbf{C}) = \begin{bmatrix} 1 & 3 \\ 4 & 5 \end{bmatrix} \begin{bmatrix} 1 + 4 & 6 + 4 & 7 + 3 \\ 4 + 2 & 7 + 3 & 2 + 5 \end{bmatrix}$$

$$= \begin{bmatrix} 1 & 3 \\ 4 & 5 \end{bmatrix} \begin{bmatrix} 5 & 10 & 10 \\ 6 & 10 & 7 \end{bmatrix}$$

$$= \begin{bmatrix} 23 & 40 & 31 \\ 50 & 90 & 75 \end{bmatrix}$$

Similarly, we find that

$$\mathbf{AB} + \mathbf{AC} = \begin{bmatrix} 1 & 3 \\ 4 & 5 \end{bmatrix} \begin{bmatrix} 1 & 6 & 7 \\ 4 & 7 & 2 \end{bmatrix} + \begin{bmatrix} 1 & 3 \\ 4 & 5 \end{bmatrix} \begin{bmatrix} 4 & 4 & 3 \\ 2 & 3 & 5 \end{bmatrix}$$

$$= \begin{bmatrix} 13 & 27 & 13 \\ 24 & 59 & 38 \end{bmatrix} + \begin{bmatrix} 10 & 13 & 18 \\ 26 & 31 & 37 \end{bmatrix}$$

$$= \begin{bmatrix} 23 & 40 & 31 \\ 50 & 90 & 75 \end{bmatrix}$$

These results serve to illustrate the general finding that $\mathbf{A}(\mathbf{B} + \mathbf{C}) = \mathbf{AB} + \mathbf{AC}$.

THE COMMUTATIVE LAW

Provided that the matrices are conformable for addition, the addition of matrices is commutative. Thus, if **A** and **B** are of the same order,

$$\mathbf{A} + \mathbf{B} = \mathbf{B} + \mathbf{A}$$

For example, if

$$\mathbf{A} = \begin{bmatrix} 1 & 3 \\ 4 & 6 \end{bmatrix} \quad \text{and} \quad \mathbf{B} = \begin{bmatrix} 4 & 7 \\ 1 & 2 \end{bmatrix}$$

it can be verified that $\mathbf{A} + \mathbf{B} = \mathbf{B} + \mathbf{A}$.

On the other hand, the multiplication of matrices is *not* generally commutative, which implies that **AB** *does not always equal* **BA**. As seen earlier, the product **BA** *may not exist* even though **AB** is defined. As a consequence, the potential for the product matrix **AB** to equal the product matrix **BA** exists *only* when **A** and **B** are square and of the same order. That matrix multiplication is not generally commutative is illustrated by the following example where we let

$$\mathbf{A} = \begin{bmatrix} 3 & 7 \\ 1 & 0 \end{bmatrix} \quad \text{and} \quad \mathbf{B} = \begin{bmatrix} 1 & 6 \\ 2 & 0 \end{bmatrix}$$

In this case, the product **AB** is given by

$$\begin{bmatrix} 3 & 7 \\ 1 & 0 \end{bmatrix} \begin{bmatrix} 1 & 6 \\ 2 & 0 \end{bmatrix} = \begin{bmatrix} 17 & 18 \\ 1 & 6 \end{bmatrix}$$

On the other hand, the product **BA** is given by

$$\begin{bmatrix} 1 & 6 \\ 2 & 0 \end{bmatrix} \begin{bmatrix} 3 & 7 \\ 1 & 0 \end{bmatrix} = \begin{bmatrix} 9 & 7 \\ 6 & 14 \end{bmatrix}$$

Here, it will be observed that the product matrix **AB** is not equal to **BA** and, even though it is *possible* for **BA** to equal **AB**, matrix multiplication is not generally commutative.

However, matrix multiplication is commutative in two cases that are of special importance. First, when dealing with a square matrix of order r and an identity matrix of the same order, we find that

$$\mathbf{I}_r\mathbf{A}_r = \mathbf{A}_r\mathbf{I}_r = \mathbf{A}_r$$

The second situation in which matrix multiplication is commutative involves the product of a square matrix of order r by the null matrix of the same order. In this case,

$$\mathbf{A}_r\mathbf{O}_r = \mathbf{O}_r\mathbf{A}_r = \mathbf{O}_r$$

Problems for Solution

1. Suppose we are provided with the following information concerning the number of physicians by specialty and age in each of two regions:

Region I

Age	Medical Specialty		
	General Practice	Surgery	Internal Medicine
30–39	400	60	190
40–49	800	240	30
50–59	110	52	70
60–69	30	3	6

Region II

Age	Medical Specialty		
	General Practice	Surgery	Internal Medicine
30–39	212	83	364
40–49	350	612	87
50–59	62	316	79
60–69	14	4	12

Using matrix notation, determine the number of physicians in these two regions by age and type of medical specialty.

2. Using the information provided in Problem 1, determine whether Region I has a greater or a fewer number of physicians by age and medical specialty than Region II.

3. Suppose we are provided with the following information concerning the number of restricted activity days per person per year experienced by 1,000 males grouped by age and family income:

Age	Family Income		
	Under $6,000	$6,000–$14,999	$15,000 and Over
10–29	20	17	18
30–49	35	29	21
50–69	42	37	36

Also suppose that the corresponding data for females are as follows:

Age	Family Income		
	Under $6,000	$6,000–$14,999	$15,000 and Over
10–29	12	9	3
30–49	28	27	15
50–69	37	25	27

Using matrix notation determine:

a. The number of disability days per person per year for males and females by age and family income.
b. Whether males experienced a greater or a lesser number of disability days per person per year by age and family income than females.

4. Referring to the data in Table 17-4, use matrix notation to determine the total revenue anticipated if the fee schedule that will prevail during the period is as follows:

Component of Care	Charge per Unit
Room and Board	$120
Laboratory Examination	$30
Operative Procedures	$110

5. Assume that our institution provides services S_1, S_2, and S_3 in the medical management of diagnostic conditions M_1, M_2, M_3, and M_4. On the basis of historical records, suppose further that management has estimated the total service requirements by diagnosis for the coming period as follows:

Condition	Service		
	S_1	S_2	S_3
M_1	1,200	1,300	20
M_2	800	600	400
M_3	500	480	150
M_4	0	370	230

Also suppose that the number of staffhours of labor of type L_1, L_2, L_3, and L_4 are employed in producing each unit of service S_1, S_2, and S_3 as follows:

Service	Labor			
	L_1	L_2	L_3	L_4
S_1	1.3	.5	0	.1
S_2	2.1	1.2	2.3	1.0
S_3	0	0	1.5	2.6

What is the total amount of labor by type of labor that we expect to employ during the coming period?

6. If the wages paid to L_1, L_2, L_3, and L_4 are \$7.00, \$6.00, \$9.00, and \$4.00 per hour, respectively, what is the total wage bill in Problem 5 above?
7. Suppose that the foregone income for each bed disability day by age and sex is as follows

Age	Male	Female
10–29	\$24	\$20
30–49	58	32
50–69	60	36

Use these data to estimate the income foregone by the two groups described in Problem 3.

8. Suppose that the following data represent the amount of supplies of type R_1, R_2, and R_3 acquired to produce each unit of service S_1, S_2, S_3, and S_4

Supply Item	Service			
	S_1	S_2	S_3	S_4
R_1	0	2.0	1.0	0
R_2	3.0	4.0	0	0
R_3	0	0	2.0	3.0

Suppose that the following data represent the total volume of S_1, S_2, S_3, and S_4 provided during a given period.

Service	Volume
S_1	180
S_2	200
S_3	600
S_4	700

How much of each supply item was required during the period?

9. Suppose that the following data represent the distribution of the volume of services S_1, S_2, S_3, and S_4 among conditions M_1, M_2, M_3, and M_4.

Service	Condition				Total
	M_1	M_2	M_3	M_4	
S_1	50	20	30	80	
S_2	60	10	30	100	
S_3	260	40	175	125	
S_4	30	60	0	110	

Use the resource coefficients presented in Problem 8 to calculate the amount of each supply item required to treat patients presenting conditions M_1, M_2, M_3, and M_4.

10. Suppose that our institution uses labor of type L_1 and L_2 to provide services S_1, S_2, S_3, and S_4 to patients presenting conditions M_1 and M_2. Further, suppose that the number of staffhours of L_1 and L_2 labor required to provide a *unit* of each service is as follows.

Labor	Service			
	S_1	S_2	S_3	S_4
L_1	.2	0	.1	.5
L_2	0	.2	.3	1.4

Also assume that the total volume of service provided to patients presenting conditions M_1 and M_2 is given by the following data.

Service	M_1	M_2
S_1	1,780 units	120 units
S_2	0 units	312 units
S_3	76 units	82 units
S_4	100 units	0 units

Use these data to determine the amount of labor of type L_1 and L_2 that is required to provide service to patients presenting conditions M_1 and M_2.

Chapter 18

Additional Matrix Operations

Objectives

After completing this chapter, you should be able to:

1. Understand and use linear transformations;
2. Find the transpose of a matrix or a vector;
3. Perform the operations of addition, subtraction, and multiplication on transposed matrices;
4. Partition matrices;
5. Perform the operation of multiplication on partitioned matrices.

Chapter Map

The sections comprising this chapter may be summarized as follows:

Section Number	Required Reading	Optional Reading	Generic Development	Application to Management	Fundamental Principles	Complex Material
	(R)	(O)	(G)	(A)	(F)	(C)
18.1	x		x	x	x	
18.2	x		x	x	x	
18.3		x	x	x	x	
18.4		x	x	x		x
18.5		x		x	x	
Appendix		x	x			x

Having described matrix addition, subtraction, and multiplication, we consider several other matrix operations before examining the matrix algebra counterpart to division. Specifically, this chapter is devoted to the use of matrix algebra when addressing problems requiring a linear transformation of one variable into another as well as the techniques of finding the transpose of a matrix, partitioning matrices, and multiplying partitioned matrices.

18.1 LINEAR TRANSFORMATIONS(R,A,G,F)

Suppose we are interested in a system of linear equations that assumes the form

$$y_1 = a_{11}x_1 + a_{12}x_2$$
$$y_2 = a_{21}x_1 + a_{22}x_2$$

In this system of equations, the variables y_1 and y_2 are expressed as weighted sums of the variables x_1 and x_2 where the assigned weights are given by the coefficients a_{11}, a_{12}, a_{21}, and a_{22}. Such a *weighted sum* is called a *linear combination* of the x's. The term linear may be used to describe the system of equations since none of the x's are raised to a power greater than one and the system is free of interactive terms (i.e., no term contains more than one element of x).

Also observe that by letting

$$\mathbf{y} = \begin{bmatrix} y_1 \\ y_2 \end{bmatrix}; \quad \mathbf{A} = \begin{bmatrix} a_{11} & a_{12} \\ a_{21} & a_{22} \end{bmatrix} \quad \text{and} \quad \mathbf{x} = \begin{bmatrix} x_1 \\ x_2 \end{bmatrix}$$

we may express this system of equations in the form

$$\mathbf{y} = \mathbf{A}\mathbf{x} \tag{18.1}$$

In this case, the matrix **A** contains the coefficients that transform the elements of **x** into **y** and it is in this regard that **A** is said to represent a linear transformation of **x** into **y**.

Linear transformations in this form may be used to approximate many of the processes by which health care is delivered. For example, assume that a known and invariant amount of service is required by each patient who is treated for the conditions M_1, M_2, and M_3. Representing these services by S_1, S_2, and S_3, suppose further that the relationship between the per capita use of each procedure and the three diagnoses is given by the data in Table 18-1. Letting $M_1{*}$, $M_2{*}$, and $M_3{*}$ represent the number of patients who receive treatment for morbidities M_1, M_2, and M_3, we may use the techniques described earlier to determine the amount of each service required by a patient population of known size and diagnostic distribution. In this situation, we find that the system of linear equations given by

$$S_1{*} = 2.0\,(M_1{*}) + 3.0\,(M_2{*}) + 1.0\,(M_3{*})$$

$$S_2{*} = 4.0\,(M_1{*}) + 6.0\,(M_2{*}) + 8.0\,(M_3{*})$$

$$S_3{*} = 7.0\,(M_1{*}) + 9.0\,(M_2{*}) + 5.0\,(M_3{*})$$

provides a mechanism by which the patient population may be transformed into the service requirements represented by $S_1{*}$, $S_2{*}$, and $S_3{*}$.

Using matrix notation, we may express the system of linear equations in the form

$$\mathbf{s} = \mathbf{Tm} \tag{18.2}$$

where

$$\mathbf{s} = \begin{bmatrix} S_1{*} \\ S_2{*} \\ S_3{*} \end{bmatrix} \qquad \mathbf{T} = \begin{bmatrix} 2.0 & 3.0 & 1.0 \\ 4.0 & 6.0 & 8.0 \\ 7.0 & 9.0 & 5.0 \end{bmatrix} \qquad \text{and} \quad \mathbf{m} = \begin{bmatrix} M_1{*} \\ M_2{*} \\ M_3{*} \end{bmatrix}$$

In this specification, the coefficients contained in the matrix **T** transform a patient population of known or assumed size and diagnostic distribution into the quantity and composition of required services.

Consider next the transformation of the service mix into the corresponding set of resource requirements. To simplify the illustration, assume that labor of type R_1 and consumable supplies of type R_2 are the only resources required to provide services S_1, S_2, and S_3. Table 18-2 presents the assumed resource requirements per unit of service. In this table, the labor component is measured in terms of staff hours of use while the required amount of consumable supplies is measured in terms of physical units. On the basis

Table 18-1 The Per Capita Use of Service by Morbidity*

	Morbidity		
Service	M_1	M_2	M_3
S_1	2.0	3.0	1.0
S_2	4.0	6.0	8.0
S_3	7.0	9.0	5.0

*The entries in this table refer to the number of units of service required by each patient treated for the specified conditions.

Table 18-2 Resource Requirements Per Unit of Service

	Service		
Resource	S_1	S_2	S_3
R_1	2.0	1.5	2.5
R_2	1.0	2.0	3.0

of these data, we may express the transformation of the column vector **s** into the total amount of required resources by

$$R_1{}^* = 2.0\,(S_1{}^*) + 1.5\,(S_2{}^*) + 2.5\,(S_3{}^*)$$

$$R_2{}^* = 1.0\,(S_1{}^*) + 2.0\,(S_2{}^*) + 3.0\,(S_3{}^*)$$

where

$$R_1{}^* = \text{total amount of required labor}$$

$$R_2{}^* = \text{total amount of required supplies.}$$

Letting

$$\mathbf{r} = \begin{bmatrix} R_1{}^* \\ R_2{}^* \end{bmatrix}; \qquad \mathbf{U} = \begin{bmatrix} 2.0 & 1.5 & 2.5 \\ 1.0 & 2.0 & 3.0 \end{bmatrix} \quad \text{and, as before, } \mathbf{s} = \begin{bmatrix} S_1{}^* \\ S_2{}^* \\ S_3{}^* \end{bmatrix}$$

we may express this system of linear equations by

$$\mathbf{r} = \mathbf{Us} \tag{18.3}$$

Equation 18.2 shows that we may substitute **Tm** for **s** and obtain

$$\mathbf{r} = \mathbf{UTm} \tag{18.4}$$

In this equation, the product matrix **UT** transforms the column vector **m** into the column vector **r**. Stated differently, Equation 18.4 transforms a patient population of known or assumed size and diagnostic distribution into the required amount of the resources R_1 and R_2. Returning to our numerical example, it can be verified that

$$\mathbf{r} = \begin{bmatrix} 27.5\ (M_1{}^*) & 37.5\ (M_2{}^*) & 26.5\ (M_3{}^*) \\ 31.0\ (M_1{}^*) & 42.0\ (M_2{}^*) & 32.0\ (M_3{}^*) \end{bmatrix}$$

Thus, under the simplifying assumptions introduced earlier, these linear transformations may be used to estimate $R_1{}^*$ and $R_2{}^*$ from a known or assumed number of patients who present the conditions M_1, M_2, and M_3.

Now, suppose that we also know or estimate the factor prices P_1 and P_2 that are associated with resources R_1 and R_2, respectively. Letting

$$\mathbf{p}' = [P_1 \quad P_2]$$

we find that

$$c = \mathbf{p}'\mathbf{UTm} \tag{18.5}$$

yields the total direct costs of managing the conditions M_1, M_2, and M_3. Such an expression is of rather obvious value when developing the budget of a health care institution. Given that the budgetary unit of analysis is the patient and the related condition, Equation 18.5 may be used to estimate the (scalar) cost c irrespective of the sizes of **U**, **T**, **m**, and **p**, provided that these terms are conformable for multiplication.

In general terms, this example illustrates the result that, if a linear transformation may be expressed in the form

$$\mathbf{y} = \mathbf{Ax}$$

and if we are also able to express **x** by

$$\mathbf{x} = \mathbf{Bz}$$

then, it follows that

$$\mathbf{y} = \mathbf{ABz}$$

The last expression allows us to transform the vector **z** into **y** by the product matrix **AB**.

18.2 THE TRANSPOSE OF A MATRIX(R,G,A,F)

There are many situations in which data that have been extracted from a table and formed into a matrix must be rearranged slightly. When considering the rearrangement of the matrix, it is important to note that the specification of which variable is to define the rows, and hence which variable is to define the columns, is an arbitrary decision. It is possible to interchange the rows and the columns without suffering a loss of information.

As an example, consider the matrix **T**, which was defined in the previous section by

$$\mathbf{T} = \begin{bmatrix} 2.0 & 3.0 & 1.0 \\ 4.0 & 6.0 & 8.0 \\ 7.0 & 9.0 & 5.0 \end{bmatrix}$$

As will be recalled, the rows of this matrix were defined by the services S_1, S_2, and S_3 while the columns were defined by the morbidities M_1, M_2, and M_3. If we wish to define the rows by diagnostic categories and the columns by the components of service, we find that the matrix **T** becomes

$$\begin{bmatrix} 2.0 & 4.0 & 7.0 \\ 3.0 & 6.0 & 9.0 \\ 1.0 & 8.0 & 5.0 \end{bmatrix}$$

It should be noted that, even though the elements of these two matrices are the same, the rearranged matrix and the original matrix **T** are not equal. Rather, the rows of matrix **T** have become the columns of the rearranged matrix. Stated differently, the columns of the rearranged matrix are the rows of the matrix **T**. When matrices are related in this fashion, each is said to be the transpose of the other.

In general, the *transpose of matrix* **A** is the matrix whose rows are the columns of **A**. As seen previously, when interchanging the rows and columns of a matrix, it is necessary to retain the order of the elements appearing in each of the original rows (or columns). For purposes of future reference, we shall use a prime to designate the transpose of a matrix. Thus, if we let

$$\mathbf{A} = \begin{bmatrix} 1 & 3 & 2 \\ 7 & 8 & 4 \end{bmatrix}$$

we find that the transpose of **A** is given by

$$\mathbf{A}' = \begin{bmatrix} 1 & 7 \\ 3 & 8 \\ 2 & 4 \end{bmatrix}$$

In this example, the matrix **A** is of order 2×3 but $\mathbf{A}'$ is of order 3×2. In general, if **A** is of order $r \times c$, then $\mathbf{A}'$ will be of order $c \times r$. When specifying the order of a transposed matrix, it is convenient to employ the notation $(\mathbf{A}')_{r \times c}$ or the notation $(\mathbf{A}_{r \times c})'$. As might be surmised from this discussion, the transpose of a transposed matrix is the original matrix. For example, if we let

$$\mathbf{A} = \begin{bmatrix} 7 & 8 & 12 \\ 4 & 1 & 0 \end{bmatrix}$$

then

$$\mathbf{A}' = \begin{bmatrix} 7 & 4 \\ 8 & 1 \\ 12 & 0 \end{bmatrix}$$

As can be verified, the transpose of $\mathbf{A}'$, which is written $(\mathbf{A}')'$, is the matrix **A**.

As implied earlier, the operation of transposition is required when management wishes to interchange the rows and columns of a matrix. In addition, it also is frequently necessary to add or multiply transposed matrices. These operations are described in the Appendix at the end of this chapter.

Similar to the discussion above, the transpose of a row vector is a column vector and vice versa. For example, if we let

$$\mathbf{a} = \begin{bmatrix} 1 \\ 0 \\ 2 \end{bmatrix}$$

then

$$\mathbf{a}' = [1 \quad 0 \quad 2]$$

Observe that this notation is consistent with our earlier work where we denoted the transpose of a matrix by the superscript prime.

18.3 PARTITIONING MATRICES(O,G,A,F)

To illustrate the technique of *partitioning matrices,* suppose that we divide the output of the hospital into stay-specific services, which are measured by the number of days of care, and ancillary services, which are measured in terms of the frequency with which each of these procedures is provided. Also suppose that the data in Table 18-3 reflect the use of these services by patients who were hospitalized with the conditions M_1, M_2, M_3, and M_4 during the previous period. As this table demonstrates, we assume that stay-specific services are provided to patients occupying private, semiprivate, and ward accommodations. Similarly, we assume that the laboratory services provided by the hospital are composed of the examinations E_1, E_2, and E_3 and that the only operative procedures performed by the institution are OP_1, OP_2, and OP_3. We may extract the data presented in Table 18-3 and construct the matrix

$$\mathbf{D} = \begin{bmatrix} 1{,}600 & 1{,}000 & 0 & 260 & 900 & 400 & 100 & 0 & 0 \\ 0 & 20{,}000 & 8{,}000 & 800 & 150 & 320 & 0 & 20 & 0 \\ 0 & 10{,}000 & 9{,}000 & 370 & 560 & 730 & 10 & 0 & 30 \\ 0 & 6{,}000 & 7{,}000 & 430 & 730 & 750 & 400 & 0 & 0 \end{bmatrix}$$

Table 18-3 Use of Service by Diagnosis

	Stay-Specific Services			Ancillary Services					
				Laboratory			Operative Procedures		
Diagnosis	Private	Semi-Private	Ward	E_1	E_2	E_3	OP_1	OP_2	OP_3
M_1	1,600	1,000	0	260	900	400	100	0	0
M_2	0	20,000	8,000	800	150	320	0	20	0
M_3	0	10,000	9,000	370	560	730	10	0	30
M_4	0	6,000	7,000	430	730	750	400	0	0

Suppose we are interested in examining the use of stay-specific services as opposed to the use of ancillary services. In this case, we may divide the matrix **D** into the two components of interest and obtain

$$\mathbf{D} = [\mathbf{D}_1 \quad \mathbf{D}_2]$$

where

$$\mathbf{D}_1 = \begin{bmatrix} 1{,}600 & 1{,}000 & 0 \\ 0 & 20{,}000 & 8{,}000 \\ 0 & 10{,}000 & 9{,}000 \\ 0 & 6{,}000 & 7{,}000 \end{bmatrix}$$

and

$$\mathbf{D}_2 = \begin{bmatrix} 260 & 900 & 400 & 100 & 0 & 0 \\ 800 & 150 & 320 & 0 & 20 & 0 \\ 370 & 560 & 730 & 10 & 0 & 30 \\ 430 & 730 & 750 & 400 & 0 & 0 \end{bmatrix}$$

Thus, $\mathbf{D}_1$ may be viewed as a matrix of order 4×3 while $\mathbf{D}_2$ may be viewed as a matrix of order 4×6. When presented in this form, the matrix **D** is called a partitioned matrix while the matrices $\mathbf{D}_1$ and $\mathbf{D}_2$ are referred to as *submatrices* of **D**.

We also might be interested in the use of stay-specific services, laboratory examinations, and operative procedures. In this case, the matrix **D** is partitioned into the submatrices $\mathbf{D}_1$, $\mathbf{D}_2$, and $\mathbf{D}_3$. (Note that this $\mathbf{D}_2$ is not the same as the previous one.) Following this procedure, we find that

$$\mathbf{D} = [\mathbf{D}_1 \quad \mathbf{D}_2 \quad \mathbf{D}_3]$$

where $\mathbf{D}_1$ is defined as before while

$$\mathbf{D}_2 = \begin{bmatrix} 260 & 900 & 400 \\ 800 & 150 & 320 \\ 370 & 560 & 730 \\ 430 & 730 & 750 \end{bmatrix} \quad \text{and} \quad \mathbf{D}_3 = \begin{bmatrix} 100 & 0 & 0 \\ 0 & 20 & 0 \\ 10 & 0 & 30 \\ 400 & 0 & 0 \end{bmatrix}$$

As a further example of how we might partition the matrix **D**, suppose we are interested in the use of stay-specific services, laboratory examinations, and operative procedures by patients hospitalized with conditions M_1 and M_2 as opposed to the use of these services by patients hospitalized with conditions M_3 and M_4. In this case, we partition the matrix **D** into the six submatrices that are given by

$$\mathbf{D} = \begin{bmatrix} \mathbf{D}_{11} & \mathbf{D}_{12} & \mathbf{D}_{13} \\ \mathbf{D}_{21} & \mathbf{D}_{22} & \mathbf{D}_{23} \end{bmatrix}$$

where $\mathbf{D}_{1j}$ and $\mathbf{D}_{2j}$ for $j = 1$, 2, and 3 are submatrices of order 2×3.

Consider next the partitioning of a matrix into submatrices that have the same number of rows and columns. In general, if we have a matrix of order $r \times k$, we may form four submatrices, as follows

$$\mathbf{A} = \begin{bmatrix} \mathbf{B}_{p \times m} & \mathbf{C}_{p \times (k-m)} \\ \mathbf{D}_{(r-p) \times m} & \mathbf{E}_{(r-p) \times (k-m)} \end{bmatrix}$$

In this example, **B**, **C**, **D**, and **E** are submatrices whose orders are shown as subscripts.

18.4 MULTIPLYING PARTITIONED MATRICES(O,G,A,C)

If two matrices are partitioned so that the submatrices are conformable for multiplication, the product matrix **AB** may be expressed in partitioned form so that the submatrices of the product matrix **AB** are functions of the submatrices comprising **A** and **B**. For example, let

$$\mathbf{A} = \begin{bmatrix} \mathbf{A}_{11} & \mathbf{A}_{12} \\ \mathbf{A}_{21} & \mathbf{A}_{22} \end{bmatrix} \quad \text{and} \quad \mathbf{B} = \begin{bmatrix} \mathbf{B}_{11} \\ \mathbf{B}_{21} \end{bmatrix}$$

The product matrix **AB** is then given by

$$\mathbf{AB} = \begin{bmatrix} \mathbf{A}_{11}\mathbf{B}_{11} + \mathbf{A}_{12}\mathbf{B}_{21} \\ \mathbf{A}_{21}\mathbf{B}_{11} + \mathbf{A}_{22}\mathbf{B}_{21} \end{bmatrix}$$

provided that the appropriate submatrices are conformable for multiplication and the results are conformable for addition.

Thus, when matrices have been partitioned appropriately, the submatrices comprising their product are obtained by treating the submatrices of each as an element. Referring to our example, we see that the first term appearing in the product matrix **AB** is given by $\mathbf{A}_{11}\mathbf{B}_{11} + \mathbf{A}_{12}\mathbf{B}_{12}$, which illustrates the latter point. Further, the product $\mathbf{A}_{11}\mathbf{B}_{11}$ is derived by the normal process of matrix multiplication.

Let us now return to Table 18-3 and consider the revenue generated by providing ancillary services only. Suppose that the fees charged for each operative procedure and laboratory examination are as follows

Laboratory Examination	Fee Per Examination	Operative Procedure	Fee Per Procedure
E_1	\$10	OP_1	\$40
E_2	\$20	OP_2	\$60
E_3	\$30	OP_3	\$80

We may express this fee schedule in the form

$$\mathbf{f}' = [10 \quad 20 \quad 30 \quad 40 \quad 60 \quad 80]$$

which may be partitioned as follows

$$\mathbf{f}' = [\mathbf{e}' \quad \mathbf{g}']$$

where $\mathbf{e}' = [10 \quad 20 \quad 30]$ and $\mathbf{g}' = [40 \quad 60 \quad 80]$. Extracting data from Table 18-3 concerning the use of laboratory examinations and operative procedures, we may form the matrix

$$\mathbf{S} = \begin{bmatrix} 260 & 800 & 370 & 430 \\ 900 & 150 & 560 & 730 \\ 400 & 320 & 730 & 750 \\ 100 & 0 & 10 & 400 \\ 0 & 20 & 0 & 0 \\ 0 & 0 & 30 & 0 \end{bmatrix}$$

where the rows of the matrix **S** are defined by the services E_1, E_2, E_3, OP_1, OP_2, and OP_3 while the columns are defined by the conditions M_1, M_2, M_3, and M_4. The matrix **S** may then be partitioned into four submatrices of order 3×2 as follows

$$\mathbf{S} = \begin{bmatrix} \mathbf{T} & \mathbf{U} \\ \mathbf{V} & \mathbf{W} \end{bmatrix}$$

Observe that the submatrix **T** represents the use of laboratory examinations by patients hospitalized with morbidities M_1 and M_2 while the submatrix **V** represents the use of operative procedures by patients hospitalized with these conditions. The submatrices **U** and **W** reflect the use of these services by patients hospitalized with conditions M_3 and M_4.

As can be verified, the revenue earned by providing service to these patients is given by

$$\mathbf{f}'\mathbf{S} = [\mathbf{e}' \quad \mathbf{g}'] \begin{bmatrix} \mathbf{T} & \mathbf{U} \\ \mathbf{V} & \mathbf{W} \end{bmatrix}$$

Thus, the first element of the product **fS** corresponds to the revenues earned by providing ancillary services to patients hospitalized with conditions M_1 and M_2 while the second refers to the income generated by providing these services to patients hospitalized with conditions M_3 and M_4.

Performing the required calculations, we find that

$$\mathbf{e}'\mathbf{T} = [10 \quad 20 \quad 30] \begin{bmatrix} 260 & 800 \\ 900 & 150 \\ 400 & 320 \end{bmatrix} = [32{,}600 \quad 20{,}600]$$

and

$$\mathbf{g}'\mathbf{V} = [40 \quad 60 \quad 80] \begin{bmatrix} 100 & 0 \\ 0 & 20 \\ 0 & 0 \end{bmatrix} = [4{,}000 \quad 1{,}200]$$

Hence,

$$\begin{aligned} \mathbf{e}'\mathbf{T} + \mathbf{g}'\mathbf{V} &= [\ 4{,}000 \quad 1{,}200] + [32{,}600 \quad 20{,}600] \\ &= [36{,}600 \quad 21{,}800] \end{aligned}$$

Following a similar procedure, it can be verified that

$$\mathbf{e}'\mathbf{U} + \mathbf{g}'\mathbf{W} = [39{,}600 \quad 57{,}400]$$

These calculations suggest that the product **fS** is given by

$$\mathbf{f}'\mathbf{S} = [36{,}600 \quad 21{,}800 \quad 39{,}600 \quad 57{,}400]$$

These results indicate that, by providing ancillary services to patients hospitalized with conditions M_1 and M_2, the hospital earned revenues of \$36,600 and \$21,800, respectively. Similarly, the income generated by providing ancillary services to patients hospitalized with conditions M_3 and M_4 amounted to \$39,600 and \$57,400, respectively.

Note that there is a great similarity between the analyses of section 18.1 and 18.4. The first deals with costs and the second deals with revenues. In section 18.4 the fee schedule $\mathbf{f}'$ is analogous to the product $\mathbf{p}'\mathbf{U}$ of section 18.1 ($\mathbf{p}'\mathbf{U}$ is also a row vector). The vector $\mathbf{f}'$ gives revenue per unit of service, while $\mathbf{p}'\mathbf{U}$ gives cost per unit of service. In section 18.4 the matrix **S** shows the services provided to *all patients*, for all conditions. **S** has the same shape as the matrix **T** in section 18.1 but **T** shows the services *per patient*, and **S** could be obtained by multiplying **T** by a diagonal matrix having the elements of **m** along the diagonal.

$$\text{i.e.} \qquad \mathbf{S} = \mathbf{T} \cdot \text{Diag}(\mathbf{m})$$

Finally $\mathbf{f}'\mathbf{S}$ is a vector whose elements must be summed to obtain total revenues, whereas c (i.e. $\mathbf{p}'\mathbf{UTm}$) is a scalar which gives total cost.

18.5 APPLICATIONS TO HEALTH CARE MANAGEMENT[O,A,F]

Linear transformations, the technique of transposition, and the multiplication of partitioned matrices may be employed in health care management, as the foregoing discussion shows clearly. As seen earlier, linear transformations of the form

$$c = \mathbf{p}'\mathbf{UTm}$$

are of considerable value to the process of budget development. Obviously, this formulation also may be used when determining the actual costs of operations that, when compared with the standard of performance as expressed by the budget, provides one of the most important elements in the internal control system of the institution. In this case, a comparison of current activity with the standard of operational performance provides the data base on which management may identify productive units that require the implementation of remedial action.

Also note that, for conceptual reasons, it frequently is necessary to express a set of data in one form but, before performing mathematical operations, the data must be rearranged into a more suitable form. As an example, consider the direct apportionment method of allocating the costs of general service centers to centers providing direct patient care. Let the numbers 1, · · ·, 5 correspond to the general service centers of the institution and the numbers 6, · · ·, 10 correspond to the direct patient care centers. The direct apportionment method may be represented by a system of linear equations of the form

$$T_6 = a_{16}z_1 + a_{26}z_2 + a_{36}z_3 + a_{46}z_4 + a_{56}z_5 + c_6$$

$$T_7 = a_{17}z_1 + a_{27}z_2 + a_{37}z_3 + a_{47}z_4 + a_{57}z_5 + c_7$$

$$T_8 = a_{18}z_1 + a_{28}z_2 + a_{38}z_3 + a_{48}z_4 + a_{58}z_5 + c_8$$

$$T_9 = a_{19}z_1 + a_{29}z_2 + a_{39}z_3 + a_{49}z_4 + a_{59}z_5 + c_9$$

$$T_{10} = a_{110}z_1 + a_{210}z_2 + a_{310}z_3 + a_{410}z_4 + a_{510}z_5 + c_{10}$$

In this formulation, the term a_{ij} corresponds to an apportionment coefficient by which the costs of center i are transferred to center j. Also observe that

1. $0 \leq a_{ij} \leq 1$ and $\Sigma_{j=6}^{10} a_{ij} = 1$;
2. z_i corresponds to the total costs originally assigned to general service center i;
3. c_j corresponds to the total costs originally assigned to direct patient care center j.

Letting

$$\mathbf{t} = \begin{bmatrix} T_6 \\ T_7 \\ T_8 \\ T_9 \\ T_{10} \end{bmatrix}; \qquad \mathbf{z} = \begin{bmatrix} z_1 \\ z_2 \\ z_3 \\ z_4 \\ z_5 \end{bmatrix}; \qquad \mathbf{c} = \begin{bmatrix} c_6 \\ c_7 \\ c_8 \\ c_9 \\ c_{10} \end{bmatrix}$$

and

$$\mathbf{A} = \begin{bmatrix} a_{16} & a_{17} & a_{18} & a_{19} & a_{110} \\ a_{26} & a_{27} & a_{28} & a_{29} & a_{210} \\ a_{36} & a_{37} & a_{38} & a_{39} & a_{310} \\ a_{46} & a_{47} & a_{48} & a_{49} & a_{410} \\ a_{56} & a_{57} & a_{58} & a_{59} & a_{510} \end{bmatrix}$$

The system of linear equations may be expressed in the form

$$\mathbf{t} = \mathbf{A}'\mathbf{z} + \mathbf{c}$$

Observe that, in order to express the system of linear equations in matrix form, it is necessary to find the transpose of **A**.

Appendix 18A

The Addition and Multiplication of Transformed Matrices[O,F,C]

We examine here the matrix operations of addition and multiplication as applied to transposed matrices. As seen earlier, we might express the sum of two matrices, **A** and **B**, by

$$\mathbf{A} + \mathbf{B} = \mathbf{C}$$

Retaining the notation introduced earlier, we let $\mathbf{A}'$ and $\mathbf{B}'$ represent the transpose of the matrices **A** and **B**, respectively. When performing the operation of addition on the transposed matrices, we find that

$$\mathbf{A}' + \mathbf{B}' = \mathbf{C}'$$

Thus, the transpose of a sum of matrices is equal to the sum of the two matrices that have been transposed.

Concerning the product of transposed matrices, we assert that

$$(\mathbf{AB})' = \mathbf{B}'\mathbf{A}'$$

As an example of this statement, let

$$\mathbf{A} = \begin{bmatrix} 0 & 1 & 1 \\ 2 & 0 & 1 \end{bmatrix} \quad \text{and} \quad \mathbf{B} = \begin{bmatrix} 10 & 2 \\ 3 & 5 \\ 1 & 3 \end{bmatrix}$$

As before, we find that

$$\mathbf{AB} = \begin{bmatrix} 0 & 1 & 1 \\ 2 & 0 & 1 \end{bmatrix} \begin{bmatrix} 10 & 2 \\ 3 & 5 \\ 1 & 3 \end{bmatrix} = \begin{bmatrix} 4 & 8 \\ 21 & 7 \end{bmatrix}$$

and

$$(\mathbf{AB})' = \begin{bmatrix} 4 & 21 \\ 8 & 7 \end{bmatrix}$$

Given that

$$\mathbf{B}' = \begin{bmatrix} 10 & 3 & 1 \\ 2 & 5 & 3 \end{bmatrix} \quad \text{and} \quad \mathbf{A}' = \begin{bmatrix} 0 & 2 \\ 1 & 0 \\ 1 & 1 \end{bmatrix}$$

the product $\mathbf{B}'\mathbf{A}'$ is found to be

$$\begin{bmatrix} 10 & 3 & 1 \\ 2 & 5 & 3 \end{bmatrix} \begin{bmatrix} 0 & 2 \\ 1 & 0 \\ 1 & 1 \end{bmatrix} = \begin{bmatrix} 4 & 21 \\ 8 & 7 \end{bmatrix}$$

Hence, the transpose of the product matrix is the product of the transposed matrices **A** and **B** with the order of the matrices reversed. We may extend this result to accommodate the multiplication of more than two matrices. For example, $(\mathbf{ABCDE})' = \mathbf{E}'\mathbf{D}'\mathbf{C}'\mathbf{B}'\mathbf{A}'$, provided that the matrices are conformable for multiplication.

Problems for Solution

1. Suppose the activity of our institution is limited to the medical management of diagnostic conditions M_1, M_2, M_3, and M_4. Also suppose that, on the basis of historical data, it is expected that 6,000, 8,000, 12,000, and 1,000 patients will be admitted during the coming period with conditions M_1, M_2, M_3, and M_4, respectively. Also suppose that the service requirements per patient have been estimated as follows:

	Service			
Condition	S_1	S_2	S_3	S_4
M_1	2.0	4.0	3.0	0
M_2	0	1.0	2.0	3.0
M_3	0	0	3.0	7.0
M_4	8.0	9.0	0	1.0

Further assume that the labor and the supply requirements per unit of service are as follows

	Labor (in Staff Hours)			
Service	L_1	L_2	L_3	L_4
S_1	1.2	2.1	3.0	0
S_2	6.0	5.8	4.5	3.0
S_3	.5	.2	0	0
S_4	1.0	0	1.2	0

	Supplies (in Physical Units)			
Service	CS_1	CS_2	CS_3	CS_4
S_1	1.0	2.0	0	1.0
S_2	2.0	3.0	4.0	0
S_3	1.0	6.0	0	3.0
S_4	0	0	3.0	2.0

Using matrix notation, determine the total labor and supply requirements for the coming period.

2. Suppose that the management of the institution described in Problem 1 estimates that the factor prices of L_1, L_2, L_3, L_4, CS_1, CS_2, CS_3, and CS_4 that will prevail during the coming period are as follows:

Labor	Wage Rate	Supply	Price/Unit
L_1	\$6.00	CS_1	\$3.00
L_2	\$7.50	CS_2	\$ .50
L_3	\$8.00	CS_3	\$ .10
L_4	\$9.00	CS_4	\$6.50

Employ these data to determine the expected cost of providing care to the patient population.

3. Referring to the institution in Problems 1 and 2, assume that the following data represent the actual operating results of the current period. The size and diagnostic mix of the patient population during the current period is as follows:

Condition	Number of Patients
M_1	4,000
M_2	7,000
M_3	9,000
M_4	1,200

The number of services per patient by type of service and diagnostic condition are as follows:

	Service			
Condition	S_1	S_2	S_3	S_4
M_1	3.0	3.0	4.0	0
M_2	0	2.0	3.0	4.0
M_3	0	0	4.0	6.0
M_4	9.0	9.0	0	2.0

The labor requirements per unit of service by type of labor and service are as follows:

	Labor (in Staff Hours)			
Service	L_1	L_2	L_3	L_4
S_1	1.4	2.0	3.1	0
S_2	6.5	6.0	4.6	3.1
S_3	.6	.3	0	0
S_4	1.1	0	1.3	0

The supply requirements per unit of service by type of supply item and service are as follows:

	Consumable Supplies (in Physical Units)			
Service	CS_1	CS_2	CS_3	CS_4
S_1	2.0	1.0	0	2.0
S_2	2.0	3.0	4.0	0
S_3	2.0	5.0	0	4.0
S_4	0	0	4.0	3.0

If the factor prices for L_1, L_2, L_3, L_4, CS_1, CS_2, CS_3, and CS_4 are as follows:

Labor	Wage Rate	Supply	Cost/Unit
L_1	6.50	CS_1	3.00
L_2	8.00	CS_2	.60
L_3	9.00	CS_3	.50
L_4	10.00	CS_4	7.00

use these data to determine the cost of operations during the current period. What is the cost variance associated with labor? Consumable supplies? What is the total cost variance for the period?

4. If

$$\mathbf{A} = \begin{bmatrix} 1 & 7 \\ 12 & 3 \\ 6 & 2 \end{bmatrix}; \quad \mathbf{B} = \begin{bmatrix} 8 & 9 & 6 & 4 \\ 17 & 12 & 3 & 6 \\ 4 & 7 & 0 & 1 \\ 9 & 0 & 1 & 0 \end{bmatrix}; \quad \mathbf{C} = \begin{bmatrix} 1 & 6 & 9 & 7 \\ 13 & 15 & 21 & 0 \end{bmatrix}$$

Find $\mathbf{A}'$, $\mathbf{B}'$, and $\mathbf{C}'$.

5. Letting

$$\mathbf{A} = \begin{bmatrix} 1 & 0 & 3 \\ 6 & 1 & 7 \\ 8 & 2 & 5 \end{bmatrix} \quad \text{and} \quad \mathbf{B} = \begin{bmatrix} 3 & 7 & 6 \\ 2 & 0 & 7 \\ 1 & 3 & 6 \end{bmatrix}$$

Find $\mathbf{AA}'$, $\mathbf{BB}'$, $\mathbf{AB}'$, and $\mathbf{B}'\mathbf{A}$. Also, verify that

$$(\mathbf{A} + \mathbf{B})(\mathbf{A} + \mathbf{B})' = \mathbf{AA}' + \mathbf{BA}' + \mathbf{AB}' + \mathbf{BB}'$$

6. Suppose that the following data reflect the number of services by component of service and by type of payer:

Responsible Party	Laboratory Procedures (# units)	Radiological Services (# units)	Operative Procedures (# units)	Room & Board (days of care)
Self Pay	600	300	200	800
Blue Cross	1,900	1,200	890	12,000
Medicare	1,700	970	780	1,900
Medicaid	980	630	550	430

Also assume that the fees charged by the hospital apply to all parties, and that these charges are as follows:

Component	Fee
Room & Board	$200/day
Laboratory Procedures	$30/procedure
Radiological Services	$20/procedure
Operative Procedure	$170/procedure

What is the revenue earned by providing laboratory and radiological services to patients for whom care is financed by private and public funds, respectively? What is the revenue earned by providing room and board as well as operative procedures to patients for whom care is financed by private and public funds, respectively?

Chapter 19

The Determinant and the Inverse of a Matrix

Objectives

After completing this chapter, you should be able to:

1. Define the determinant of a matrix;
2. Evaluate the determinant for matrices of order 2 and 3;
3. Use the determinant to find the inverse;
4. Use the inverse to solve systems of linear equations.

Chapter Map

The sections comprising this chapter may be summarized as follows:

Section Number	Required Reading	Optional Reading	Generic Development	Application to Management	Fundamental Principles	Complex Material
	(R)	(O)	(G)	(A)	(F)	(C)
19.1	x	x	x	x	x	x
19.1.1	x		x	x	x	
19.1.2		x	x			x
19.2	x	x	x	x	x	x
19.2.1	x		x	x	x	
19.2.2		x	x			x
19.3		x	x		x	
19.4		x		x	x	
Appen-dix		x	x			x

The matrix operations of addition, subtraction, and multiplication have been discussed. However, we have not considered division because this operation is not defined in its usual sense in matrix algebra. Rather than "dividing" by matrix **A**, we multiply by a matrix that is referred to as the inverse of **A**. Thus, one of the major objectives of this chapter is to describe methods of determining the inverse of a matrix. First, however, it is necessary to examine the determinant of a matrix which, in turn, plays a role in finding the inverse of the matrix. After examining matrix inversion, we then consider the use of the inverse in a management setting.

19.1 THE DETERMINANT OF A MATRIX(R,O,G,A,F,C)

Although the literature concerning determinants is vast, in this chapter we focus on the elementary methods of evaluation. In general, a determinant is a scalar value that is nothing more than the sum of selected products of the elements of the matrix from which it is derived. Each of the sums of the selected products is multiplied by $+1$ or -1 according to well-defined rules that will be described later.

In general, determinants are defined for square matrices only. In fact, determinants of nonsquare matrices are not defined and do not exist. When dealing with a square matrix **A**, the determinant of **A** is denoted by $|\mathbf{A}|$, $\|\mathbf{A}\|$, or det $(\mathbf{A})$. In this text, however, we employ the notation $|\mathbf{A}|$ to refer to the determinant of the matrix. The determinant of a matrix of order **n** is referred to as an nth order determinant and the determinant of a scalar is the number itself. The process of obtaining the value that corresponds to $|\mathbf{A}|$ is referred to as *evaluating* the determinant or *expanding* the determinant. The reason for this terminology will become clear later.

19.1.1 Second-Order Determinants(R,G,A,F)

Consider first the determinant of a square matrix of order 2. The value of the determinant of such a matrix is obtained by subtracting the product of the off-diagonal elements from the product of the diagonal elements. Thus, if we let

$$\mathbf{A} = \begin{bmatrix} a_{11} & a_{12} \\ a_{21} & a_{22} \end{bmatrix}$$

the determinant of **A** is given by

$$|\mathbf{A}| = \begin{vmatrix} a_{11} & a_{12} \\ a_{21} & a_{22} \end{vmatrix} = a_{11}a_{22} - a_{12}a_{21} \quad \textbf{(19.1)}$$

(Notice that a determinant is shown with straight vertical bars, while a matrix uses square brackets.)

As an example of Equation 19.1, let

$$\mathbf{A} = \begin{bmatrix} 21 & 12 \\ 1 & 2 \end{bmatrix}$$

from which we find that

$$|\mathbf{A}| = 21(2) - 1(12) = 30$$

Second-order determinants can play a direct role in many of the statistical analyses that might be performed by the health administrator. For example, the standard deviation may be used by management to describe the dispersion exhibited by a set of data. In addition, the standard deviation is required when estimating averages and performng many statistical tests of significance.

Any standard statistics text will reveal that the results of

$$s^2 = \frac{n(\Sigma x_i^2) - (\Sigma x_i)^2}{n(n - 1)} \quad \textbf{(19.2.1)}$$

and

$$s = \sqrt{\frac{n(\Sigma x_i^2) - (\Sigma x_i)^2}{n(n - 1)}} \quad \textbf{(19.2.2)}$$

represent the variance and standard deviation of a sample consisting of the values $x_1, \cdots, x_i, \cdots, x_n$, respectively. For example, suppose that the lengths of stay associated with a sample of five patients are 2, 4, 7, 10, and 2 days. Letting x_i represent the length of stay experienced by patient i, we find that the calculations required for the computation of s^2 and s are as follows

x_i	x_i^2
2	4
4	16
7	49
10	100
2	4
$\Sigma x_i = 25$	$\Sigma x_i^2 = 173$

Given that $n = 5$, we simply substitute appropriately into Equation 19.2.1 and obtain

$$s^2 = \frac{5(173) - (25)^2}{5(4)}$$

$$= 12$$

which implies that the sample standard deviation is given by

$$s = \sqrt{12}$$

$$\cong 3.46 \text{ days}$$

To many students, Equations 19.1.1 and 19.1.2 are quite complex, formidable, and difficult to remember. However, these equations may be simplified by employing matrix notation. When considering a sample consisting of the values $x_1, \cdots, x_i, \cdots, x_n$, the sample variance and standard deviation may be expressed in the form

$$s^2 = \lambda|\mathbf{X}'\mathbf{X}| \qquad \textbf{(19.3.1)}$$

and

$$s = \sqrt{\lambda|\mathbf{X}'\mathbf{X}|} \qquad \textbf{(19.3.2)}$$

respectively. In this formulation, we let the scalar $\lambda = 1/n(n - 1)$ and

$$\mathbf{X} = \begin{bmatrix} 1 & x_1 \\ \cdot & \cdot \\ \cdot & \cdot \\ \cdot & \cdot \\ 1 & x_i \\ \cdot & \cdot \\ \cdot & \cdot \\ \cdot & \cdot \\ 1 & x_n \end{bmatrix}$$

As a result, the product matrix $\mathbf{X}'\mathbf{X}$ is given by

$$\mathbf{X}'\mathbf{X} = \begin{bmatrix} n & \Sigma x_i \\ \Sigma x_i & \Sigma x_i^2 \end{bmatrix}$$

from which we find that

$$\lambda|\mathbf{X}'\mathbf{X}| = \frac{n\Sigma x_i^2 - (\Sigma x_i)^2}{n(n-1)}$$

which is identical to Equation 19.2.1.

Referring to our numerical example, we find that the scalar $\lambda = 1/5(4)$ while the matrix $\mathbf{X}$ is given by

$$\mathbf{X} = \begin{bmatrix} 1 & 2 \\ 1 & 4 \\ 1 & 7 \\ 1 & 10 \\ 1 & 2 \end{bmatrix}$$

As a result, we find that

$$\mathbf{X}'\mathbf{X} = \begin{bmatrix} 1 & 1 & 1 & 1 & 1 \\ 2 & 4 & 7 & 10 & 2 \end{bmatrix} \begin{bmatrix} 1 & 2 \\ 1 & 4 \\ 1 & 7 \\ 1 & 10 \\ 1 & 2 \end{bmatrix} = \begin{bmatrix} 5 & 25 \\ 25 & 173 \end{bmatrix}$$

Thus, employing Equation 19.3.1, we obtain

$$\begin{aligned} s^2 &= \frac{1}{5(4)} \begin{vmatrix} 5 & 25 \\ 25 & 173 \end{vmatrix} \\ &= \frac{5(173) - (25)^2}{5(4)} \\ &= 12 \end{aligned}$$

These calculations suggest that, as before, the sample standard deviation is approximately 3.46 days. Without providing formal proof, we simply assert that one-way analysis of variance, two-way analysis of variance without interaction, and two-way analysis of variance with interaction is greatly simplified by the general notation $|\mathbf{A}'\mathbf{A}|$.

19.1.2 The Determinant of Higher Order Matrices(O,G,C)

When deriving the determinant and the inverse of higher order matrices, managers should use high-speed electronic computers to perform the required calculations. However, to maximize the usefulness of the computer, it is necessary to provide a basic understanding of at least one process by which the determinant of higher order matrices is evaluated. In this section, then, we consider the process of evaluating the determinant of a matrix of order 3×3.

In deriving the simultaneous solution to a system of linear equations, suppose that we require the determinant of the 3×3 matrix

$$\mathbf{A} = \begin{bmatrix} 1 & 2 & 0 \\ 4 & 2 & 5 \\ 2 & 1 & 1 \end{bmatrix} \tag{19.4}$$

Basing our calculations on the first row (i.e., elements 1, 2, and 0), we assert that

$$\begin{aligned} |\mathbf{A}| &= 1(+1) \begin{vmatrix} 2 & 5 \\ 1 & 1 \end{vmatrix} + 2(-1) \begin{vmatrix} 4 & 5 \\ 2 & 1 \end{vmatrix} + 0(1) \begin{vmatrix} 4 & 2 \\ 2 & 1 \end{vmatrix} \\ &= 1(2 - 5) - 2(4 - 10) + 0(4 - 4) \\ &= -3 + 12 = 9 \end{aligned}$$

Here, the determinant of **A** is calculated by the sum of the signed products of the elements 1, 2, 0 and the determinants

$$\begin{vmatrix} 2 & 5 \\ 1 & 1 \end{vmatrix}, \quad \begin{vmatrix} 4 & 5 \\ 2 & 1 \end{vmatrix}, \quad \text{and} \quad \begin{vmatrix} 4 & 2 \\ 2 & 1 \end{vmatrix}$$

respectively. Each of these determinants is obtained by eliminating the row and the column in which the element appears. For example, the element $a_{11} = 1$ is multiplied by the determinant

$$\begin{vmatrix} 2 & 5 \\ 1 & 1 \end{vmatrix}$$

which contains the terms that remain after eliminating the elements appearing in the first row and the first column. Similarly, if we ignore the -1, the element a_{12} or 2 is multiplied by the determinant

$$\begin{vmatrix} 4 & 5 \\ 2 & 1 \end{vmatrix}$$

which contains the elements that remain after eliminating the first row and the second column of the matrix. Determinants obtained by eliminating the row and the column in which the element appears are called the minors of **A**. For example, the determinant

$$\begin{vmatrix} 2 & 5 \\ 1 & 1 \end{vmatrix}$$

is called the minor of element a_{11} while the determinant

$$\begin{vmatrix} 4 & 5 \\ 2 & 1 \end{vmatrix}$$

is the minor of element a_{12}.

The sign attached to the products of each element and the corresponding minor are determined according to the following rule. Recall that the notation a_{ij} refers to the element that appears in row i and column j of **A**. The product of a_{ij} and its minor are multiplied by $(-1)^{i+j}$ where the power of -1 is the sum of the two subscripts. For example, the element 1 that appears in row 1 and column 1 is a_{11}. Thus, the product of a_{11} and its minor is multiplied by $(-1)^{1+1}$ or $+1$. Similarly, the product of a_{12}, which in our example is 2, and its minor are multiplied by $(-1)^{1+2}$ or -1.

To generalize these findings, let us consider the 3×3 matrix

$$\mathbf{A} = \begin{bmatrix} a_{11} & a_{12} & a_{13} \\ a_{21} & a_{22} & a_{23} \\ a_{31} & a_{32} & a_{33} \end{bmatrix}$$

and denote the minor of element a_{ij} by $|\mathbf{M}_{ij}|$ where $\mathbf{M}_{ij}$ is a submatrix of $\mathbf{A}$ that is obtained by eliminating row i and column j. For example, $|\mathbf{M}_{11}|$ is the minor of the element a_{11}, which is found to be

$$|\mathbf{M}_{11}| = \begin{vmatrix} a_{22} & a_{23} \\ a_{32} & a_{33} \end{vmatrix}$$

Using a similar procedure, we may derive $|\mathbf{M}_{12}|$ and $|\mathbf{M}_{13}|$ which, of course, are the minors of the elements a_{12} and a_{13}, respectively. Using this notation, we find that the evaluation of the determinant may be expressed by

$$|\mathbf{A}| = a_{11}(-1)^{1+1}|\mathbf{M}_{11}| + a_{12}(-1)^{1+2}|\mathbf{M}_{12}| + a_{13}(-1)^{1+3}|\mathbf{M}_{13}| \qquad \textbf{(19.5.1)}$$

As can be verified, the details of Equation 19.5.1 are as follows

$$|\mathbf{A}| = a_{11}(-1)^{1+1}\begin{vmatrix} a_{22} & a_{23} \\ a_{32} & a_{33} \end{vmatrix} + a_{12}(-1)^{1+2}\begin{vmatrix} a_{21} & a_{23} \\ a_{31} & a_{33} \end{vmatrix} + a_{13}(-1)^{1+3}\begin{vmatrix} a_{21} & a_{22} \\ a_{31} & a_{32} \end{vmatrix} \qquad \textbf{(19.5.2)}$$

After performing the indicated calculations, we find that

$$|\mathbf{A}| = a_{11}(a_{22}a_{33} - a_{32}a_{23}) - a_{12}(a_{21}a_{33} - a_{31}a_{23}) + a_{13}(a_{21}a_{32} - a_{31}a_{22}) \qquad \textbf{(19.5.3)}$$

In general, when calculations are based on the elements of a row, the nth order determinant may be expressed by

$$|\mathbf{A}| = \sum_{j=1}^{n} a_{ij}(-1)^{i+j}|\mathbf{M}_{ij}| \quad \text{for any } i \tag{19.6.1}$$

On the other hand, when calculations are based on the elements of a column, the nth order determinant is found by

$$|\mathbf{A}| = \sum_{i=1}^{n} a_{ij}(-1)^{i+j}|\mathbf{M}_{ij}| \quad \text{for any } j \tag{19.6.2}$$

Although we have demonstrated the process of evaluating the determinant using the elements of the first row, the technique may be applied to any row or column. To illustrate, let us return to the matrix

$$\mathbf{A} = \begin{bmatrix} 1 & 2 & 0 \\ 4 & 2 & 5 \\ 2 & 1 & 1 \end{bmatrix}$$

that was introduced earlier. Basing our calculations on the third row, we find that

$$\begin{aligned} |\mathbf{A}| &= 2(-1)^{3+1} \begin{vmatrix} 2 & 0 \\ 2 & 5 \end{vmatrix} + 1(-1)^{3+2} \begin{vmatrix} 1 & 0 \\ 4 & 5 \end{vmatrix} + 1(-1)^{3+3} \begin{vmatrix} 1 & 2 \\ 4 & 2 \end{vmatrix} \\ &= 2(10 - 0) - 1(5 - 0) + 1(2 - 8) \\ &= 20 - 5 - 6 = 9 \end{aligned}$$

which agrees with our earlier result.

Consider next the evaluation of the minor using the elements appearing in the third column. In this case we find that

$$|\mathbf{A}| = 0(-1)^{1+3}\begin{vmatrix} 4 & 2 \\ 2 & 1 \end{vmatrix} + 5(-1)^{2+3}\begin{vmatrix} 1 & 2 \\ 2 & 1 \end{vmatrix} + 1(-1)^{3+3}\begin{vmatrix} 1 & 2 \\ 4 & 2 \end{vmatrix}$$

$$= 0 - 5(1 - 4) + 1(2 - 8)$$

$$= 15 - 6$$

$$= 9$$

which also agrees with our earlier results. Thus, the evaluation of the minor using the elements appearing in any row or column leads to the *same* result. As a consequence, irrespective of which row or column is used, the value of the determinant is the same. Also, once the row or column has been selected and the sign attached to the first product has been determined, the signs of the following products alternate from minus to plus and from plus to minus.

The evaluation of the determinant of matrices of order greater than 3 is simply an extension of the procedure described in this section. Here, however, the process is recursive in that the minor $|\mathbf{M}_{ij}|$ is in turn expanded by the procedure outlined above.

Also note that the selection of a row or column that contains one or more zeros reduces the number of calculations required to evaluate the determinant. For example, using the first row of

$$\mathbf{A} = \begin{bmatrix} 3 & 0 & 0 \\ 6 & 9 & 4 \\ 2 & 1 & 5 \end{bmatrix}$$

we find that the determinant is given by

$$|\mathbf{A}| = a_{11}(-1)^{1+1}|\mathbf{M}_{11}|$$

since $a_{12}(-1)^{1+2}|\mathbf{M}_{12}|$ and $a_{13}(-1)^{1+3}|\mathbf{M}_{13}|$ are both zero. Performing these calculations, we find that

$$|\mathbf{A}| = 3(-1)^2\begin{vmatrix} 9 & 4 \\ 1 & 5 \end{vmatrix}$$

$$= 3(45 - 4)$$

$$= 123$$

At this point, the reader should verify that equivalent results are obtained when any other row or column is used to evaluate the determinant of the matrix.

19.2 THE INVERSE OF A MATRIX(R,O,G,A,F,C)

As mentioned earlier, an understanding of the techniques of evaluating the determinant is necessary to matrix inversion, which plays a role in the matrix algebra counterpart to division. We usually denote the inverse of a matrix, say the matrix **A**, by $\mathbf{A}^{-1}$, which is read "the inverse of **A**," "**A** inverse," or "**A** to the minus one." We define the *inverse of matrix* **A** as that matrix whose product with **A** is the identity matrix. This definition may be expressed in the form

$$\mathbf{A}\mathbf{A}^{-1} = \mathbf{I} \qquad \textbf{(19.7)}$$

To illustrate the meaning of the term inverse, suppose we determine that the cost of providing 500 x-rays is \$1,500. We also might be interested in determining the average or unit cost of producing x-rays. On the basis of this information, we find that the cost per unit is \$1,500/500 or \$3. Such a calculation involves the solution of an equation that assumes the form

$$500x = \$1{,}500$$

where x may be regarded as a scalar value. One method of solving this expression is to multiply both sides of the equation by 1/500, which yields

$$1/500(500)x = (1/500)\$1{,}500$$
$$x = \$3$$

We may develop a general notation that applies to this sort of problem by considering the scalars a, b, and x. Here, it will be observed that the equation

$$ax = b \qquad (a \neq 0) \qquad \textbf{(19.8.1)}$$

may be solved for x by

$$x = \frac{1}{a}b = a^{-1}b \qquad \textbf{(19.8.2)}$$

where a^{-1} is the inverse of a. Observe that in solving for x we have multiplied by the *inverse* of a.

19.2.1 The Inverse of a Matrix of Order 2[(R,G,A,F)]

We may extend this solution to accommodate a set of simultaneous equations. As an example, economic theory suggests that, in the absence of collective bargaining and other confounding factors, the market mechanism forces the price of a factor input to a level that results in an equality between the supply of the resource and the demand for the input. Assuming we may apply this theory to the health industry, suppose we express the market demand curve for registered nurses by

$$x_1 = 14{,}000 - .5x_2 \qquad \textbf{(19.9.1)}$$

where x_2 is the salary paid to registered nurses and x_1 is the corresponding quantity. Such an expression suggests that, as the salary paid to registered nurses increases, the demand for this resource declines while the obverse also is true. Similarly, let

$$x_1 = 2{,}000 + .1x_2 \qquad \textbf{(19.9.2)}$$

represent the supply curve for registered nurses. This expression suggests that, as the salary paid to registered nurses increases, the corresponding quantity also increases. This situation may be portrayed graphically, as in Figure 19-1, where the market demand for registered nurses is equal to supply only when $x_2 = x_2^*$ and $x_1 = x_1^*$. Our objective then is to determine the values of x_1 and x_2 that simultaneously satisfy Equations 19.9.1 and 19.9.2.

These equations may be solved simultaneously by substitution. Using this procedure, we may substitute $2{,}000 + .1x_2$ for x_1 in Equation 19.9.1 and obtain

$$\begin{aligned} 2{,}000 + .1x_2 &= 14{,}000 - .5x_2 \\ .6x_2 &= 12{,}000 \\ x_2 &= \$20{,}000 \end{aligned}$$

Hence, x_1 may be found by

$$\begin{aligned} x_1 &= 2{,}000 + .1(20{,}000) \\ &= 4{,}000 \end{aligned}$$

registered nurses.

Figure 19-1 Supply and Demand for Registered Nurses

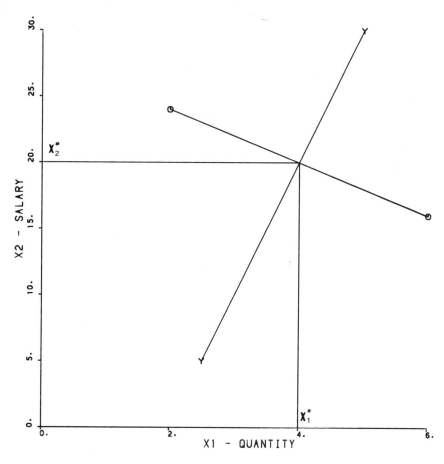

Thus, when dealing with a set of simple equations, the method of substitution is a satisfactory technique for determining values that simultaneously satisfy the expressions. However, when dealing with systems that involve many equations and variables, obtaining a simultaneous solution using the methods that have been described can be tedious and perhaps prohibitive.

Returning to our example, let us rearrange Equations 19.9.1 and 19.9.2 slightly so that we obtain

$$x_1 + .5x_2 = 14{,}000$$

$$10x_1 - x_2 = 20{,}000$$

We may now write these equations in the form

$$\begin{bmatrix} 1 & .5 \\ 10 & -1 \end{bmatrix} \begin{bmatrix} x_1 \\ x_2 \end{bmatrix} = \begin{bmatrix} 14{,}000 \\ 20{,}000 \end{bmatrix} \tag{19.10}$$

Letting

$$\mathbf{x} = \begin{bmatrix} x_1 \\ x_2 \end{bmatrix}, \quad \mathbf{A} = \begin{bmatrix} 1 & .5 \\ 10 & -1 \end{bmatrix} \quad \text{and} \quad \mathbf{b} = \begin{bmatrix} 14{,}000 \\ 20{,}000 \end{bmatrix}$$

we find that the system of equations may be expressed in the form

$$\mathbf{Ax} = \mathbf{b} \tag{19.11.1}$$

Observe that we may solve for $\mathbf{x}$ by premultiplying both sides of the expression by $\mathbf{A}^{-1}$ which yields

$$\mathbf{A}^{-1}\mathbf{Ax} = \mathbf{A}^{-1}\mathbf{b} \tag{19.11.2}$$

and since $\mathbf{A}^{-1}\mathbf{A} = \mathbf{I}$, we have

$$\boxed{\mathbf{x} = \mathbf{A}^{-1}\mathbf{b}} \tag{19.12}$$

Hence, in order to find the simultaneous solution for this system of equations, it is necessary to calculate $\mathbf{A}^{-1}$. Given that the matrix $\mathbf{A}$ is of order 2, we find that $\mathbf{I}$ is also of order 2 and is given by

$$\mathbf{I} = \begin{bmatrix} 1 & 0 \\ 0 & 1 \end{bmatrix}$$

Letting

$$\mathbf{A}^{-1} = \begin{bmatrix} a_{11}^* & a_{12}^* \\ a_{21}^* & a_{22}^* \end{bmatrix}; \quad \mathbf{A} = \begin{bmatrix} a_{11} & a_{12} \\ a_{21} & a_{22} \end{bmatrix} \quad \text{and} \quad \mathbf{I} = \begin{bmatrix} 1 & 0 \\ 0 & 1 \end{bmatrix}$$

and substituting into Equation 19.7 we obtain

$$\begin{bmatrix} a_{11} & a_{12} \\ a_{21} & a_{22} \end{bmatrix} \begin{bmatrix} a_{11}^* & a_{12}^* \\ a_{21}^* & a_{22}^* \end{bmatrix} = \begin{bmatrix} 1 & 0 \\ 0 & 1 \end{bmatrix} \tag{19.13}$$

Our problem is to find the values of $\mathbf{A}^{-1}$ such that Equation 19.13 is satisfied. Expanding this expression, we find that

$$a_{11}a_{11}^* + a_{12}a_{21}^* = 1 \tag{19.14.1}$$

$$a_{21}a_{11}^* + a_{22}a_{21}^* = 0 \tag{19.14.2}$$

$$a_{11}a_{12}^* + a_{12}a_{22}^* = 0 \tag{19.14.3}$$

$$a_{21}a_{12}^* + a_{22}a_{22}^* = 1 \tag{19.14.4}$$

If we now multiply Equation 19.14.1 by a_{21} and Equation 19.14.2 by a_{11}, we obtain

$$a_{21}a_{11}a_{11}^* + a_{21}a_{12}a_{21}^* = a_{21}$$
$$a_{21}a_{11}a_{11}^* + a_{11}a_{22}a_{21}^* = 0$$

If we now solve for a_{21}^*, we find that

$$a_{21}^* = -\frac{a_{21}}{a_{11}a_{22} - a_{12}a_{21}}$$

Referring to Equation 19.1, we find that $a_{11}a_{22} - a_{12}a_{21}$ is the determinant of $\mathbf{A}$. As a result, we may write

$$a_{21}^* = -\frac{a_{21}}{|\mathbf{A}|} \tag{19.15.1}$$

If we replicated this procedure for the terms a_{12}^*, a_{11}^*, and a_{22}^* we would find that

$$a_{12}^* = -\frac{a_{12}}{|\mathbf{A}|} \tag{19.15.2}$$

$$a_{11}^* = \frac{a_{22}}{|\mathbf{A}|} \tag{19.15.3}$$

$$a_{22}^* = \frac{a_{11}}{|\mathbf{A}|} \tag{19.15.4}$$

Thus, the inverse of a 2 × 2 matrix is found to be

$$\mathbf{A}^{-1} = \frac{1}{|\mathbf{A}|} \begin{bmatrix} a_{22} & -a_{12} \\ -a_{21} & a_{11} \end{bmatrix} \tag{19.16}$$

where the elements of the inverse of **A** are given by Equations 19.15.1 through 19.15.4.

Returning to our example, recall that the matrix **A** was found to be

$$\mathbf{A} = \begin{bmatrix} 1 & .5 \\ 10 & -1 \end{bmatrix}$$

From Equation 19.1, we find that

$$\begin{aligned} |\mathbf{A}| &= 1(-1) - 10(.5) \\ &= -6 \end{aligned}$$

and the inverse of **A** is found to be

$$\mathbf{A}^{-1} = \frac{1}{-6} \begin{bmatrix} -1 & -\frac{1}{2} \\ -10 & 1 \end{bmatrix} = \begin{bmatrix} \frac{1}{6} & \frac{1}{12} \\ \frac{10}{6} & -\frac{1}{6} \end{bmatrix}$$

That $\mathbf{A}^{-1}\mathbf{A} = \mathbf{I}$ is verified as follows

$$\begin{bmatrix} \frac{1}{6} & \frac{1}{12} \\ \frac{10}{6} & -\frac{1}{6} \end{bmatrix} \begin{bmatrix} 1 & \frac{1}{2} \\ 10 & -1 \end{bmatrix} = \begin{bmatrix} 1 & 0 \\ 0 & 1 \end{bmatrix}$$

The solution for the set of equations

$$x_1 + .5x_2 = 14{,}000$$
$$10x_1 - x_2 = 20{,}000$$

is found by

$$\mathbf{x} = \mathbf{A}^{-1}\mathbf{b}$$

$$= \begin{bmatrix} \frac{1}{6} & \frac{1}{12} \\ \frac{10}{6} & -\frac{1}{6} \end{bmatrix} \begin{bmatrix} 14{,}000 \\ 20{,}000 \end{bmatrix}$$

$$= \begin{bmatrix} \frac{14{,}000}{6} + \frac{20{,}000}{12} \\ \frac{140{,}000}{6} - \frac{20{,}000}{6} \end{bmatrix}$$

$$= \begin{bmatrix} 4{,}000 \\ 20{,}000 \end{bmatrix}$$

As before, we find that $x_1 = 4{,}000$ and $x_2 = 20{,}000$. These findings imply that, when demand is equal to supply, 4,000 registered nurses will be employed at a salary of $20,000 per year.

Application of the Inverse in Regression Analysis

In addition to its importance in deriving the solution to systems of equations, the inverse can also play an important role in statistics in estimating the parameters of equations that assume the general form

$$y = b_0 + b_1 x$$

Here, the coefficient b_0 represents the y intercept (i.e., the value of y when $x = 0$) while b_1 corresponds to the slope (the change in y resulting from a unit change in x) of the equation. When estimating the parameters of such an equation we usually employ regression analysis, which requires paired observations on the variables x and y. Assuming that we have n data points, we might represent our information in the form $(x_1, y_1), \cdots, (x_i, y_i), \cdots, (x_n, y_n)$ where (x_i, y_i) corresponds to the values of x and y for observation i. As may be verified by consulting any standard statistics text, the *normal equations* for least squares regression analysis assume the form

$$b_0 n + b_1 \Sigma x_i = \Sigma y_i \tag{19.17.1}$$

$$b_0 \Sigma x_i + b_1 \Sigma x_i^2 = \Sigma x_i y_i \tag{19.17.2}$$

We may now let

$$\mathbf{b} = \begin{bmatrix} b_0 \\ b_1 \end{bmatrix}; \qquad \mathbf{X} = \begin{bmatrix} 1 & x_1 \\ \cdot & \cdot \\ \cdot & \cdot \\ \cdot & \cdot \\ 1 & x_i \\ \cdot & \cdot \\ \cdot & \cdot \\ \cdot & \cdot \\ 1 & x_n \end{bmatrix}$$

and

$$\mathbf{y} = \begin{bmatrix} y_1 \\ \cdot \\ \cdot \\ \cdot \\ y_i \\ \cdot \\ \cdot \\ \cdot \\ y_n \end{bmatrix}$$

Employing this notation, we find that

$$\mathbf{X}'\mathbf{X} = \begin{bmatrix} n & \Sigma x_i \\ \Sigma x_i & \Sigma x_i^2 \end{bmatrix}$$

while

$$\mathbf{X}'\mathbf{y} = \begin{bmatrix} \Sigma y_i \\ \Sigma x_i y_i \end{bmatrix}$$

Hence, the normal equations given by 19.17.1 and 19.17.2 may be expressed in the form

$$\mathbf{X}'\mathbf{X}\mathbf{b} = \mathbf{X}'\mathbf{y}$$

from which we find that

$$\mathbf{b} = (\mathbf{X}'\mathbf{X})^{-1}\mathbf{X}'\mathbf{y} \qquad \textbf{(19.17.3)}$$

Thus, in order to calculate the coefficients b_0 and b_1 we simply multiply $\mathbf{X}'\mathbf{y}$ by the inverse of $\mathbf{X}'\mathbf{X}$.

To illustrate the use of Equation 19.17.3, suppose that management wants to estimate the parameters of a total cost function that assumes the form

$$TC = b_0 + b_1 V$$

In this formulation, we let

TC represent total cost,
b_0 represent fixed cost component of total cost, and
b_1 represent the variable component of total cost where V represents the volume of service as measured by days of care.

Also assume that the following data are available to management:

TC_i (in dollars)	V_i (days of care)
600	50
620	65
730	72
740	84
760	90
790	110

In this example, we find that

$$\mathbf{X} = \begin{bmatrix} 1 & 50 \\ 1 & 65 \\ 1 & 72 \\ 1 & 84 \\ 1 & 90 \\ 1 & 110 \end{bmatrix} \quad \text{and} \quad \mathbf{y} = \begin{bmatrix} 600 \\ 620 \\ 730 \\ 740 \\ 760 \\ 790 \end{bmatrix}$$

from which we obtain

$$\mathbf{X}'\mathbf{X} = \begin{bmatrix} 6 & 471 \\ 471 & 39{,}165 \end{bmatrix}$$

and

$$\mathbf{X}'\mathbf{y} = \begin{bmatrix} 4{,}240 \\ 340{,}320 \end{bmatrix}$$

On the basis of these calculations, we find the inverse of $\mathbf{X}'\mathbf{X}$ as follows:

$$(\mathbf{X}'\mathbf{X})^{-1} = \frac{1}{13{,}149} \begin{bmatrix} 39{,}165 & -471 \\ -471 & 6 \end{bmatrix}$$

$$= \begin{bmatrix} 2.9785535 & -.0358202 \\ -.0358202 & .0004563 \end{bmatrix}$$

Referring to Equation 19.17.3, we obtain

$$\mathbf{b} = \begin{bmatrix} 2.9785535 & -.0358202 \\ -.0358202 & .0004563 \end{bmatrix} \begin{bmatrix} 4{,}240 \\ 340{,}320 \end{bmatrix}$$

$$= \begin{bmatrix} 438.74 \\ 3.41 \end{bmatrix}$$

These results suggest that the fixed cost component of our unit is approximately \$438.74 and that total costs increase by \$3.41 on each occasion that an additional unit of care is provided.

19.2.2 The Inverse of Higher Order Matrices(O,G,C)

Having described the technique for finding the inverse of a 2×2 matrix, we turn now to the problem of calculating the inverse of higher order matrices. As seen in Section 19.1.2, the determinant of a matrix of order n may be evaluated by

$$|\mathbf{A}| = \sum_{j=1}^{n} a_{ij}(-1)^{i+j}|\mathbf{M}_{ij}|$$

for any row i where $|\mathbf{M}_{ij}|$ is the minor of element a_{ij}. We now define the product $(-1)^{i+j}|\mathbf{M}_{ij}|$ as the cofactor of a_{ij}. Thus, the cofactor of a_{ij} is the product of its minor $|\mathbf{M}_{ij}|$ and the factor $(-1)^{i+j}$. Letting γ_{ij} (the Greek letter gamma) represent the cofactor of a_{ij}, we may write

$$\gamma_{ij} = (-1)^{i+j}|\mathbf{M}_{ij}| \qquad \textbf{(19.18)}$$

To illustrate the use of Equation 19.18, we return to Equation 19.4, where the matrix **A** was defined by

$$\mathbf{A} = \begin{bmatrix} 1 & 2 & 0 \\ 4 & 2 & 5 \\ 2 & 1 & 1 \end{bmatrix}$$

and calculate the corresponding cofactors as follows. Beginning with the first column and using Equation 19.18, we find that the cofactors of the elements a_{11}, a_{21} and a_{31} are

$$\gamma_{11} = (-1)^{1+1}\begin{vmatrix} 2 & 5 \\ 1 & 1 \end{vmatrix}; \qquad \gamma_{21} = (-1)^{2+1}\begin{vmatrix} 2 & 0 \\ 1 & 1 \end{vmatrix};$$

$$\gamma_{31} = (-1)^{3+1}\begin{vmatrix} 2 & 0 \\ 2 & 5 \end{vmatrix}$$

or -3, -2, and 10, respectively. Similarly, the cofactors of the elements in the second column are

$$\gamma_{12} = (-1)^{1+2} \begin{vmatrix} 4 & 5 \\ 2 & 1 \end{vmatrix}; \qquad \gamma_{22} = (-1)^{2+2} \begin{vmatrix} 1 & 0 \\ 2 & 1 \end{vmatrix};$$

$$\gamma_{32} = (-1)^{3+2} \begin{vmatrix} 1 & 0 \\ 4 & 5 \end{vmatrix}$$

or 6, 1, and −5, respectively. Finally, the cofactors of the elements in the third column are

$$\gamma_{13} = (-1)^{1+3} \begin{vmatrix} 4 & 2 \\ 2 & 1 \end{vmatrix}; \qquad \gamma_{23} = (-1)^{2+3} \begin{vmatrix} 1 & 2 \\ 2 & 1 \end{vmatrix};$$

$$\gamma_{33} = (-1)^{3+3} \begin{vmatrix} 1 & 2 \\ 4 & 2 \end{vmatrix}$$

or 0, 3, and −6, respectively. We may now form the matrix of cofactors as follows

$$\mathbf{G} = \begin{bmatrix} \gamma_{11} & \gamma_{12} & \gamma_{13} \\ \gamma_{21} & \gamma_{22} & \gamma_{23} \\ \gamma_{31} & \gamma_{32} & \gamma_{33} \end{bmatrix} = \begin{bmatrix} -3 & 6 & 0 \\ -2 & 1 & 3 \\ 10 & -5 & -6 \end{bmatrix}$$

At this point in the analysis, it is necessary to find the transpose of **G** that is given by

$$\mathbf{G}' = \begin{bmatrix} \gamma_{11} & \gamma_{21} & \gamma_{31} \\ \gamma_{12} & \gamma_{22} & \gamma_{32} \\ \gamma_{13} & \gamma_{23} & \gamma_{33} \end{bmatrix}$$

Since $|\mathbf{A}| = \Sigma a_{ij}\gamma_{ij}$ (summing over j or i for any chosen value of i or any chosen value of j) and it can be shown that $\Sigma a_{ij}\gamma_{ik} = 0$ for $j \neq k$, the product $\mathbf{G}'\mathbf{A}$ yields

$$\begin{bmatrix} \gamma_{11} & \gamma_{21} & \gamma_{31} \\ \gamma_{12} & \gamma_{22} & \gamma_{32} \\ \gamma_{13} & \gamma_{23} & \gamma_{33} \end{bmatrix} \begin{bmatrix} a_{11} & a_{12} & a_{13} \\ a_{21} & a_{22} & a_{23} \\ a_{31} & a_{32} & a_{33} \end{bmatrix} = \begin{bmatrix} |\mathbf{A}| & 0 & 0 \\ 0 & |\mathbf{A}| & 0 \\ 0 & 0 & |\mathbf{A}| \end{bmatrix}$$

To illustrate this important result, let us refer to our numeric example where we find that

$$\mathbf{G}' = \begin{bmatrix} -3 & -2 & 10 \\ 6 & 1 & -5 \\ 0 & 3 & -6 \end{bmatrix}$$

which, when postmultiplied by **A**, yields

$$\mathbf{G}'\mathbf{A} = \begin{bmatrix} -3 & -2 & 10 \\ 6 & 1 & -5 \\ 0 & 3 & -6 \end{bmatrix} \begin{bmatrix} 1 & 2 & 0 \\ 4 & 2 & 5 \\ 2 & 1 & 1 \end{bmatrix} = \begin{bmatrix} 9 & 0 & 0 \\ 0 & 9 & 0 \\ 0 & 0 & 9 \end{bmatrix}$$

Here, it will be observed that $|\mathbf{A}| = 9$ appears as the diagonal element of the product matrix $\mathbf{G}'\mathbf{A}$ while the other elements are zero. In other words, the sum of the products between the elements of a row (or column) and their corresponding cofactors yields the determinant of the matrix. Conversely, the sum of the products of the elements in a row (or column) and the cofactors of the elements appearing in a different row (or column) is zero. As a result we may assert that

$$|\mathbf{A}| = \sum_{j=1}^{n} \gamma_{ij} a_{ij} = \sum_{j=1}^{n} a_{ij} \gamma_{ij} \tag{19.19.1}$$

which is simply a restatement of Equation 19.6.1, and that

$$\Sigma a_{ij} \gamma_{kj} = 0 \qquad \text{for } k \neq i \tag{19.19.2}$$

Returning to our example, recall that

$$\mathbf{G}'\mathbf{A} = \begin{bmatrix} 9 & 0 & 0 \\ 0 & 9 & 0 \\ 0 & 0 & 9 \end{bmatrix}$$

If we now multiply the product matrix $\mathbf{G}'\mathbf{A}$ by $1/|\mathbf{A}|$ we obtain

$$\left(\frac{1}{|\mathbf{A}|}\mathbf{G}'\right)\mathbf{A} = \begin{bmatrix} 1 & 0 & 0 \\ 0 & 1 & 0 \\ 0 & 0 & 1 \end{bmatrix}$$

which, of course, is the identity matrix. Since $\mathbf{A}^{-1}\mathbf{A} = \mathbf{I}$ and

$$\left(\frac{1}{|\mathbf{A}|}\mathbf{G}'\right)\mathbf{A} = \mathbf{I},$$

the inverse of **A** is given by

$$\mathbf{A}^{-1} = \frac{1}{|\mathbf{A}|}\mathbf{G}' \qquad \textbf{(19.20)}$$

where, as before, $|\mathbf{A}|$ is the determinant of **A** and $\mathbf{G}'$ is the transposed matrix of cofactors. That the technique described may be used to find the inverse of a 2 × 2 matrix is demonstrated in the Appendix at the end of this chapter.

In summary, the derivation of $\mathbf{A}^{-1}$ using general notation is as follows. Letting

$$\mathbf{A} = \begin{bmatrix} a_{11} & a_{12} & a_{13} \\ a_{21} & a_{22} & a_{23} \\ a_{31} & a_{32} & a_{33} \end{bmatrix}$$

we determine the cofactors of a_{ij} by

$$\gamma_{ij} = (-1)^{i+j}|\mathbf{M}_{ij}|$$

which yields

$$\begin{bmatrix} \gamma_{11} & \gamma_{12} & \gamma_{13} \\ \gamma_{21} & \gamma_{22} & \gamma_{23} \\ \gamma_{31} & \gamma_{32} & \gamma_{33} \end{bmatrix}$$

The transpose of this matrix is given by

$$\begin{bmatrix} \gamma_{11} & \gamma_{21} & \gamma_{31} \\ \gamma_{12} & \gamma_{22} & \gamma_{32} \\ \gamma_{13} & \gamma_{23} & \gamma_{33} \end{bmatrix}$$

which, when multiplied by $1/|\mathbf{A}|$, yields $\mathbf{A}^{-1}$ as follows

$$\mathbf{A}^{-1} = \frac{1}{|\mathbf{A}|} \begin{bmatrix} \gamma_{11} & \gamma_{21} & \gamma_{31} \\ \gamma_{12} & \gamma_{22} & \gamma_{32} \\ \gamma_{13} & \gamma_{23} & \gamma_{33} \end{bmatrix} = \frac{1}{|\mathbf{A}|}\mathbf{G}'$$

Finally, the expression

$$\mathbf{A}^{-1}\mathbf{A} = \mathbf{I}$$

should be verified each time $\mathbf{A}^{-1}$ is calculated.

19.3 THE EXISTENCE OF THE INVERSE(O,G,F)

Essentially two conditions must be satisfied by the matrix **A** in order for $\mathbf{A}^{-1}$ to exist. First, matrix inversion is possible only when dealing with square matrices, which implies that $\mathbf{A}^{-1}$ exists only when **A** is square. Second, since the inverse of **A** is given by the product of the reciprocal of the determinant and the transpose of the matrix of cofactors, $\mathbf{A}^{-1}$ exists only when $|\mathbf{A}|$ is nonzero. If $|\mathbf{A}|$ is zero, the scalar $1/|\mathbf{A}|$ is not defined and hence $\mathbf{A}^{-1}$ does not exist.

When the value of the determinant of a square matrix is found to be zero, the matrix is said to be *singular*, and when the determinant of a square matrix is found to be nonzero, the matrix is said to be *nonsingular*. As a result, $\mathbf{A}^{-1}$ exists for square nonsingular matrices.

19.4 APPLICATIONS TO HEALTH CARE MANAGEMENT(O,A,F)

Throughout this chapter we have referred to the use of the inverse in solving a system of linear equations. To provide an additional illustration of this technique, consider a departmental unit, such as the laboratory or radiology, that provides a set of standardized services. If we view the department as consisting of n functional units, we may let a_{ij} correspond to the portion of the capacity in unit i required to produce a unit of service j. Similarly, we let b_i represent the total capacity of unit i and x_j represent the frequency with which service j is provided. We may now represent the use of available capacity by

$$\begin{bmatrix} a_{11} & \cdots & a_{1j} & \cdots & a_{1n} \\ \vdots & & \vdots & & \vdots \\ a_{i1} & \cdots & a_{ij} & \cdots & a_{in} \\ \vdots & & \vdots & & \vdots \\ a_{n1} & \cdots & a_{nj} & \cdots & a_{nn} \end{bmatrix} \begin{bmatrix} x_1 \\ \vdots \\ x_j \\ \vdots \\ x_n \end{bmatrix} = \begin{bmatrix} b_1 \\ \vdots \\ b_j \\ \vdots \\ b_n \end{bmatrix}$$

which may be expressed in the form

$$\mathbf{Ax} = \mathbf{b}$$

Under the assumption that available capacity is neither underused nor overused, the quantities of service that satisfy this relationship are given by

$$\mathbf{x} = \mathbf{A}^{-1}\mathbf{b}$$

This illustration may be modified slightly to accommodate the entire hospital enterprise. For example, we might assume that the output of the hospital may be represented by the number and mix of patients who are medically managed through the provision of diagnostic and therapeutic services. Assume that the hospital produces services $s_1, \cdots, s_i, \cdots, s_x$ to patients presenting the diagnostic conditions $M_1, \cdots, M_j, \cdots, M_x$. If a_{ij} represents the amount of service i required in the management of each patient presenting diagnosis j, d_j represents the number of patients presenting diagnosis j, and $\hat{s}_i$ corresponds to the maximum attainable rate of producing s_i, we find that

$$\begin{array}{ccccccc} a_{11}d_1 + \cdots & + a_{1j}d_j + & \cdots & + a_{1x}d_x & = & \hat{s}_1 \\ \vdots & \vdots & & \vdots & & \vdots \\ a_{i1}d + \cdots & + a_{ij}d_j + & \cdots & + a_{ix}d_x & = & \hat{s}_i \\ \vdots & \vdots & & \vdots & & \vdots \\ a_{x1}d + \cdots & + a_{xj}d_j + & \cdots & + a_{xx}d_x & = & \hat{s}_x \end{array}$$

Letting

$$\mathbf{A} = \begin{bmatrix} a_{11} & \cdots & a_{1j} & \cdots & a_{1x} \\ \vdots & & \vdots & & \vdots \\ a_{i1} & \cdots & a_{ij} & \cdots & a_{ix} \\ \vdots & & \vdots & & \vdots \\ a_{x1} & \cdots & a_{xj} & \cdots & a_{xx} \end{bmatrix}; \qquad \mathbf{d} = \begin{bmatrix} d_1 \\ \vdots \\ d_j \\ \vdots \\ d_x \end{bmatrix}; \qquad \hat{\mathbf{s}} = \begin{bmatrix} \hat{s}_1 \\ \vdots \\ \hat{s}_i \\ \vdots \\ \hat{s}_x \end{bmatrix}$$

we find that this system of equations may be expressed by

$$\mathbf{Ad} = \hat{\mathbf{s}}$$

Here **A** corresponds to the matrix consisting of the quantity of each service provided per patient while **d** is a column vector representing the quantity and composition of patients and $\hat{\mathbf{s}}$ corresponds to the maximum amount of each service that can be provided with the existing capacity. Similar to our earlier work, we find that the solution that satisfies this equation is given by

$$\mathbf{d} = \mathbf{A}^{-1}\hat{\mathbf{s}}$$

Under the assumption that the system of linear equations is correct, this result yields the quantity and composition of patients who can be medically managed without either overusing or underusing hospital capacity.

Consider next a slightly different situation. Assume that departments $B_1, \cdots, B_i, \cdots, B_x$ provide services $s_1, \cdots, s_i, \cdots, s_x$, respectively. Suppose further that

$$\begin{aligned} ATC_1 &= f(s_1) \\ &\vdots \\ ATC_i &= f(s_i) \\ &\vdots \\ ATC_x &= f(s_x) \end{aligned}$$

represent the functional relation between unit costs and the volume of service provided. Referring to Chapter 10, the conditions

$$\begin{array}{ll} f'(s_1^*) = 0; & f''(s_1^*) > 0 \\ \vdots & \vdots \\ f'(s_i^*) = 0; & f''(s_i^*) > 0 \\ \vdots & \vdots \\ f'(s_x^*) = 0; & f''(s_x^*) > 0 \end{array}$$

indicate that $s_1^*, \cdots, s_i^*, \cdots, s_x^*$ represent the rates of activity that minimize the unit costs of operating departments $B_1, \cdots, B_i, \cdots, B_x$, respectively.

As before, if we let a_{ij} represent the amount of service i required in the management of each patient presenting diagnosis j, and s_i^* correspond to the rate of activity that minimizes the cost per service, we may form the system of linear equations

$$\begin{array}{ccccc} a_{11}d_1^* + \cdots & + a_{1j}d_j^* + \cdots & + a_{1x}d_x^* & = & s_1^* \\ \vdots & \vdots & \vdots & & \vdots \\ a_{i1}d_1^* + \cdots & + a_{ij}d_j^* + \cdots & + a_{ix}d_x^* & = & s_i^* \\ \vdots & \vdots & \vdots & & \vdots \\ a_{x1}d_1^* + \cdots & + a_{xj}d_j^* + \cdots & + a_{xx}d_x^* & = & s_x^* \end{array}$$

Letting

$$\mathbf{s}^* = \begin{bmatrix} s_1^* \\ \vdots \\ s_i^* \\ \vdots \\ s_x^* \end{bmatrix} \quad \text{and} \quad \mathbf{d}^* = \begin{bmatrix} d_1^* \\ \vdots \\ d_j^* \\ \vdots \\ d_x^* \end{bmatrix}$$

we find that this system of equations may be expressed in the form

$$\mathbf{Ad}^* = \mathbf{s}^*$$

where, as before, **A** corresponds to the matrix consisting of the quantity of each service provided per patient. The solution that satisfies this system of equations is given by

$$\mathbf{d}^* = \mathbf{A}^{-1}\mathbf{s}^*$$

In this case, notice that the elements of the column vector **d*** correspond to the quantity and composition of patients who minimize the unit costs of operating departments $B_1, \cdots, B_i, \cdots, B_x$. Assuming that 80 to 90 percent of all admissions are elective, the administrator might monitor and control the size and mix of patients managed per period so as to approximate the column vector **d*** and thereby increase the percentage of time that the departments of the hospital are operated at minimum average costs. Notice that these results provide the basis for ensuring that the services required by the community are provided at minimum cost.

Appendix 19A

Using Cofactors to Find the Inverse of a Matrix of Order 2[(O,G,C)]

To illustrate the use of cofactors to find the inverse of a matrix of order 2, let us return to the system of equations

$$x_1 + .5x_2 = 14{,}000$$
$$10x_1 - x_2 = 20{,}000$$

where, as before, we let

$$\mathbf{A} = \begin{bmatrix} 1 & .5 \\ 10 & -1 \end{bmatrix}$$

Applying Equation 12.17, the cofactors of the elements a_{11}, a_{12}, a_{21}, and a_{22} are given by

$$\gamma_{11} = (-1)^{1+1}|-1| = -1$$
$$\gamma_{12} = (-1)^{1+2}|10| = -10$$
$$\gamma_{21} = (-1)^{2+1}|.5| = -.5$$
$$\gamma_{22} = (-1)^{2+2}|1| = 1$$

Thus, the matrix of cofactors is given by

$$\begin{bmatrix} \gamma_{11} & \gamma_{12} \\ \gamma_{21} & \gamma_{22} \end{bmatrix} = \begin{bmatrix} -1 & -10 \\ -.5 & 1 \end{bmatrix}$$

while the transpose of this matrix is seen to be

$$\begin{bmatrix} \gamma_{11} & \gamma_{21} \\ \gamma_{12} & \gamma_{22} \end{bmatrix} = \begin{bmatrix} -1 & -.5 \\ -10 & 1 \end{bmatrix}$$

Further, the product of the transpose of the matrix of cofactors and the matrix **A** is given by

$$\begin{bmatrix} \gamma_{11} & \gamma_{21} \\ \gamma_{12} & \gamma_{22} \end{bmatrix} \begin{bmatrix} a_{11} & a_{12} \\ a_{21} & a_{22} \end{bmatrix}$$

which, in terms of our example, is seen to be

$$\begin{bmatrix} -1 & -.5 \\ -10 & 1 \end{bmatrix} \begin{bmatrix} 1 & .5 \\ 10 & -1 \end{bmatrix} = \begin{bmatrix} -6 & 0 \\ 0 & -6 \end{bmatrix}$$

Since the determinant of **A** was found to be -6 earlier, we find that

$$\begin{bmatrix} \gamma_{11} & \gamma_{21} \\ \gamma_{12} & \gamma_{22} \end{bmatrix} \begin{bmatrix} a_{11} & a_{12} \\ a_{21} & a_{22} \end{bmatrix} = \begin{bmatrix} \gamma_{11}a_{11} + \gamma_{21}a_{21} & \gamma_{11}a_{12} + \gamma_{21}a_{22} \\ \gamma_{12}a_{11} + \gamma_{22}a_{21} & \gamma_{12}a_{12} + \gamma_{22}a_{22} \end{bmatrix}$$

which reduces to

$$\begin{bmatrix} |\mathbf{A}| & 0 \\ 0 & |\mathbf{A}| \end{bmatrix}$$

This result allows us to assert that

$$\frac{1}{|\mathbf{A}|} \begin{bmatrix} \gamma_{11} & \gamma_{21} \\ \gamma_{12} & \gamma_{22} \end{bmatrix} \begin{bmatrix} a_{11} & a_{12} \\ a_{21} & a_{22} \end{bmatrix} = \frac{1}{|\mathbf{A}|} \begin{bmatrix} |\mathbf{A}| & 0 \\ 0 & |\mathbf{A}| \end{bmatrix} = \begin{bmatrix} 1 & 0 \\ 0 & 1 \end{bmatrix}$$

Given that $\mathbf{A}^{-1}\mathbf{A} = \mathbf{I}$, the inverse of matrix **A** may be calculated by

$$\mathbf{A}^{-1} = \frac{1}{|\mathbf{A}|} \begin{bmatrix} \gamma_{11} & \gamma_{21} \\ \gamma_{12} & \gamma_{22} \end{bmatrix}$$

Thus,

> the inverse of the matrix **A** is obtained by transposing the matrix of cofactors and multiplying the resulting matrix by the reciprocal of the determinant.

Problems for Solution

1. Given that

$$\mathbf{A} = \begin{bmatrix} 1 & 7 & 6 \\ 3 & 10 & 4 \\ 6 & 9 & 7 \end{bmatrix}; \quad \mathbf{B} = \begin{bmatrix} 12 & 0 & 6 \\ 1 & 7 & 6 \\ 0 & 3 & 7 \end{bmatrix}; \quad \mathbf{C} = \begin{bmatrix} 3 & 1 & 7 \\ 6 & 2 & 9 \\ 15 & 20 & 21 \end{bmatrix}$$

Find $|\mathbf{A}|$, $|\mathbf{B}|$, $|\mathbf{C}|$ using the first row, the second column and the third row.

2. Letting

$$\mathbf{X}' = \begin{bmatrix} 1 & 1 & 1 & 1 \\ 7 & 2 & 9 & 8 \end{bmatrix}$$

find $|\mathbf{X}'\mathbf{X}|$.

3. Suppose that our hospital provides services S_1, S_2, and S_3 to patients presenting conditions M_1, M_2, and M_3. Also suppose that the following data reflect the quantity and composition of services that are required by each patient presenting one of the three conditions.

Condition	Service		
	S_1	S_2	S_3
M_1	0	1.0	2.0
M_2	2.0	3.0	1.0
M_3	0	0	2.0

If

$$\hat{S}_1 = 800{,}000$$
$$\hat{S}_2 = 100{,}000$$
$$\hat{S}_3 = 900{,}000$$

represents the maximum obtainable rates of producing S_1, S_2, and S_3 what is the quantity and composition of patients who can be medically managed under the condition that the institution uses its full productive potential?

4. Referring to the institution described in Problem 3, suppose that the labor requirements per unit of service are as follows:

Service	Labor Per Unit of Service		
	L_1	L_2	L_3
S_1	2.5	0	3.0
S_2	1.0	2.5	0
S_3	3.0	1.0	2.7

Assuming that services S_1, S_2, and S_3 are produced at the rates $\hat{S}_1$, $\hat{S}_2$, and $\hat{S}_3$, respectively, what is the quantity and composition of labor that will be required to sustain this rate of operation?

5. Again, referring to the institution in Problem 3, suppose that the supply requirements per unit of service are as follows:

Service	Supply Requirements Per Unit of Service			
	CS_1	CS_2	CS_3	CS_4
S_1	1.0	0	3.0	4.0
S_2	3.0	6.0	8.0	0
S_3	10.0	1.0	3.0	2.0

Find the quantity and composition of consumable supplies required to sustain the rate of activity given by $\hat{S}_1$, $\hat{S}_2$, $\hat{S}_3$.

6. Given that

$$\mathbf{A} = \begin{bmatrix} 1 & 6 & 9 \\ 0 & 3 & 10 \\ 12 & 14 & 5 \end{bmatrix}; \quad \mathbf{B} = \begin{bmatrix} 3 & 6 & 12 \\ 0 & 2 & 7 \\ 9 & 6 & 0 \end{bmatrix}; \quad \mathbf{C} = \begin{bmatrix} 9 & 6 \\ 4 & 0 \end{bmatrix}$$

find $\mathbf{A}^{-1}$, $\mathbf{B}^{-1}$, and $\mathbf{C}^{-1}$; show that $\mathbf{A}^{-1}\mathbf{A}$, $\mathbf{B}^{-1}\mathbf{B}$, and $\mathbf{C}^{-1}\mathbf{C}$ equal $\mathbf{I}$.

7. Referring to Problem 4 of Chapter 10, suppose that the following data reflect the quantity and composition of service required by each patient admitted with conditions M_1, M_2, and M_3.

Condition	Service/Case		
	S_1	S_2	S_3
M_1	2.0	0	1.0
M_2	1.0	2.0	1.0
M_3	3.0	1.0	0

What is the size and diagnostic mix of patients that results in each of the departments operating at minimum unit costs?

Solutions to Selected Problems

Chapter 4

1. a. 1, 2, 3, 4, 5
   b. 7, 8, 9, 10, 11
   c. −6, −5, −4, −3, −2
   d. 7, 11, 13, 17, 19
   e. 3/8, 6/8, 9/8, 12/8, 15/8
2. a. 3/8, 3/4, 9/8, 3/2, 15/8
   b. 3/8, 3/4, 1 1/8, 1 1/2, 1 7/8
   c. .375, .75, 1.125, 1.5, 1.875
   d. .38, .75, 1.13, 1.50, 1.88
3. a. *Associative Law*

$$3 + 7 + 5 + 9 = (3 + 7) + (5 + 9) = 10 + 14 = 24$$
$$= 3 + 7 + 14 = 3 + 21 = 24$$
$$= 10 + 5 + 9 = 15 + 9 = 24$$

Commutative Law

$$3 + 7 + 5 + 9 = (7 + 3) + (9 + 5) = 10 + 14 = 24$$
$$= 7 + 9 + 3 + 5 = 16 + 8 = 24$$

etc.

b. *Associative Law*

$$3 \times 7 \times 5 \times 9 = (3 \times 7) \times (5 \times 9) = 21 \times 45 = 945$$
$$= 21 \times 5 \times 9 = 105 \times 9 = 945$$

Commutative Law

$$3 \times 7 \times 5 \times 9 = 3 \times 5 \times 9 \times 7 = 15 \times 63 = 945$$
$$= 3 \times 9 \times 7 \times 5 = 27 \times 35 = 945$$

etc.

4. a. *Correct answer*

$$3 - 7 - 5 - 9 = (-4) - 5 - 9 = (-9) - 9 = -18$$

Incorrect answers

$$3 - 7 - 5 - 9 = (3 - 7) - (5 - 9) = (-4) - (-4) = 0$$
$$3 - 7 - 5 - 9 = (3 - 5) - (9 - 7) = (-2) - (2) = -4$$

etc.

b. *Correct answer*

$$3 \div 7 \div 5 \div 9 = \left(\frac{3}{7}\right) \div 5 \div 9 = \left(\frac{3}{35}\right) \div 9 = \frac{3}{315} \simeq .0095238$$

Incorrect answers

$$3 \div 7 \div 5 \div 9 = \left(\frac{3}{7}\right) \div \left(\frac{5}{9}\right) = \left(\frac{3}{7}\right) \times \left(\frac{9}{5}\right) = \frac{27}{35} \simeq .7714285$$
$$3 \div 7 \div 5 \div 9 = \left(\frac{3}{5}\right) \div \left(\frac{9}{7}\right) = \left(\frac{3}{5}\right) \times \left(\frac{7}{9}\right) = \frac{21}{45} \simeq .4666\dot{6}$$

etc.

5. a. $(m + n)(a + b) = m(a + b) + n(a + b) = ma + mb + na + nb$

$(3 + 4)(7 + 10) = 7 \times 17 = 119$

$3(7 + 10) + 4(7 + 10) = (3 \times 17) + (4 \times 17) = 51 + 68 = 119$

$(3 \times 7) + (3 \times 10) + (4 \times 7) + (4 \times 10) = 21 + 30 + 28 + 40 = 119$

b. $(6 - 2)(5 - 8) = 4 \times (-3) = -12$

$6(5 - 8) - 2(5 - 8) = (-18) - (-6) = -12$

$$\begin{aligned} 6(5) + 6(-8) - 2(5) - 2(-8) &= 30 - 48 - 10 + 16 \\ &= -18 - 10 + 16 \\ &= -28 + 16 \\ &= -12 \end{aligned}$$

6. a. $\frac{3}{4} + \frac{2}{4} + \frac{6}{4} = \frac{11}{4} = 2\frac{3}{4}$

b. $\frac{3}{16} + \frac{7}{16} = \frac{10}{16} = \frac{5}{8}$

c. $\frac{5}{9} + \frac{3}{4} = \frac{20}{36} + \frac{27}{36} = \frac{47}{36} = 1\frac{11}{36}$

d. $$\begin{aligned} \frac{5}{6} + \frac{3}{5} + \frac{6}{7} &= \frac{5 \times 35}{210} + \frac{3 \times 42}{210} + \frac{6 \times 30}{210} \\ &= \frac{175 + 126 + 180}{210} \\ &= \frac{481}{210} \\ &= 2\,\frac{61}{210} \end{aligned}$$

7. a. $\frac{1}{2} \times \frac{3}{4} = \frac{3}{8}$

b. $\frac{5}{6} \times \frac{3}{11} = \frac{5}{22}$

c. $\frac{4}{3} \times \frac{3}{4} = 1$

d. $\frac{7}{5} \times \frac{14}{15} = \frac{98}{75}$

e. $\frac{1}{2} \div \frac{3}{4} = \frac{1}{2} \times \frac{4}{3} = \frac{2}{3}$

f. $\frac{5}{6} \div \frac{3}{11} = \frac{5}{6} \times \frac{11}{3} = \frac{55}{18} = 3\,\frac{1}{18}$

g. $\frac{4}{3} \div \frac{3}{4} = \frac{4}{3} \times \frac{4}{3} = \frac{16}{9} = 1\,\frac{7}{9}$

h. $\frac{7}{5} \div \frac{14}{15} = \frac{7}{5} \times \frac{15}{14} = \frac{3}{2} = 1\,\frac{1}{2}$

8. a. x—variable and term
   b. 17—constant and term
   c. $13x$—term composed of constant 13 and variable x
   d. $4xyz$—term composed of constant 4 and three variables x, y and z
   e. $15 + 12mn$—binomial expression composed of 2 terms: 15 and $12mn$
   f. $3a^2 + 4ab + 5b^2$—trinomial expression composed of three terms: $3a^2$, $4ab$, and $5b^2$
9. a. z^5
   b. m^3n^2
   c. a^4b^5
   d. c^{-1}
   e. y^4x^{-1}
   f. $2a$
   g. $125a^6$
   h. 4

Chapter 6

1. a. $x_2 = 1 - \frac{4}{5}x_1 \qquad (x_1, x_2) = \left(\frac{5}{13}, \frac{9}{13}\right)$

   $x_2 = \frac{6}{7} - \frac{3}{7}x_1$

   b. $x_2 = -\frac{4}{3} + \frac{7}{3}x_1 \qquad (x_1, x_2) = \left(-\frac{7}{29}, -\frac{55}{29}\right)$

   $x_2 = -\frac{5}{2} - \frac{5}{2}x_1$

   c. $x_2 = \frac{1}{3} - \frac{2}{3}x_1 \qquad (x_1, x_2) = (-1, 1)$

   $x_2 = \frac{5}{2} + \frac{3}{2}x_1$

   d. $x_2 = \frac{15}{7} - \frac{5}{7}x_1 \qquad (x_1, x_2) = \left(\frac{3}{8}, \frac{15}{8}\right)$

   $x_2 = 2 - \frac{1}{3}x_1$

e. $x_2 = 8 - 2x_1 \qquad (x_1, x_2) = \left(\frac{5}{3}, \frac{14}{3}\right)$

$x_2 = 6 - \frac{4}{5}x_1$

Graphic Solutions for Problem 1

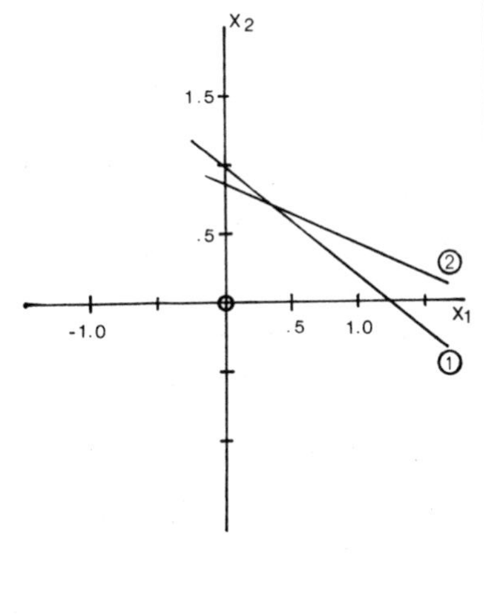

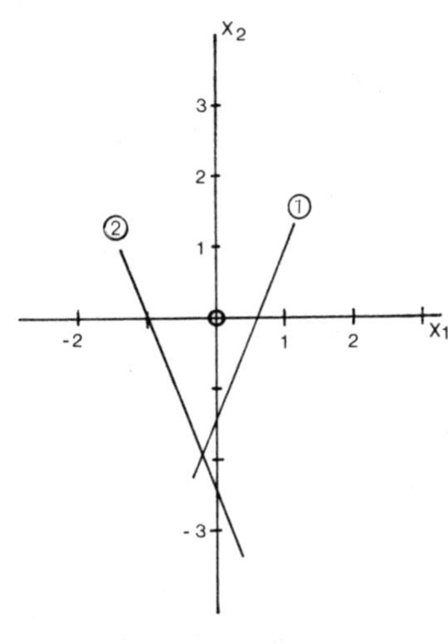

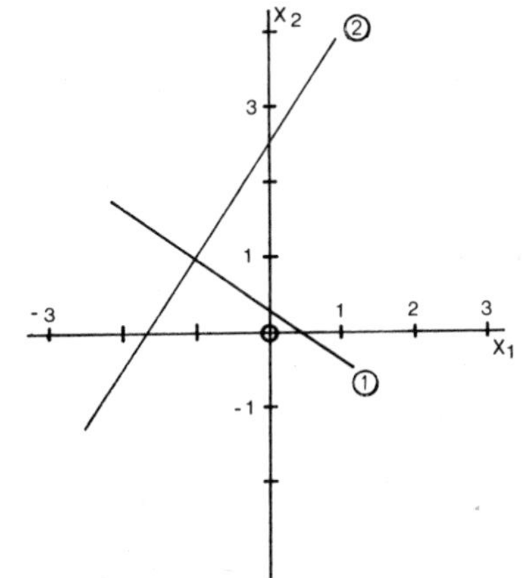

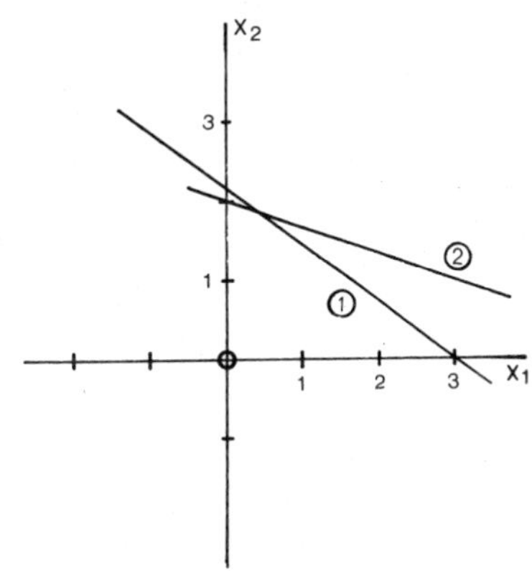

Chapter 7

1. a. $3x^2 + 5x + 1 = 2 + 4x$

$$3x^2 + x - 1 = 0$$

$$x = \frac{-1 \pm \sqrt{1 + 12}}{6}$$

$$x \cong -.768 \text{ or } .434$$

b. $.5x^2 - 2x + 14 = 3 + 6x$

$$.5x^2 - 8x + 11 = 0$$

$$x = \frac{8 \pm \sqrt{64 - 22}}{1}$$

$$x \cong 1.519 \text{ or } 14.481$$

c. $2x^2 - x + 10 = 3 + 2x$

$$2x^2 - 3x + 7 = 0$$

$$x = \frac{3 \pm \sqrt{9 - 56}}{4}$$

Roots are imaginary. There is no intersection.

d. $x^2 - 3x + 5 = 2x - 1.25$

$$x^2 - 5x + 6.25 = 0$$

$$x = \frac{5 \pm \sqrt{25 - 25}}{2}$$

$$x = 2.5$$

There is only one point where the straight line intersects (touches) the parabola.

Graphic Solution for Problem 1

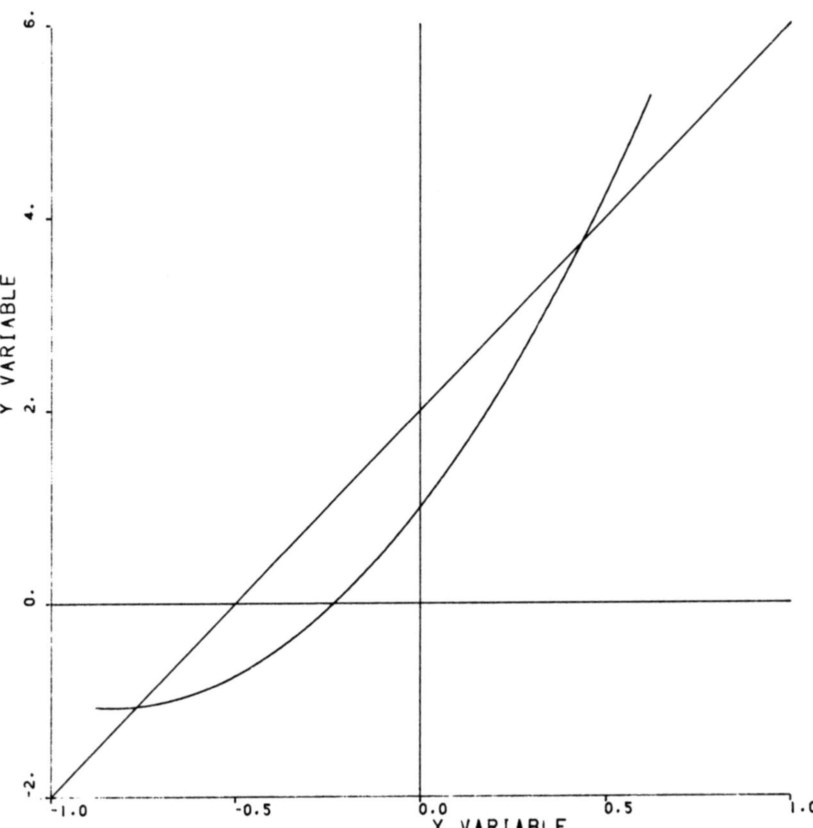

Graphic Solution for Problem 1

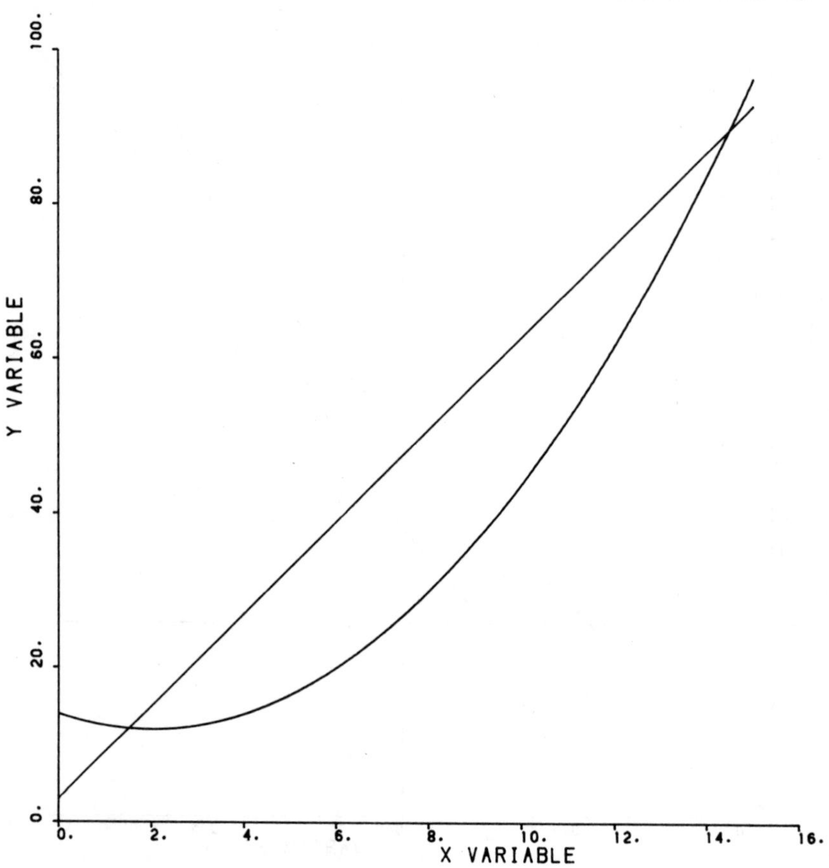

Chapter 8

2. \$2,000 @ 15% (rounded to nearest dollar)

Years:	2	4	6	8	10
Simple:	2,600	3,200	3,800	4,400	5,000
Compounded:	2,645	3,498	4,626	6,118	8,091

3. $12,000 discounted at 11% (rounded to nearest dollar)

Years:	2	4	6	8	10
Present Value:	9,739	7,905	6,416	5,207	4,226

4. Monthly Rate: 1.1% 1.7%
   Annual Rates
   Nominal: 13.2% 20.4%
   Effective: 14.03% 22.42%

5. Annual net cash flows ($000—Years 0 to 10)

   (−1, 388 76 150 136 244 192 315 377 435 435 435)

Discount Rate:	2%	6%	12%	20%
Net Present Value:	1,058.18	514.22	−37.01	−480.53

   With increasing discount rates, the project looks more and more unattractive.

Chapter 10

1. $f'(x) = -60 + .2x$

   $f'(300) = 0$

   $f''(x) = .2$

   ∴ a hospital with a complement of 300 beds results in a minimum average cost of $81,200/bed.

2. $f'(x) = 12 - .6x$

   $\therefore f'(20) = 0$

   $f''(x) = -6$

   As a result, the maximum of service per input is 130 units.

3. a. $f'(x) = 6.5 - 8x + \frac{3}{2}x^2$

   b. $f'(x) = \frac{1}{5}x - 5$

4. a. $f'(x_1) = .24x - 48$

   $f'(200) = 0$

   $f''(x_1) = .24$

   Minimum costs of $200/unit are obtained when 200 units are provided.

   b. $f'(x_2) = .4x - 16$

   $f'(400) = 0$

   $f''(x_2) = .04$

   Minimum costs of $200/unit are obtained when 400 units of service are provided.

   c. $f'(x_3) = .064 - 12$

   $\therefore f'(200) = 0$

   $f''(x_3) = .06$

   Minimum costs of $300/unit are obtained when 200 units of service are provided.

Chapter 11

2. $\partial y/\partial L = .5\left(\frac{1240.66}{20}\right)$

   $\cong 31.0$ units

3. $\partial u/\partial A = .001(E^2) + .0001(I)$

   $\partial u/\partial A = .001(7)^2 + .0001(4{,}000)$

   $\cong .45$ units

6. $$\frac{\partial y}{\partial L} = .2\left(\frac{174.11}{50}\right)$$
$$\cong .70 \text{ units}$$
$$\partial y/\partial L = .2\left(\frac{693.14}{50}\right)$$
$$= 2.77 \text{ units}$$

10. $\partial R/\partial P = -.98R/P \qquad \partial R/\partial C = .69R/C \qquad \partial R/\partial I = .30R/I$

$\partial R/\partial A = .39R/A$ and $\partial R/\partial O = -.20R/O$

R is positively related to A, C, and I, and negatively related to P and O.

Chapter 12

1. $$TR = \int_0^{600} 100dx$$
$$= 100(600) - 100(0)$$
$$TR = \$60{,}000$$

2. $$TC = \int (.2x + 5)dx$$
$$= .1x^2 + 5x + 200$$

a. $TC_{800} = (.1(800)^2 + 5(800) + 200) - (200)$
$= \$68{,}000$

b. $TC_{2000} = (.1(2000)^2 + 5(2000) + 200) - (200)$
$= \$410{,}000$

c. $TC_{600} = (.1(600)^2 + 5(600) + 200) - (200)$
$= \$39{,}000$

4. a. $F(x) = \frac{1}{4}x^4 - \frac{7}{3}x^3 + \frac{3}{2}x^2 - 10x$

b. $F(x) = \frac{2}{15}x^5 + x^3 - 4x + c$

c. $F(x) = -\frac{3}{4}x^2 - 2x^3 + 4x + c$

7. a. $R_{200,000} = .01(200)^3$
$= \$80,000$

b. $R_{700,000} = .01(700)^3$
$= \$3,430,000$

c. $R_{100,000} = .01(100)^3$
$= \$10,000$

8. a. $1,008
b. $25,040
c. $36,048

11. a. approximately $ 400
b. approximately $1,200
c. approximately $1,600

Chapter 13

1.

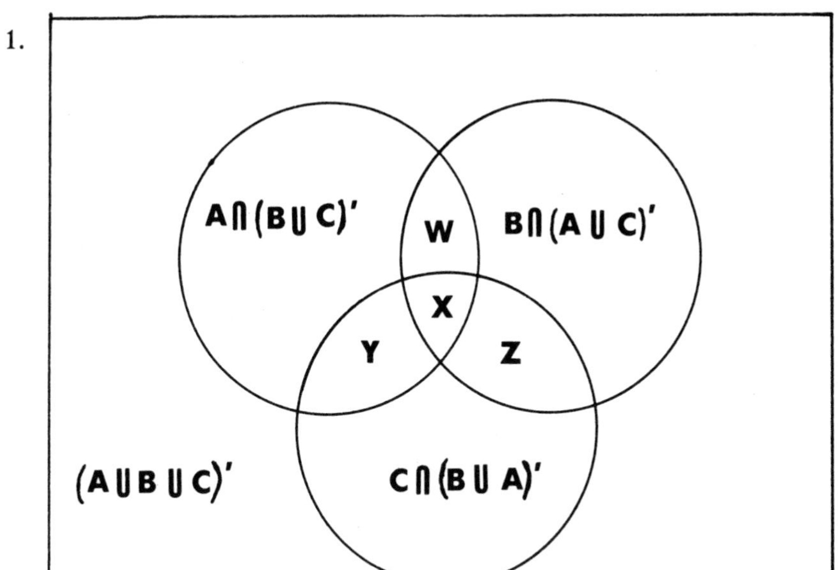

W = A ∩ B ∩ C'
X = A ∩ B ∩ C

Y = A ∩ B' ∩ C
Z = B ∩ C ∩ A'

2.

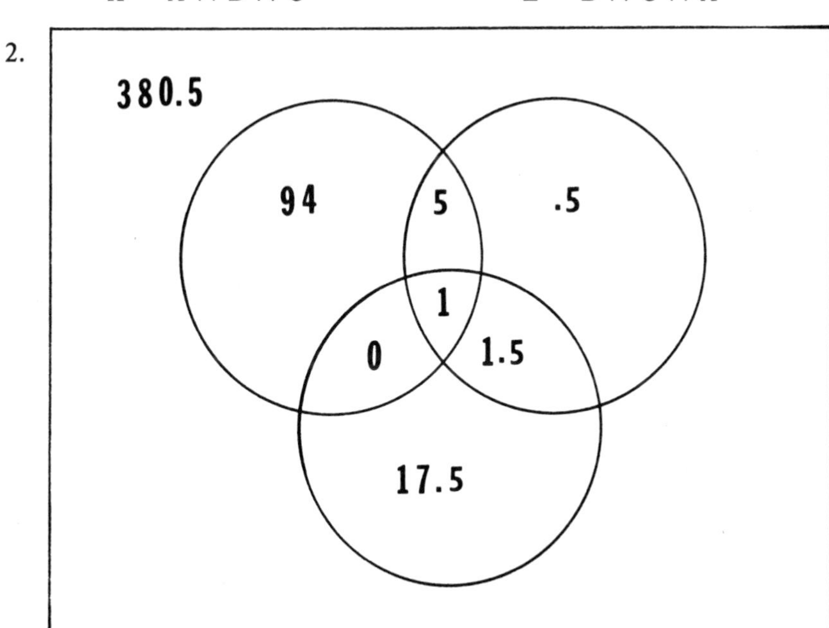

3.

	Emergencies (B)			Nonemergencies (B')			
	Admitted (C)	Not Admitted (C')	Sub Total	Admitted (C)	Not Admitted (C')	Sub Total	TOTAL
Emergency Visit (A)	1	5	6	0	94	94	100
No Emergency Visit (A')	1.5	.5	2	17.5	380.5	398	400
TOTAL	2.5	5.5	8	17.5	474.5	492	500

4.

$A \cap B \cap C$	$A \cap B \cap C'$	$A \cap B$	$A \cap B' \cap C$	$A \cap B' \cap C'$	$A \cap B'$	A
$A' \cap B \cap C$	$A' \cap B \cap C'$	$A' \cap B$	$A' \cap B' \cap C$	$A' \cap B' \cap C'$	$A' \cap B'$	A'
$B \cap C$	$B \cap C'$	B	$B' \cap C$	$B' \cap C'$	B'	U

5. Left half = B
   Bottom half = A'
   Top Right Quarter = $A \cap B'$

6.

	Admitted			Not Admitted			
	Emergency	Nonemergency	Sub Total	Emergency	Nonemergency	Sub Total	TOTAL
Emergency Visit	1	0	1	5	94	99	100
No Emergency Visit	1.5	17.5	19	.5	380.5	381	400
TOTAL	2.5	17.5	20	5.5	474.5	480	500

7. a. {a, c, e, g}
   c. {i, j, k, l, m, n, o, p, q, r, s, t, u, v, w}
   e. {m, p, q}
   g. U
   i. {i, j, k, l, n, o, r, s, t, u, v, w}
   k. Ø
   m. see (i) $A' \cap B' = (A \cup B)'$
   o. {b, d, f, h, i, j, k, l, m, n, o, p, q, r, s, t, u, v, w, x, y, z}

Chapter 14

1. .70
3. $P(A_2 \cap B_1) = .33$

   $P(A_2 \cap B_2) = .27$

   $P(A_2 \cup B_2) = .63$

   $P(A_2|B_1) \cong .47 \neq .60$

   $\therefore$ A_2 and B_1 are not independent

5. $P(A) = .50$

   $P(B) = .40$

   $P(A \cap B) = .10$

   $P(B|A) = .20$

6. 2.75 days
7. 3.1 visits
10. 3.25 units

Chapter 15

1. a. .1546
   b. .4286
   c. .3974
   d. .0395
3. a. .1292
   b. .1186

c. .3520
d. .0700

6. a. .0197
b. .3897
c. .1446

Chapter 16

1. a. $\mathbf{A} = \begin{bmatrix} 50 & 95 & 62 & 78 \\ 72 & 107 & 96 & 81 \\ 12 & 23 & 72 & 186 \end{bmatrix}$

b. $\sum_{j=1}^{4} a_{1j} = 285$ visits between 8 and 12 a.m.;

$\sum_{j=1}^{4} a_{2j} = 356$ visits between 1 and 5 p.m.;

$\sum_{j=1}^{4} a_{3j} = 293$ visits between 6 and 8 p.m.;

$\sum_{i=1}^{3} a_{i1} = 134$ visits by persons in the range 5–14 years;

$\sum_{i=1}^{3} a_{i2} = 225$ visits by persons in the range 15–24 years;

$\sum_{i=1}^{3} a_{i3} = 230$ visits by persons in the range 25–34 years;

$\sum_{i=1}^{3} a_{i4} = 345$ visits by persons in the range 35–44 years;

$\sum_{j=1}^{4} \sum_{i=1}^{3} a_{ij} = \sum_{i=1}^{3} \sum_{j=1}^{4} a_{ij} = 934$ visits by all persons and at all times

2. $\mathbf{a}'\mathbf{B} = [10 \quad 20 \quad 30] \begin{bmatrix} 500 & 600 \\ 50 & 90 \\ 70 & 300 \end{bmatrix}$

$$= [8{,}100 \quad 16{,}800]$$

The total costs of producing these services in April and May were $8,100 and $16,800, respectively.

4. Matrices **A**, **B**, **D**, and **E** are square matrices,
   Matrix **D** is an upper triangular matrix,
   Matrix **B** is a lower triangular matrix.

Chapter 17

1. $$\begin{bmatrix} 400 & 60 & 190 \\ 800 & 240 & 30 \\ 110 & 52 & 70 \\ 30 & 3 & 6 \end{bmatrix} + \begin{bmatrix} 212 & 83 & 364 \\ 350 & 612 & 87 \\ 62 & 316 & 79 \\ 14 & 4 & 12 \end{bmatrix} = \begin{bmatrix} 612 & 143 & 554 \\ 1{,}150 & 852 & 117 \\ 172 & 368 & 149 \\ 44 & 7 & 18 \end{bmatrix}$$

2. $$\begin{bmatrix} 400 & 60 & 190 \\ 800 & 240 & 30 \\ 110 & 52 & 70 \\ 30 & 3 & 6 \end{bmatrix} - \begin{bmatrix} 212 & 83 & 364 \\ 350 & 612 & 87 \\ 62 & 316 & 79 \\ 14 & 4 & 12 \end{bmatrix} = \begin{bmatrix} 188 & -23 & -174 \\ 450 & -372 & -57 \\ 48 & -264 & -9 \\ 16 & -1 & -6 \end{bmatrix}$$

   Region I has more general practitioners than Region II; however Region II has more surgeons and specialists in internal medicine than Region I.

7. The rates per 1,000 males, by income, are as follows:

$$[24 \quad 58 \quad 60] \begin{bmatrix} 20 & 17 & 18 \\ 35 & 29 & 21 \\ 42 & 37 & 36 \end{bmatrix} = [\$5{,}030 \quad \$4{,}310 \quad \$1{,}746]$$

   The rates per 1,000 females, by income, are as follows:

$$[20 \quad 32 \quad 36] \begin{bmatrix} 12 & 9 & 3 \\ 28 & 27 & 15 \\ 37 & 25 & 27 \end{bmatrix} = [\$2{,}468 \quad \$1{,}944 \quad \$1{,}512]$$

8.

$$\begin{bmatrix} 0 & 2.0 & 1.0 & 0 \\ 3.0 & 4.0 & 0 & 0 \\ 0 & 0 & 2.0 & 3.0 \end{bmatrix} \begin{bmatrix} 180 \\ 200 \\ 600 \\ 700 \end{bmatrix} = \begin{bmatrix} 1{,}000 \\ 1{,}340 \\ 3{,}300 \end{bmatrix}$$

9.

$$\begin{bmatrix} 0 & 2.0 & 1.0 & 0 \\ 3.0 & 4.0 & 0 & 0 \\ 0 & 0 & 2.0 & 3.0 \end{bmatrix} \begin{bmatrix} 50 & 20 & 30 & 80 \\ 60 & 10 & 30 & 100 \\ 260 & 40 & 175 & 125 \\ 30 & 60 & 0 & 110 \end{bmatrix}$$

$$= \begin{bmatrix} 380 & 60 & 235 & 325 \\ 390 & 100 & 210 & 640 \\ 610 & 260 & 350 & 580 \end{bmatrix}$$

Chapter 18

1. a. The volume of each service is given by $\mathbf{m}'\mathbf{T}'$

$$[6{,}000 \quad 8{,}000 \quad 12{,}000 \quad 1{,}000] \begin{bmatrix} 2.0 & 4.0 & 3.0 & 0 \\ 0 & 1.0 & 2.0 & 3.0 \\ 0 & 0 & 3.0 & 7.0 \\ 8.0 & 9.0 & 0 & 1.0 \end{bmatrix} =$$

$$[20{,}000 \quad 41{,}000 \quad 70{,}000 \quad 109{,}000]$$

b. The number of staff hours by type of labor are given by $\mathbf{m}'\mathbf{T}'\mathbf{U}_1{}'$

$$[20{,}000 \quad 41{,}000 \quad 70{,}000 \quad 109{,}000] \begin{bmatrix} 1.2 & 2.1 & 3.0 & 0 \\ 6.0 & 5.8 & 4.5 & 3.0 \\ .5 & .2 & 0 & 0 \\ 1.0 & 0 & 1.2 & 0 \end{bmatrix} =$$

$$[414{,}000 \quad 293{,}800 \quad 375{,}300 \quad 123{,}000]$$

c. The supply requirements, by type of supply, are as follows: $\mathbf{m}'\mathbf{T}'\mathbf{U}_2'$

$$[20{,}000 \quad 41{,}000 \quad 70{,}000 \quad 109{,}000]\begin{bmatrix} 1.0 & 2.0 & 0 & 1.0 \\ 2.0 & 3.0 & 4.0 & 0 \\ 1.0 & 6.0 & 0 & 3.0 \\ 0 & 0 & 3.0 & 2.0 \end{bmatrix} =$$

$$[172{,}000 \quad 583{,}000 \quad 491{,}000 \quad 448{,}000]$$

2. The labor costs are given by $\mathbf{p}'\mathbf{U}_1\mathbf{Tm}$

$$[\$6.00 \quad \$7.50 \quad \$8.00 \quad \$9.00]\begin{bmatrix} 414{,}000 \\ 293{,}800 \\ 375{,}300 \\ 123{,}000 \end{bmatrix} = \$8{,}796{,}900$$

The supply costs are given by $\mathbf{p}'\mathbf{U}_2\mathbf{Tm}$

$$[\$3.00 \quad \$.50 \quad \$.10 \quad \$6.50]\begin{bmatrix} 172{,}000 \\ 583{,}000 \\ 491{,}000 \\ 448{,}000 \end{bmatrix} = \$3{,}768{,}600$$

Hence, the total costs are \$12,565,500 (i.e., \$8,796,900 + \$3,768,600).

4.

$$\mathbf{A}' = \begin{bmatrix} 1 & 12 & 6 \\ 7 & 3 & 2 \end{bmatrix}; \qquad \mathbf{B}' = \begin{bmatrix} 8 & 17 & 4 & 9 \\ 9 & 12 & 7 & 0 \\ 6 & 3 & 0 & 1 \\ 4 & 6 & 1 & 0 \end{bmatrix}$$

$$\mathbf{C}' = \begin{bmatrix} 1 & 13 \\ 6 & 15 \\ 9 & 21 \\ 7 & 0 \end{bmatrix}$$

Chapter 19

1. $|\mathbf{A}| = -143$; $|\mathbf{B}| = 390$; $|\mathbf{C}| = 225$

2. $\mathbf{X}'\mathbf{X} = \begin{bmatrix} 4 & 26 \\ 26 & 198 \end{bmatrix}$

6. $\mathbf{A}^{-1} = \begin{bmatrix} -.46125 & .35424 & .12177 \\ .44280 & -.38007 & -.03690 \\ -.13284 & .21402 & .01107 \end{bmatrix}$

$$\mathbf{B}^{-1} = \begin{bmatrix} -1.16667 & 2 & .5 \\ 1.75 & -3 & -.58333 \\ -.5 & 1 & .16667 \end{bmatrix}$$

$$\mathbf{C}^{-1} = \begin{bmatrix} 0 & .25 \\ .16667 & -.375 \end{bmatrix}$$

Glossary

A

Absolute Maximum—2; 4—The highest value a function can assume.

Absolute Value—2—The distance of a given number from zero as measured on the real number line; the absolute value of x is written in the form $|x|$, and its value is always positive.

Adjacent Side—7.3,.4—See right-triangle.

Algebraic Expression—4.5—A group of one or more terms that are to be added or subtracted, as indicated by the signs joining them.

Analysis of Variance—A statistical technique used to show that several groups of observations do, or do not, all have the same mean.

Annuity—8—A series of periodic (annual or monthly) payments made in fulfillment of a (financial) contract.

Antiderivative—9.3; 12.2—A function whose derivative is a specified or given function. The function $F(x)$ is an antiderivative of $f(x)$ if $F'(x) = f(x)$. This, when evaluated for values of x between a and b ($a \leq x \leq b$), gives the area underneath the curve, $f(x)$, between a and b.

Antidifferentiation—12—The process of finding an antiderivative.

Antilogarithm (Antilog)—4.6—A number which has a given and specific logarithm. The number b is the antilog of a if $a = \log(b)$.

Arithmetic Mean—14.4—The sum of all numbers in a sample divided by the number of observations in the sample ($\bar{x} = \Sigma x/n$).

Associative Law—4.3; 17.6—One of the basic laws of arithmetic and algebra, which states that a series of (only) additions or (only) multiplications (which must be performed on one pair of numbers at a time) can start from the left and work to the right, or proceed in the reverse direction. *See also:* commutative, distributive

Asymptote—A line or axis which a curve approaches but never reaches.

Note: For more extensive information and background, most terms in this glossary are keyed to the text. This keying is shown by the number(s) immediately following the term. The references are either to chapter, or to chapter and subsection(s). For any one term, references to different chapters are separated by semicolons.

Average—The arithmetic mean.

Axiom—4.3—A basic law in mathematics that is stated without proof, and forms the basis for proving the truth (or falsehood) of other mathematical statements.

B

Bar Chart—3.3—A graphic portrayal of data in which the categories of the independent variable are represented by bars, and the value of the dependent variable is given by the length of the bar for each category. *See also:* class interval

Base—4.6—In a power the base is the repeated factor. *See also:* logarithm

Basic Outcome—13.2—The simplest possible result of an experiment. One element or point in a sample space. Only one basic outcome can occur in any trial or experiment, and one *must* occur. *See also:* event

Binary Relationship—5.2—A relationship involving *two* variables.

Binomial—4.5—An algebraic expression with two terms, eg. $ax + b$.

Boolean Algebra—An algebra of variables which can have only two states, usually represented by (0, 1) or by (no, yes) or (off, on).

Break-Even Point—7.6; 7.7—The volume of service for which all costs are recovered in the form of revenue.

C

Characteristic—4.6—The number in a common logarithm which locates the decimal point in the antilogarithm.

Chi-Square—A special statistical distribution, based on the sum of the squares of normally distributed variables, that is used to test for a strong (or weak) relationship between variables used to define a contingency table.

Circle Diagram—3.6—A graphic portrayal of data where sectors of the circle represent different categories of the independent variable, and the relative size of each sector gives the relative value of the dependent variable. It is also called a *pie* chart.

Class Interval—3.3—The range of values associated with a given class, grouping, or category of the independent variable, where the categories can be themselves expressed numerically (eg. grouping patients by age). *See also:* bar chart

Coefficient—4.5—A term is the product of one or more variables with a number (or a constant). The number is referred to as the coefficient.

Cofactor—19-A—The cofactor of any element of a square matrix is the determinant of the sub-matrix obtained by eliminating the row and column containing the element. The cofactor is used in evaluating determinants or inverses of matrices of order 3 or greater.

Column Vector—16—A matrix consisting of a single column.

Common Denominator—4.4—In order to add two (or more) fractions they must all have the same (ie. a common) denominator.

Common Logarithm—4.6—A logarithm with base 10.

Commutative Law—4.3; 17.6—One of the basic laws of arithmetic and algebra which states that the order of a pair of numbers (or factors) being added (or multiplied) is unimportant (ie. $xy = yx$). *See also:* associative, distributive

Complement—13—The complement of a set A is comprised of all elements in the universal set, U, which are not contained in set A.

Compound Interest—8.1; 8.3—Interest which is paid on principal plus previously accumulated interest.

Compounding Interval—8.1—The length of time between dates at which interest payable is calculated. The standard interval (for comparison purposes) is one year, although intervals of a day, and one, three, and six months are commonly used.

Conditional Probability—14.2,.3—The probability that event A will occur given that event B is known or assumed to have occurred.

Conformable (Matrix)—17.2,.4—In order to perform matrix arithmetic matrices must be conformable. For addition or subtraction of two matrices conformability requires the same dimensions. For multiplication conformability requires that the number of columns of the first matrix be equal to the number of rows of the second.

Constant—4.5—A number, or a letter used to represent a fixed number in an algebraic expression. By convention the early letters of the alphabet are used for constants, while the latter letters are used for variables. (eg. In $y = a + bx$, the constants are a and b, while the variables are x and y.) *See also:* variable

Contingency Table—13.7—A table used to display the number of observations in a survey or experiment which have each of the possible combinations of values of two or more variables of interest. For example in analyzing the population of patients in a hospital it may be desirable to know if there is a relationship between age and sex. A two dimensional contingency table would be used to show the numbers of males and females for each age group. *See also:* chi-square, relationship

Continuous Distribution—15.2—A distribution in which the states of the process can be represented by any point on the real number line. *See also:* discrete distribution

Continuous Function—10.2,.3—The function $y = f(x)$ is continuous at the point a if it has a single, finite value at that point, and there are no breaks or sudden changes at, or close to, that point. The technical requirements are expressed as:

1. $\lim_{x \to a} f(x)$ exists
2. $f(a)$ exists
3. $\lim_{x \to a} f(x) = f(a)$

Continuous Sample Space—14.2; 15.1—A sample space where the basic outcomes can have any real value (ie. the real number line).

Continuous Variable—2.2; 4.3—A variable which may assume any value on the real number line.

Coordinate—3.5—In a Cartesian (rectangular grid) system of drawing graphs, any point is located by its distance from the origin measured along, or parallel to, each axis. In a two-dimensional system, with X and Y axes, point i is represented by two coordinates (x_i, y_i).

Cosine—7.5—A trigonometric function of the specified angle (θ) in a right triangle, defined as the ratio of the lengths of the adjacent side and the hypotenuse, $\cos(\theta) = b/c$. *See also:* sine

Cross-Product—11.4—A term created by multiplying two variables.

Cube Root—4; 7.3—The variable y is the cube root of x ($y = \sqrt[3]{x}$), if $y^3 = x$. For example $\sqrt[3]{27} = 3$, because $3 \times 3 \times 3 = 27$. *See also:* square root

Cumulative Frequency Distribution—15.2—A frequency distribution where the categories are arranged in a natural order and the successive category frequencies, f_i, are added to each other so that the cumulative frequencies $cf_1 = f_1$, $cf_2 = f_1 + f_2$, $cf_3 = f_1 + f_2 + f_3$, and so on.

Curvilinear Function—See non-linear curve.

D

Definite Integral—12.4—An antiderivative evaluated between a specific lower limit, a, and a specific upper limit, b.

DeMorgan's Laws—13.6—Laws of set operations for unions and intersections.

Denominator—4.4—The bottom number in a fraction. (eg. b in a/b).

Density Function—See distribution.

Dependent Variable—2.2; 5—A variable whose value is determined by one or more independent variables.

Derivative—9.2; 10—The rate of change of the dependent variable of a specified function as related to changes in one or more of the independent variables. If $y = f(x)$ then the derivative of y with respect to x is represented by dy/dx or by $f'(x)$.

Determinant—19.1—A scalar value (single number) which results from a specific set of operations on a square matrix. The determinant of the matrix **A** is written $|\mathbf{A}|$ or det (**A**).

Deterministic (Process or Model)—An experiment or physical process (or a model of one) where the inputs (independent variables) or starting conditions fix the outcome. If the inputs are known the outcome can be known or calculated. *See also:* random variable, stochastic

Diagonal (of a square matrix)—16.3—The *main diagonal* consists of all elements in the straight line from the top left corner to the bottom right corner. That is all a for $i = 1, 2, 3, \ldots, n$.

Diagonal Matrix—16.3—A square matrix for which all elements *off* the main diagonal are zero.

Differential Calculus—9.2—The branch of calculus which deals with finding rates of change of specified functions. *See also:* integral calculus

Differentiation—9.2; 10—The process of finding the derivative.

Dimension of a Matrix—16.3—The number of rows and columns of a matrix. The dimension of matrix **A** which consists of m rows and n columns is written in the form $\mathbf{A}_{m \times n}$.

Discontinuous Function—10.3—If a function is not continuous at $x = a$, the function is said to be discontinuous. *See also:* continuous function

Discount Rate—8.4—The interest rate used in carrying out present value calculations.

Discounting—8.4—The process of finding a present value.

Discrete Distribution—14—A distribution in which the states of the process can be represented on the real number line by the integer points only. The states are separate and distinct, as opposed to the continuous distribution where the states blend into each other.

Disjoint Set—13.4—Two sets are disjoint (mutually exclusive) if they have no common points.

Distribution—3.3; 14—(1) A function, $f(x)$, which represents the frequency or probability of occurrence of a specific state of a physical or logical process which can assume only discrete values of x. (eg. frequency of x-ray films acceptable ($x = 1$) or not acceptable ($x = 0$); frequency of visits by patient x; frequency of times employee x is late; frequency of reportable incidents on ward x).

Distribution—15.2—(2) A function $f(x)$ which represents the first derivative of the probability of occurrence of a specific value of x, where x is a continuous variable. In this situation $f(x)$ is called a *probability density function.* Since no physical process can actually assume *any* real value, the theoretical distribution must be integrated between specific limits to find the probability of occurrence of a value of x between the specified limits. Important theoretical distributions include the normal, the t, the exponential, the poisson, the chi-square, the F, and others.

Distributive Law—2.3—One of the basic laws of arithmetic and algebra, which states that an expression multiplied by a factor is equivalent to the sum of the terms of the expression each multiplied by the factor. For example $a(b + c) = ab + ac$.

Domain—5—The set of all possible values of x in a function, $y = f(x)$. *See also:* range

E

***e*—4.6**—The base of natural logarithms with a value of approximately 2.71828. *See also:* logarithm

Effective Annual Rate—8.1,.3—The annual compounding interest rate which gives the same effect as the specified rate and compounding interval. Given a monthly rate of one percent the effective annual rate is 12.68 percent rather than 12 percent, the nominal annual rate. The size of the discrepancy between effective and nominal rates increases with the size of the monthly rate.

Elasticity Coefficient—The percentage change in one variable resulting from a one percent change in another.

Element—13.2; 16.2,.3—A single number in a matrix or vector, or a single number or object in a set.

Elimination Method—6.3—The process of solving a system of equations in which variables are successively eliminated.

Empty Set—13.2—A set consisting of no members; also called the null set.

Enumeration—13.2—Listing all the members of a set in order to define the set.

Equality—5.2,.3—The state of being equal. *See also:* inequality

Equation—5.3—A mathematical statement that two expressions are equal.

Equivalent Fractions—4.4—Fractions which represent the same real number. A fraction can be transformed into an equivalent fraction by multiplying both the numerator and denominator by the same, arbitrarily chosen integer.

Evaluating a Determinant—19.1—The process of finding the determinant.

Event—14.2—A subset of a sample space, consisting of 0, 1 or more basic outcomes. An event can occur only if one of its basic outcomes occurs. A single basic outcome *must* be part of at least one event, and may be in several. *See also:* conditional probability

Expanding an Expression—4.5; 7.6—Applying the distributive law, by multiplying all factors in each term containing parentheses to remove as many parentheses as possible from the expression. *See also:* factoring

Explicit Form of an Equation—5.3,.4—An equation in which only the dependent variable appears (usually) on the left hand side, with coefficient and exponent equal to one. On the right hand side there is an expression involving constants and (possibly) some power of the independent variable. *See also:* implicit form

Exponent—4.6—In the power a^b the exponent b indicates the number of times the factor a is repeated.

Exponential Curve—4.6,.7—A curve generated by an equation containing a power of some base with the independent variable as the exponent. (eg. $y = ab^x$). *See also:* logarithm

Expression—4.5—See algebraic expression.

Extreme Values—9.1; 10.5—Maximum or minimum values of a function. *See also:* relative maximum

F

***F* Tables**—Statistical tables used for testing the significance of an analysis of variance.

Factor—4.5—(1) An integer multiplied by other integers to give a specified number. (2) A number, letter, term or expression multiplied by other factors to obtain a more complex term or expression.

Factor Inputs—The land, labor and capital required to produce specified goods or services.

Factoring an Expression—7.6—Finding a set of factors which when multiplied reproduce the original expression. *See also:* expanding an expression

Feasible Region—5.3—The set of all physically possible and allowable combinations of values of variables representing the operating conditions of a system. For example a hospital's bed occupancy cannot drop below zero nor rise above one hundred percent.

Finite—2.2—Something which has an upper limit or maximum, and a minimum or lower limit.

Finite Interval—2.2—A region of the real number line with specific lower and upper limits.

Finite Population (Set)—13.2—One where the members of the set can be completely counted.

First Derivative—See derivative.

Fixed Cost—A cost which does not vary as the rate of activity (output or services produced) is increased or decreased.

Flexible Budget—A budget portraying the cost implications associated with different levels or rates of activity.

Fraction—2.2—A number formed by dividing two integers. (eg. a/b where a and b are integers.)

Frequency Distribution—3.3—A distribution where the dependent variable represents a count of items within each category of the independent variable. *See also:* bar chart

Function—5.2—A rule of correspondence between two (or more) variables in which, for each value of x, the independent variable(s), there is one and only one value of y, the dependent variable. *See also:* relationship

Future Value—8.1—Either an intrinsic value, or an accumulated monetary value that an investment or project is expected to have at a specified future date. *See also:* present value

G

Gaussian Distribution—See normal distribution.

Geometric Mean—4.6—The geometric mean of n numbers is the nth root of their product.

Graph—3—A pictorial method of representing data.

Growth Rate—8.1,.3—The effective annual rate at which a population, or the value of an investment, is growing.

H

Histogram—3.3—A bar chart.

Hypotenuse—See right-triangle.

I

Identity Matrix—17.5—A diagonal matrix, **I**, of any size, in which all diagonal elements are equal to one. eg. for size 3

$$\mathbf{I} = \begin{bmatrix} 1 & 0 & 0 \\ 0 & 1 & 0 \\ 0 & 0 & 1 \end{bmatrix}$$

Imaginary Number—2.2—A number which includes $\sqrt{-1}$.

Implicit Form of an Equation—5.3,.4—An equation in which the dependent variable may appear in a single term expression with coefficient and/or exponent not equal to one, or may appear in an expression of two or more terms.

Indefinite Integral—9.3; 12.3—An antiderivative which is not evaluated for specific limits. *See also:* definite integral

Independent Events—14.3—Event A is independent of event B if the probability of its occurrence does not change if event B occurs, and vice versa. *See also:* conditional probability

Independent Variable—5.2,.3—Any variable whose value is chosen, known or assumed. In $y = f(x)$, x is the independent variable.

Inequality—5.3—A statement in which two expressions, A and B, are related so that $A < B$, $A > B$ or $A \neq B$.

Infeasible Region—5.3—The opposite of feasible region. That portion of the solution space which is either forbidden or physically impossible.

Infinite—2.2—Without limit or bounds. Increasing indefinitely. *See also:* finite

Intangible—8.5,.6—In analysis of investment projects any benefit or cost which is impossible to measure or estimate accurately is called an intangible.

Integer—2.2—Any of the numbers $\ldots -3, -2, -1, 0, 1, 2, 3 \ldots$

Integral Calculus—9.3; 12—The branch of calculus which deals with finding areas under curves defined by specified functions. *See also:* differential calculus

Intercept—5.3—The point at which a line crosses an axis, most commonly the y intercept.

Interest—8.1—An amount of money charged by a lender and paid by a borrower for the use of the principal sum of money loaned. It compensates the lender for being unable to use the principal for other purposes.

Interest Rate—8.1—A fraction (usually expressed as a percentage) used to determine the amount of interest to be repaid. A period of time for calculating the interest is specified, usually a day, month or year. *See also:* compound interest; compounding interval; discount rate; present value; simple interest

Intersection—13.4—(1) The intersection of sets A and B is the set of elements that are contained in both sets. (2) The point at which two lines cross.

Invariant—Unchanging. *See also:* Fixed cost, variable cost

Inverse Matrix—19.2—The inverse of matrix $\mathbf{A}$ is a matrix whose product with $\mathbf{A}$ is the identity matrix. In matrix algebra there is no division operation; its place is taken by inversion. The inverse of $\mathbf{A}$ is represented by $\mathbf{A}^{-1}$.

$$\mathbf{A}\mathbf{A}^{-1} = \mathbf{A}^{-1}\mathbf{A} = \mathbf{I}$$

Inverse Operations—4.3; 19.2—Operations which are the opposite of each other, or undo each other. Subtraction is the inverse of addition; division is the inverse of multiplication. In matrix algebra the *inverse* operation replaces division.

Irrational Number—2.2—A real number which cannot be expressed as the ratio of two integers (eg. $\sqrt{3}$, π, e).

L

Least Squares Regression Analysis—A statistical procedure which finds an equation to best represent a set of data by minimizing the sum of the squared distances of the points from the line represented by the equation.

Limits of Summation (Integration)—12.5—The lowest and highest values of the independent variable between which the values of the function will be summed.

Line Chart—3.2—Portrayal of data formed by connecting successive points on a graph with straight lines.

Linear Combination—18.1—An expression (or equation) formed by adding two or more expressions (or equations). Before performing the addition each expression (or equation) may be multiplied by an arbitrarily selected constant.

Linear Dependence—6.4; 18.1—A situation in which one vector is a linear combination of other vectors.

Linear Equation—5.3—An equation of the form $y = a + bx$, which represents a straight line.

Linear Transformation—18.1—The process of multiplying a vector by a scalar, or a matrix by a conformable vector to obtain a vector which is a linear combination of the vectors in the original matrix.

Logarithm—4.6—If $y = a^x$, then $x = \log_a y$ implies that x is the logarithm of y to the base a. A logarithm is a real number with an integer part called the *characteristic*, and a fractional part called the *mantissa*.

Logarithmic Function—4.6—An equation of the form $y = \log_a(x)$ where x and a are greater than zero and a is not equal to one.

Logical AND—A Boolean Algebra operation on two variables which gives a result of 1 only if both variables are 1.

Logical OR—A Boolean Algebra operation on two variables which gives a result of 1 if either or both variable is a 1, and a result of 0 only if both variables are 0.

Lowest Common Denominator—4.4—In addition of fractions we need to express each fraction with the same (*common*) denominator which is taken as the smallest (*lowest*) positive integer for which each original denominator is a factor. It is usually found by first finding the set of prime factors of each denominator and creating a set of prime factors which contains each of the other sets.

M

Main Diagonal—See diagonal.

Mantissa—4.6—The portion of a common logarithm which follows the decimal point.

Marginal Cost—9.1,.2—The change in total cost that results from changing the amount of output by one unit.

Marginal Revenue—9.1,.2—The additional revenue derived by selling an additional unit of output.

Matrix—16—A rectangular array of numbers, usually designated as having m rows and n columns ($m \times n$, or m by n).

Mean—See arithmetic mean, geometric mean.

Mixed Fraction—4.4—A fraction, larger than one, expressed as a combination of a whole number and a fraction between zero and one.

Model—A representation of an experiment, an object, or a physical process used for analyzing the properties or behaviour of an existing process, or object, recreating it in miniature (eg. a toy), or planning one which may eventually be created. A model may be in the form of a drawing, in a physical form, in words, in thoughts, in a mathematical equation or in a computer program. *See also:* deterministic, stochastic, variable

Multivariate Function—6—A function containing more than one independent variable.

Mutually Exclusive Events—14.2,.3—Two events are mutually exclusive if both cannot occur simultaneously. They have no basic outcomes in common.

N

***n* Order Determinant—19.1**—The determinant of a matrix of order n.

Natural Logarithm—4.6—A logarithm with base e.

Natural Number—2.2—Any of the numbers 1, 2, 3, . . . The set of positive integers.

Net Present Value (NPV)—8.4,.5—Defined as the present value of benefits minus the present value of costs.

Nominal Annual Rate—8.3—The specified interest rate (for a period shorter than a year) is multiplied by the ratio of the length of a year to the length of the specified period to give the nominal annual rate. For example a monthly annual rate is multiplied by 12. The nominal rate is always less than the effective rate. *See also:* effective annual rate

Nonlinear Curve (Equation)—7—A curve involving a term containing a trigonometric, or exponential function or a term containing the independent variable with an exponent not equal to one.

Nonnegativity Constraint—In physical processes most variables cannot have values below zero and a mathematical statement of such a variable ($x \geq 0$) is called a nonnegativity constraint.

Nonsingular Matrix—6.4; 19.3—A square matrix is nonsingular when its determinant is nonzero.

Normal Distribution—15.3—A special, widely used, probability distribution with the form of a continuous bell shaped curve which is symmetrical about a vertical center line.

Normal Equations—In least squares regression analysis, using differential calculus to minimize the sum of the squared deviations gives a set of normal equations which are solved to obtain the coefficients of the regression equation.

Notation—1; 2.1,.2; 16.2—A system of writing mathematical statements, consisting of symbols (letters, numbers, special signs, regular script, superscripts and subscripts, boldface and italic script) and the rules for writing the statements.

***n*th Root—4.6**—The number a is the nth root of b ($a = {}^n\sqrt{b}$) if $a^n = b$. *See also:* square root, cube root

Null Matrix—17.5—A matrix consisting of elements which are all zero.

Null Set—See empty set.

Numerator—4.4—The top number in a fraction. (eg. a in a/b).

O

Observation—In a survey or experiment the basic objects (patients, accidents, hospitals, countries) for which measurements must be obtained for each variable of interest (age, height, sex, diagnosis, number of employees, number of strikes). The set of measurements for a single object is called an observation.

Ogive—3.4—Line chart portraying a cumulative distribution.

Opposite Side—See right-triangle.

Optimization—9.1,.2; 10.4,.5—The process of identifying the strategy with the best result. The process frequently requires the identification of the minimum or maximum values assumed by a function.

Order (of a matrix)—19—The number of *independent* rows or columns of a given matrix, which is the size of the largest non-singular square matrix contained in the given matrix.

Ordered Pair—5.2—For each observation the measurements for two variables are listed, always in the same specified order.

Ordinal Ranking—A listing of items in which those with the highest values appear at the head of the list.

P

Parabola—7.4,.5—A curve that results from a quadratic equation.

Parameter—5; 6; 7—An independent variable whose value can be controlled by an experimenter or manager.

Partial Derivative—11—The derivative of the dependent variable associated with a change in one of several independent variables, while holding other independent variables constant.

Partitioned Matrix—18.3,.4—A matrix which has been divided into two or more submatrices.

Perfect Squares—7.6—An expression which can be factored into the form $(a + b)^2$ or $(a - b)^2$.

Pie Chart—See circle diagram.

Polygon—3—(1) In mathematics, a geometrical figure with a large number of sides, in which the sides are straight lines. (2) A frequency distribution graph where individual points are joined by straight lines rather than using bars.

Polynomial—4.5—An algebraic expression with more than two terms.

Positive—2.2—Greater than zero. To the right of zero on the real number line.

Postmultiply—17.4—Matrix **A** is postmultiplied by **B** if their product is expressed with **B** following **A**, as in **AB**. *See also:* premultiply

Power—4.6—(1) In mathematics a power is a shorthand method of expressing the multiplication of the same factor repeated several times. For example $a \times a \times a \times a \times a$ can be expressed as a^5 where a, the repeated factor, is called the *base*, and 5, the number of factors, is called the *exponent.* (2) In statistics the power of a test measures the ability of the test to distinguish between differences of interest.

Premultiply—17.4—Matrix **A** is premultiplied by **B** if the product is expressed with **B** before **A**, as in **BA**. *See also:* postmultiply

Present Value—8.1,.4—The value today of a given amount that is to be received in the future. *See also:* discounting

Prime Number (Factor)—4.5; 7.6—A prime number is a positive integer that has only itself and one as integer factors. The first few prime numbers are 1, 2, 3, 5, 7, 11, 13, 17, 19, 23, 27, . . .

Principal (Sum)—8.1—The amount of money loaned, deposited, or invested. *See also:* interest

Probability—14.1,.2—The likelihood that a given event will occur, measured as a fraction between 0 and 1.

Probability Distribution—3.3,.4; 14; 15—A discrete distribution in which the function represents probability (or relative frequency) rather than raw frequencies, or a continuous distribution representing probability density. *See also:* distribution

Promissory Note—8.1—A contract between a borrower and a lender stating the principal of the loan, the interest rate, and the terms and conditions of repayment.

Proper Subset—13—A is a proper subset of B if every element in A is also found in B, but there is at least one element in B not found in A.

Q

Quadratic Equation—7.3,.5—An equation in which the highest exponent is two. A quadratic equation assumes the form $ax^2 + bx + c = 0$ where $a \neq 0$.

Qualitative Data—1.1—Measurements for which no ordering is possible. Examples include a patient's sex, hair color, religion, language, diagnosis, or type of lab test; a department's specialty; an employee's job classification; and so on.

Quantitative Data—1.1—Measurements for which at least ordering (ranking) is possible. Examples include a patient's age, weight, blood pressure, seriousness of diagnosis, number of lab tests; a department's number of employees; an employee's number of times late; and so on.

Queue—2.4—A waiting line.

R

Radial Chart—3.7—A graphic method of portraying measurements of different aspects of a system along axes arranged like spokes in a wheel. The point on each axis is joined to the points on the neighboring axes, creating a polygon. If the polygon is regular then the system is in balance. Lack of balance is detected easily.

Radian—7.5—An angle formed between two radius arms in a circle, when the length of the arc between their endpoints is equal to the radius. One radian is approximately 57.3 degrees.

Radius—7.5—A straight line drawn from the center of a circle to meet the circumference.

Random Variable—A variable, resulting from an experiment or physical process, whose value is not fixed in advance but may assume one of several, or perhaps many, values. For example: a person's sex is determined at conception; the number of patients arriving at an emergency department in a given hour is random. *See also:* stochastic process, deterministic process, probability, distribution

Range—5.2—(1) The difference between the highest and lowest values in a set of numbers; (2) the set of all possible values assumed by y in a functional relation, $y = f(x)$.

Rate of Return—8.1—An imputed interest rate based on the actual or expected earnings from an investment other than a loan or a bank account deposit.

Ratio—4.4—A fraction. The ratio of a to b is a/b.

Rational Number—2.2—A number which can be expressed as a ratio of two integers where the denominator is nonzero.

Real Number—2.2—Any positive or negative integer, rational or irrational number, which can be represented by a point on the real number line. Imaginary numbers are excluded.

Real Number Line—2.2—A line representing the set of all real numbers. Negative infinity is shown on the left, zero in the center, and positive infinity on the right. Numbers are increasingly larger when moving from left to right along the line.

Reciprocal—2.2—The reciprocal of any number b is $1/b$, which is sometimes written b^{-1}.

Regression Analysis—See least squares regression analysis.

Relationship—5.2—A rule of correspondence between the values assumed by two or more variables for observations or measurements which belong together. A relationship can be expressed by an equation with two or more variables, or a set of ordered pairs (triples, quadruples, . . ., or n-tuples). *See also:* binary relationship, function

Relative Frequency—3.3—Frequency of an individual category divided by the total for all categories. *See also:* distribution

Relative Maximum (Minimum)—9.1; 10—The highest (lowest) value that a function $y = f(x)$ assumes within a given neighborhood of values of x.

Relevance Index—The relative frequency of admissions or visits to a particular hospital by the residents of a specified census tract, based on all admissions or visits by residents of the tract.

Right-Triangle—7.5—A triangle in which one of the angles is a right-angle (90 degrees) and one of the others is specified, usually by the Greek letter θ. The longest side is called the *hypotenuse* (c in the figure) and does not form part of the right angle. The angle θ is formed by the hypotenuse and the *adjacent side* (b), while the other side is given the name *opposite side* (a) because of its position opposite θ.

Right-Triangle:

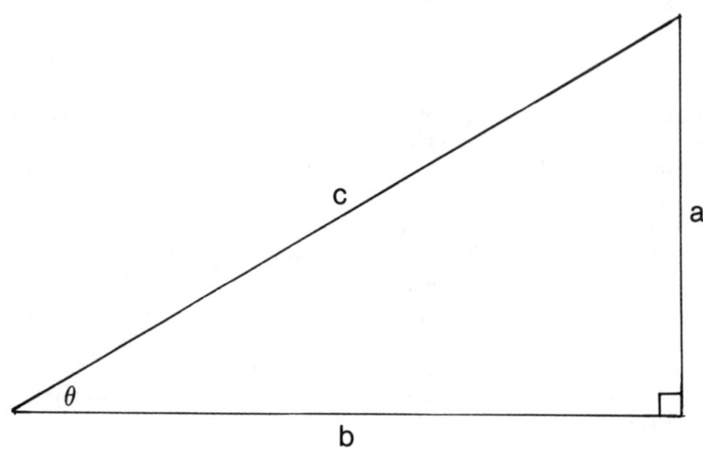

The lengths of the sides are related by the Pythagorean theorem $c^2 = b^2 + a^2$. *See also:* sine; cosine; tangent

Risk Factor—8.1—(1) That portion of the interest rate that reflects the lender's fear that the borrower might not be able to repay the principal amount when due. (2) A natural or artificial condition or event which is believed to contribute to a higher rate of incidence of a specified illness.

Roots—7.6—Values of x for which a quadratic expression in x and x^2 has the value zero.

Rounding—2.2—Adding one to the last significant digit to be retained (in a decimal fraction) if the next digit to the right is five or greater.

Row Vector—16—A matrix consisting of a single row.

S

Sample Space—14.2—The set of all possible outcomes of an experiment.

Scalar—16; 17—A scalar is any single real number.

Scattergram—3.5—A graphic portrayal of a relationship that is obtained by plotting the data as points.

Secant—9.2; 10.2—A line that intersects a curve at more than one point.

Second Derivative—10.6—The derivative of the first derivative.

Sector (of a circle)—A portion of a circle bounded by the circumference and two radius lines.

Set—13.1,.2—A collection of objects or representations of objects.

Set Equality—13.2—Occurs when two sets contain exactly the same objects.

Significant Digits—2.2—The digits in a number which are meaningful (or accurately calculated, if the number results from a series of calculations). The leftmost digits are the most significant.

Simple Interest—8.2—Interest calculated on the original principal amount, rather than being compounded. *See also:* interest, compound interest

Simultaneous Equations—6—A system of equations for which it is possible to find a set of values of all the variables such that the equations are all true at the same time.

Sine—7.5—A trigonometric function of the specified angle (θ) in a right triangle, defined as the ratio of the lengths of the opposite side and the hypotenuse, $\sin(\theta) = a/c$.

Singular Matrix—19.3—A square matrix is singular when its determinant is zero. At least one row (or column) is a linear combination of other rows (or columns).

Size of Matrix—16—See dimension of a matrix.

Slope—5.3—The steepness of a line, as expressed by the number of units change in the function for every unit change in the independent variable. This is the derivative of the function.

Slope-Intercept Form (of an Equation)—5.3—An equation of a straight line of the form $y = a + bx$, where a is the Y intercept and b is the slope of the line.

Solidus—A vertical bar (mathematical notation).

Square Matrix—16—A matrix in which the number of rows is equal to the number of columns.

Square Root—4; 7.5—The variable y is the square root of x ($y = \sqrt{x}$), if $y^2 = x$. For example $\sqrt{16} = 4$, because $4 \times 4 = 16$. *See also:* cube root

Standard Deviation—15.3—A measure of the variation or dispersion exhibited by a set of data.

Standard Normal Distribution—15.3—A normal distribution with a mean of zero and standard deviation of one.

Standard Normal Unit—15.3—Measures the distance between an observation and the mean in terms of standard deviations.

Stationary Value (of a function)—9; 10.3,.5—A point at which the slope is zero.

Stochastic (Process or Model)—(1) An experiment or physical process (or model of one) where the inputs (independent variables) or starting conditions only partially determine the outcome, which is in some aspects a random variable. (2) Such an experiment or process repeated, usually frequently. *See also:* random variable, deterministic

Straight Line Function—See linear equation.

Submatrix—See partitioned matrix.

Subscript—A small number or letter written below the normal line, used to identify a particular value of the variable it follows. *See also:* superscript

Subset—13—Set A is a subset of set B if all the elements in set A are also elements of set B.

Substitute—5.3; 7.4—To replace a variable by a specific value.

Sum of Squares—In certain statistical procedures the value of the variable is squared and summed, giving the sum of squares.

Sum of the Years Digits Method—An approach to calculating the depreciation allowance on capital equipment.

Summation—The process of adding a series of numbers.

Superscript—A small number or letter written above the line, most commonly used as an exponent in a power.

Symmetry (Symmetrical)—The property of having one half of the curve that is a mirror image of the other half.

T

t **Distribution (Table)—15**—A family of statistical distributions closely related to the normal distribution.

Tangent—7.5; 9; 10.2,.3—(1) A line that touches but does not cross a curve at a point. (2) A trigonometric function of the specified angle θ in a right triangle, defined as the ratio of the lengths of the opposite and adjacent sides, $\tan(\theta) = a/b$. (This is the slope of the line in definition (1).)

Term—4.5—The product of a number and one or more variables. Part of an algebraic expression.

Time Series—A set of observations of one or several variables which are taken at successive points in time.

Transpose—18.2—For a given matrix **A**, the transpose (represented by **A**′) is formed by writing the original rows as columns (or columns as rows).

Triangular Matrix—16; 17—A square matrix in which all of the elements above (or below) the main diagonal are zero.

Trigonometric Function—7.5—Functions relating the sides of a right-triangle to the specified angle. *See also:* sine; cosine; tangent

Trinomial—4.5—An expression with three terms.

U

Unimodal—3.3—A distribution (or curve) having one peak.

Union—13—The union of set A and set B is composed of all the elements in set A, or in set B, as well as in both sets A and B.

Universe—13—The set of all possible objects (related to a specific problem).

V

Variable—2; 4; 5; 6; 7—(1) In algebra, a letter used to represent a number in an expression or equation. *See also:* dependent variable; independent variable; constant. (2) A letter (or name) used to represent the inputs to or outcomes from an experiment or physical process. *See also:* random variable; deterministic; stochastic

Variable Cost—A cost which increases (decreases) as output is increased (decreased). *See also:* fixed cost

Venn Diagram—13.4—Diagramatic method of portraying sets and subsets.

W

Weighted Sum (Weighting)—18.1—A process of attaching the relative importance (weight) to each observation before adding. The observations are multiplied by their respective weights.

Whole Numbers—2.2—Any of the numbers 0, 1, 2, 3, 4, ...

Z

Zero Matrix—See null matrix.

Index

Note: Page numbers in italics designate tables and figures.